PHYSICAL BIOLOGY

From Atoms to Medicine

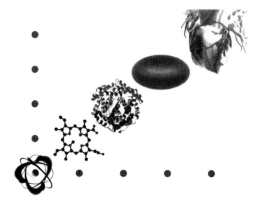

PHYSICAL BIOLOGY

From Atoms to Medicine

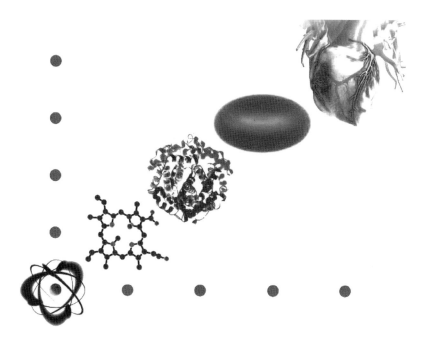

BALTIMORE · BUSTAMANTE · DOBSON · HOOD · HUSH
KOCH · KORNBERG · MACKINNON · MCCAMMON
MILLER · PARRINELLO · PHILLIPS · QUAKE · REES · THOMAS
TIRRELL · VARSHAVSKY · WHITESIDES · WOLYNES

Ahmed H. Zewail

California Institute of Technology, USA

Imperial College Press

ICP

Published by

Imperial College Press
57 Shelton Street
Covent Garden
London WC2H 9HE

Distributed by

World Scientific Publishing Co. Pte. Ltd.

5 Toh Tuck Link, Singapore 596224

USA office: 27 Warren Street, Suite 401-402, Hackensack, NJ 07601

UK office: 57 Shelton Street, Covent Garden, London WC2H 9HE

Library of Congress Cataloging-in-Publication Data
Physical biology : from atoms to medicine / editor, Ahmed H. Zewail.
    p. cm.
    Includes bibliographical references.
    ISBN-13: 978-1-84816-199-3 (hardcover : alk. paper)
    ISBN-10: 1-84816-199-9 (hardcover : alk. paper)
    ISBN-13: 978-1-84816-200-6 (pbk. : alk. paper)
    ISBN-10: 1-84816-200-6 (pbk. : alk. paper)
    1. Biophysics. 2. Biochemistry. 3. Molecular biology. I. Zewail, Ahmed H.
    QH505.P457 2008
    570--dc22

                                                    2008009371

British Library Cataloguing-in-Publication Data
A catalogue record for this book is available from the British Library.

Printed by Fuisland Offset Printing (S) Pte Ltd. Singapore

Front row (*left to right*): Peter J. Wolynes, J. Andrew McCammon, John Meurig Thomas, David Baltimore, Ahmed H. Zewail, Leroy Hood, Roderick MacKinnon, Carlos J. Bustamante. Back row (*left to right*): Christof Koch, Douglas C. Rees, Michele Parrinello, David A. Tirrell, Christopher M. Dobson, George M. Whitesides, Stephen R. Quake, Rob Phillips, Roger D. Kornberg.

Contents

Contents

Contents

Prologue

Some historians of science are inclined to classify fields and their impact with a broad brush — we are told that chemistry was the science of the 19th century, physics the science of the 20th, and that biology is the science of the 21st century. Other historians, such as the renowned Thomas Kuhn, discussed the concept of a paradigm shift and defined a separation for the impact of concepts and tools. As a practicing scientist I, in fact, see a simpler picture with less of a chasm in the progress of science. Biology without concepts and tools from physics, chemistry, and now engineering, would be severely limited. Tools are as important as concepts and, at the end, the two amalgamate to change the way we observe and the way we think.

We may recall the impact on biology of tools such as X-ray diffraction, electron microscopy, and NMR. Related and equally significant is the central role of concepts, such as chemical bonding, dynamics of transition states, and molecular recognition, in elucidating mechanisms of biological activity. From biology, the tools of RNAi, PCR, site-directed mutagenesis, and gene knockout are opening up new avenues of chemical syntheses and physical studies. Similarly, concepts of complementarity, self-assembly, and collective interactions (coherence) of molecular machines are introducing to physics and chemistry challenges and opportunities in the realm of emergence and complexity.

This volume brings about the confluence of concepts and tools and that of different disciplines to address significant problems of our time in physical biology and adjacent disciplines. Specifically, the book is structured to provide a broad perspective on the current state-of-the-art methods and concepts at the heart of chemical and biological behavior, covering the topics of visualization; theory and computation for complexity; macromolecular function, protein folding and misfolding; molecular recognition; and systems integration from cells to consciousness. The scope of tools is wide-ranging, spanning imaging, crystallography, microfluidics, single-molecule spectroscopy, and synthetic probe targeting, either molecular or by metallic particles. The perspectives are made by world leaders of physics, chemistry, and biology, and they define potential new frontiers at the interface of these disciplines, including physical, systems, and synthetic biology.

The book, which is the product of the 2007 Welch Conference on Chemical Research, is not a proceeding of reports and references. The articles are overviews, rather than reviews. They provide a panoramic view across areas of progress made, from revolutionary tools for the determination of extended structures to perspectives on 21st century biology and medicine. It was pleasing to welcome an audience of nearly a thousand people, surely reflecting the excitement about the contributions and the contributors. Personally, I wish to thank all the contributors who accepted our invitation and made this volume possible. Out of the twenty scientists invited to the conference, only John Walker was unable to travel at the time of the gathering in Houston on October 22, 2007. Two colleagues, Susan Lindquist of MIT and Julie Theriot of Stanford University, were invited but could not participate.

The Houston event would not have been possible without the generous support of The Welch Foundation and its leaders. The contribution to the organization made by the president of the foundation, Mr. Norbert Dittrich, and his staff, especially Ms. Carla Atmar, is greatly appreciated. Over the years, the Welch conferences have benefited from the wisdom of members of the Scientific Advisory Board, chaired formerly by the late Norman Hackerman and now by James L. Kinsey. As a member, I have enjoyed and benefited from the wide-ranging and stimulating discussions, which in one way or another have influenced the outcome of this endeavor. During this event we celebrated the achievements of two distinguished colleagues, Noel Hush and William Miller, who both received the 2007 Welch Award for their contributions to theoretical chemistry — their addresses are happily included in this volume.

(*From left*)
Norbert Dittrich,
Ahmed H. Zewail
and James L. Kinsey.

The timely completion of this book demanded careful orchestration among all contributors while the editor was in Pasadena or on travel. Essential to this coordination was the support of my assistants at Caltech, De Ann Lewis and Maggie Sabanpan. De Ann's special effort from the realization of the conference to the completion of the book is highly valued. Ms. Sook Cheng Lim at World Scientific made a long trip from Singapore to Pasadena and the quality of the printed volume reflects her diligence and dedication to this project. Last but not least, I wish to thank Dr. K. K. Phua for his special care and enthusiasm for a prompt publication.

Ahmed H. Zewail
Pasadena, California
December 15, 2007

chapter 1

The Preoccupations of Twenty-First-Century Biology

David Baltimore[*]

The life sciences are in transition, with the beginning of the 21st century symbolically marking the initiation of this redefinition. It is a creative, exciting repositioning because it derives from a changed perspective on what is possible. The appearance of the human genome sequence at the turn of the millennium ushered in the post-genomic era, an era of new possibilities, new challenges and new paradigms.

Until the millennium, biologists struggled to produce data. Each of us had his or her particular perspective and each toiled to flesh it out by laboriously finding relevant proteins and genes, learning how they worked and trying to link them to the larger events of biological systems. Now we are reveling in a data and materials glut. All the genes and thus all the proteins are catalogued. Gene expression profiles in particular cell types are available online and comparisons of normal and diseased tissues are appearing daily. The sequenced genome in bite size pieces is readily accessed and an increasing fraction of the genome has been knocked out in available mice. And oh, the wonderful new machines we have to work with!

Consequently, life science research has become both easier and more difficult. Easier because we no longer have to clone a gene to get access to it; we can buy it from a supplier. If one wants to know something about a

[*]Division of Biology, California Institute of Technology, 1200 E. California Blvd., MC 147-75, Pasadena, CA 91125, USA, e-mail: baltimo@caltech.edu.

property of a protein, such as its structure, or one of a related protein, one can download it and study its molecular details. But the field is egalitarian: a scientist in India or California or London has the same access to data and materials. This makes originality harder to come by. Previously, one might have had a niche in biology to oneself, maintaining the hegemony by running ahead of the pack with a big laboratory of hard workers. But today such a strategy does not work because the access to tools and materials allows one to leap frog into the fray. Thus, there is a higher premium on innovative thinking, on the creative asking of questions.

The edge today goes to people who see a new avenue of application of the ubiquitous tool bag. One way to find open avenues is to focus on recently discovered molecules whose functions are up for grabs. The post-docs in my laboratory have increasingly been seeing such an avenue in microRNAs. Only found a few years ago and still in the process of being catalogued, the biological roles of these extraordinarily powerful little strands of RNA have yet to be extensively investigated, providing the post-docs with an open field. Actually, the field is only open temporarily; it is rapidly becoming crowded as more and more investigators ask how their pet area of study might be impacted by these previously unrecognized cellular components.

To study even a single microRNA, one uses all the tools of the moment: the array technologies, the data libraries, the sequence compendia. Mountains of data can be the investigator's friend when he or she approaches them from a well-defined point of view. But the data have stories to tell of their own. These stories are not about single entities but about relationships, correlations, interactions, control mechanisms, all of which can be teased from the data themselves. In the best of circumstances, the analysis leads to a hypothesis that can be tested by other data: data that may have to be collected. In this way, the data glut gets mined for significance and pathways of knowledge are unveiled. But more importantly, the new knowledge may suggest new perturbations that can be introduced into biological systems and then interrogated for their consequences. This back and forth between data, hypotheses, experimental design, collection of new data and then verification or falsification of hypotheses is a rich and satisfying process, making contemporary biological science particularly powerful.

But this data mining and hypothesis construction is only a part of the 21st century scene. There is structural and mechanistic biochemistry, which benefits so greatly from the increased computing power we all enjoy. Protein folding, that tantalizing mystery at the core of biology, is also yielding to the power of digitalization and computation. Structural biology promises to finally deliver in this century its long-awaited fruits of precise knowledge of the molecular events that underlie the remarkable ability of linear combinatorial association of 20 amino acids to make the universe of nanomachines that carry out the processes of the biological world. And new and very expensive tools will give us insight into the aggregates of proteins that carry out the events of biology: the proteins of the synapse that send signals among neurons, the proteins of transcriptional control that give individuality to cells, the proteins of motors that distribute the contents of cells, the proteins of membranes that allow a cell to sense its environment.

It is a truism that the events of biology play out in time and time is the dimension hardest to resolve at the molecular and cellular level. It has felt like Heisenberg's uncertainty principle was at work: the better the spatial resolution, the less we know about the dynamics. In this arena, we are seeing the beginnings of a true revolution, with tools emerging that can give us spatial resolution measured even in nanmometers and still allow us to watch molecular events unfold. We can hope that the detailed events of biological catalysis will grace our textbooks in coming editions. Other techniques allow even single protein molecules or aggregates to be followed as they perambulate inside cells.

We must acknowledge that it is not only proteins that make up cellular life: lipids and carbohydrates — remembered from our textbooks but put aside by most researchers because they did not fit into the central dogma (even as modified by reverse transcription) — have stories to tell of specificity, control and interactions and these are just beginning to emerge.

They will surprise, with that surprise coming to us soon because they are drawing 21st-century attention.

What moves through the world is not proteins or carbohydrates or lipids: it is an organism, an integration of the molecules of life. In this early decade of the 21st century, there are stirrings of interest in integrative biology, usually under the rubric of systems biology and for metabolic questions, this, too, will yield over time to our computers, models and data collection. It will leave us with the toughest problem of biology facing us as the last frontier: the nervous system. Here we face integration on its largest scale, but an integration that can produce the ultimate emergent property: consciousness.

While cellular life outside of the nervous system will be unraveled and known in exquisite detail in the next few decades, the understanding of brains seems likely to tantalize biologists well into the century before a deep understanding emerges. But emerge it will and there will one day be sense to the obvious but presently incomprehensible statement that consciousness is the product of cellular behaviors.

And what will that leave us? What will be our concerns as the century ripens into older age? There will be concerns about synthesis. Just like chemists, we will prove that we understand biology by synthesizing it. And we will not just make over what evolution has produced through its slow, cumbersome four billion-year-long channeled meanderings, we will do it better. We will make things that evolution never got to. The power of gene manipulation will be not only to correct defects but also to create new capabilities. We will not be constrained by the usual constituents of cells. Carbon nanotubes and other nanostructures will interface with proteins providing new levels of control and readout. Medicine will not just be about curing but about replacing and ultimately enhancing.

As we come to the latter decades of our new century, our world will be very different. We will have harnessed the sun's energy directly, replacing the burning of stored chemical energy and watching a hot earth begin to cool back down: letting the universe slowly scour us of our poisonous gases.

Biology will have played a role there, somehow being the model for the capture and storage of the sun's energy. That will have been the conquest of the century. We will live long lives, battling disease to a standstill when it appears. We will lead enhanced lives, tightly linked to computers that will extend our range of activities and do much of the drudgery we now take on ourselves. We will still be humans of the same sort, fighting among ourselves but hopefully not with the guns that will have become so smart that no one can hope to outsmart them.

And we will continue to see vistas. Science will not come to an end because synthetic science knows no limits. There is always another structure to be made, another linkage to be created, another capability to add to our armamentarium. We will still be humans, still restless, still excited by the unrealized opportunity and committed to newness.

At least, that is the world I would hope to leave for those who come after me.

chapter 2

The World as Physics, Mathematics and Nothing Else

Alexander Varshavsky[*]

> *"There are two, and only two,*
> *kinds of statements: trivial or incorrect."*
> (Lev Landau)

The title of this chapter would not surprise readers who have encountered articles by Max Tegmark about the foundations of physics and its relation to mathematics.[1,2] Practicing scientists — most of them anyway — know that science is one. The names of different sciences, be it "chemistry," "astronomy," "immunology," "linguistics," "psychology" and so forth, are shorthands that indicate the location of a specific domain of science *vis á vis* its other domains. In part because of the multiplicity of these names, the unitary nature and hierarchic structure of science tend to be less obvious to the nonscientists. This chapter covers, briefly, a lot of territory, and introduces a simple re-definition of terms that may improve the comprehension of scientific discourses by folks outside of science. It also pauses at philosophy and its discontents, and proposes a solution.

We begin with chemistry. In the 19th century, it was a common assumption that physics and chemistry were different in a profound way. This view was shown to be untenable when quantum mechanics accounted,

[*]Division of Biology, California Institute of Technology, Pasadena, CA 91125, USA; Telephone: 626-395-3785; Fax: 626-440-9821; e-mail: avarsh@caltech.edu.

in the 1930s and later, for the properties of covalent and noncovalent interactions, and for the physical basis of the periodic table. At first, only the molecule of hydrogen (H_2) could be "explained," quantitatively, through quantum mechanics. Later, both conceptual and computational advances in a branch of physics called quantum chemistry extended this understanding to complex atoms and chemical compounds, including organic ones. In the course of these studies, there was never a reason to doubt that the receding possibility of doing exact calculations for larger molecules was caused solely by the (expected) increase in computational complexity, i.e., by nothing more mysterious than that. In other words, a statement that chemistry has been "reduced" to physics is correct in the well-defined sense of interactions that involve sufficiently small numbers of atoms, and is most likely to be also correct "in principle," for arbitrarily large molecules, their large ensembles, and their reactions.

This reductionism "in principle," as distinguished from "practice," underlies physics proper as well. The "*arrows of explanation*"[3] are far from random: they point, consistently, from complex systems, e.g., those encountered in solid-state physics, to systems with fewer and smaller components. These "arrows" eventually converge to the level of quantum mechanics of elementary particles and simple nuclei composed from them. An increase in complexity of a physical system often brings forth strikingly specific phenomena that are characteristic of larger systems and are not observed with their (sufficiently small) subsets. For instance, three molecules of hydrogen do not comprise a setting for which the concepts of temperature and pressure can be tangibly useful. But the Avogadro number or so of the same molecules more than suffices for the properties of temperature and pressure to become relevant. Examples of this kind are a legion.

High-level phenomena in the physics of many particles and the regularities — "laws" — that describe such systems are referred to as "emergent" qualities. "*More Is Different*" was the title of a 1972 paper by P. W. Anderson,[4] and more is different indeed, qualitatively so. Lev Landau was one of the first to suggest, in the 1950s (cited in Ref. 5), a general idea that at least some emergent properties of a physical system are "generic" and characteristic of distinct stable states of the matter in the sense of being quasi-independent of the underlying "microscopic" laws. The resulting hierarchies of methods and approaches are tailored to a set of emergent properties of a system, and in most cases can afford to ignore the microscopic laws altogether. The underlying "fundamental" physics is still there, but often not in a way that has explanatory power or is helpful otherwise.

The hierarchic, "embedded" relationship between physics "proper" and chemistry can be extended, through the same logic, to biology, including molecular biology (a mix of biochemistry, biophysics, genetics and cell biology), neuroscience, and other "hard" or not so hard branches of biology, for example, psychology, sociology and linguistics. All three of the latter disciplines can be argued to be complex subsets of neuroscience, as they deal with behavioral, cognitive and societal aspects of humans, the only large-brained animals that possess recursive symbolic language and thereby are capable of communications and behaviors whose range and complexities vastly exceed those of other animals. A definitive proof that plants, animals, bacteria and other living organisms are nothing else but chemical (physico-chemical) machines of vast but finite complexities is not available in a formal sense. Until about the 1960s, a view (or rather a hope) was intermittently expressed that the understanding of biological systems may require fundamentally new physical "laws."[6] There is no current support for such views.

What I am about to suggest may seem radical only if it is (incorrectly) perceived as an assertion of a "truth." Instead, I propose, below, a set of *definitions* that have (as they must) the property of self-consistency but aspire to nothing greater than that. Specifically: let us call "physics" what is traditionally denoted as "science." "Science," in this context, would be simply a synonym of "physics." This definition addresses, in one step, the often encountered confusion, largely among nonscientists, about the multiplicity of "sciences," as if they are fundamentally separate entities. They are not, and the above definition takes care of that. Physics "proper," then, would be defined as the one that deals with systems that are distinct from living organisms and encompass domains (high-energy physics, astronomy/cosmology, solid-state physics, applied physics, etc.) that are conventionally denoted as "physics." A particular subset of physics, called "chemistry," is defined as physical systems that involve large sets of atoms significantly above 0°K but at sufficiently low energies (temperatures) and pressures to enable their (reversible) interactions and formation of metastable atomic aggregates called molecules. That's chemistry. Biology, then — at least the biology known to us — is a particular subset of chemistry (and hence of physics) that involves temperatures between ~0°C and ~200°C, plenty of H_2O, carbon/nitrogen/phosphorus-based compounds, and complex (but finitely complex) entities, called cells, that have the properties of *replicators*, i.e., the ability to assimilate parts of media in which a replicator grows and to make (occasionally imperfect) copies of itself. The imperfection of copying results in *mutations* and leads to a quasi-random variation in the rate of propagation of replicator's progeny, thereby making possible the evolution, through "natural" selection (i.e., competition), among nonidentical replicators.

The extant cell-based replicators, while fundamentally similar in that they employ lipid membranes and DNA/RNA/protein-based machines, are also diverse in several respects. Numerically, most cells on Earth are prokaryotes. These cells lack a membrane-enclosed intracellular nucleus and do not (strictly) depend on multicellularity and division of labor among closely related cells. A subset of replicators called eukaryotes evolved to contain the nucleus, multiple chromosomes, and a range of other complex intracellular structures. Some eukaryotes became multicellular, with division of labor amongst the organism's cells and with intricate mechanisms that both repair the instruction tape (DNA) and recombine it with other (usually similar) DNA in germ-line cells specifically reserved for that purpose. This *genetic* recombination, largely through sexual reproduction and *meiosis*, is likely to be a form of repair/rejuvenation, i.e., the prevention of aging of a lineage of organisms related through descent. DNA recombination also contributes to the ongoing competition with other organisms, including parasites. Prokaryotes are proficient too, in their own ways, in repairing their components, particularly DNA, and in carrying out DNA recombination, a process that also contributes to DNA repair.

While the evolution that led to this state of affairs was, to a large extent, the evolution through natural selection, a quasi-neutral mutational drift, referred to as *neutral* evolution, is also capable of producing increases in organismal complexity, and may have underlain, in part, the emergence of large multicellular organisms.[7–9] All of this is chemistry of course (i.e., a subset of physics), in the realm of complex assemblies of large and small molecules that are partitioned within enclosures made of lipid membranes, and function as replicators. As one might expect, these assemblies are fragile: even in the absence of overt insults, and especially upon them, cells employ roughly a third to a half of their entire machinery and "fuel" (ATP and related compounds) to keep undoing various errors and to repair, nonstop, their critical subsystems, with an emphasis on DNA. The resulting emergency-ward pandemonium (exacerbated by randomness of molecular collisions that underlie these circuits) includes the incessant repair of the repair devices themselves. All of these processes involve, among other things, a highly active, regulated turnover (destruction followed by resynthesis) of specific intracellular proteins.[10] Despite these remarkable efforts, most cells eventually die, through predation, aging (deterioration), or programmed cell death (apoptosis).

One lineage of mammals has yielded, over the last $\sim 10^6$ years, the only species that was capable of sophisticated (as distinguished from rudimentary) imitation, a skill that may have underlain the development, over the course of human evolution, of a syntax-enabled, vocabulary-rich language. This unique capability, in conjunction with other features of (evolved) human mind, led to the primacy of humans among mammals and other large animals. It also made possible the technological civilization, and underlies its continuing development. The emergence of language, the much later invention of writing, and the still later inventions of printing (15th century) and electronic communications (20th century) have unleashed the *second species* of replicator, referred to as *memes*, by analogy with *genes*.[11,12] Memes are stories, songs, habits, religious beliefs, specific skills, and ways of doing things that can spread from person to person through imitation, either directly or through devices such as books, TV or the Internet. Memes, like DNA, can be thought of as information. DNA propagates from cell to cell, whereas memes, in a qualitatively distinct way (hence the *second* replicator) propagate from one mind to another meme-capable mind. DNA evolves without "intentional regard" for a vehicle (cell) that contains it. Similarly, memes, in their journeys from mind to mind, are free of "intentional concerns" about the minds that receive a meme and either spread it further, or hold it, or misremember ("mutate") it, or forget ("delete") it. The memetic evolution (a part of it is referred to as the evolution of culture) is much faster than a DNA-based one. Memes are thus a separate, profoundly distinct replicator. They did not exist before the emergence of human language, and now influence ("drive") the evolution of humans and their societies concurrently with DNA-based evolution, in complex and insufficiently understood ways. Potential ramifications of memes are fascinating (e.g., Ref. 12), but there is no credible reason to suspect that in discussing memes we might have left chemistry/physics behind. We did not. What we left behind are *complexity* levels of a discourse that underlies "conventional" chemistry, entering the field of a particular "high-level" chemistry, with its own "high-level" tools (they are still so-so, those tools).

Is medicine physics? Of course it is, as medicine is a branch of applied biology, and biology, as we already discussed, is chemistry/physics. The coming revolution in medicine will involve not only qualitatively better ways to do surgery, but also pharmacological (drug-based) therapies that will take into account, at last, the massive interconnectedness and redundancy of molecular circuits in living cells. Most of conventional single-compound (and even multi-compound) drugs of today are incapable of such finesse. Therefore even otherwise useful drugs exhibit undesirable side effects. Yet another problem is our continuing helplessness in containing (let alone curing) major human cancers once they spread beyond a surgeon's knife. The problem is exacerbated by genomic instability of many, possibly most, cancers. This property increases the heterogeneity of malignant cells in the course of tumor progression or anticancer treatment and is one reason for the failure of most drug-based cancer therapies.[13,14] A few relatively rare cancers can often be cured through chemotherapy but require cytotoxic treatments of a kind that cause severe side effects and are themselves carcinogenic.[15,16] Recent advances, including the use of antiangiogenic compounds and inhibitors of specific kinases, hold the promise of rational, curative therapies.[17,18] Nevertheless, major human cancers are still incurable once they have metastasized.

I recently proposed an approach to cancer therapy that involves homozygous deletions (HDs). They are present in many, possibly most, cancers, and differ from any other attribute of a cancer cell by the fact that an HD *cannot revert.*[19] Thus, a treatment that homes exclusively on cells that *lack* specific DNA sequences that are present in normal cells may be not only curative but substantially free of side effects as well. The difficulty here is that a homozygous deletion is an "absence," and therefore it cannot be a conventional molecular target. Nevertheless, an HD-specific anticancer regimen, termed *deletion-specific targeting* (DST), is feasible (on paper so far), at the price of relatively complex, "large" molecule-based designs.[19] If the complexity of DST-type strategies is unavoidable (that remains to be determined), approaches of this kind might be a harbinger of therapies to come. The virtues of simplicity notwithstanding, a complex problem, such as an assured cure of cancer, or a selective elimination of damaged (e.g., aged) mitochondria in cells of a patient, or other such feats may require commensurately sophisticated solutions. Can small compounds, with their inherently low informational content, ever enable a definitive cure of cancer that is also free of collateral damage? The notion that underlies (and motivated) the DST strategy[19] is that a curative, side effects-free treatment may require polymer-scale, multitarget, Boolean-type circuits, i.e., that simpler (smaller) drugs ultimately will not do, particularly in regard to side effects. The task at hand is to address the validity of this assumption.

"I heard a great deal, many lies and falsehoods, but the longer I lived the more I understood that there were really no lies. Whatever doesn't really happen is dreamed at night. It happens to one if it doesn't happen to another, tomorrow if not today, or a century hence if not next year. What difference can it make? Often I heard tales of which I said, 'Now this is a thing that cannot happen.' But before the year had elapsed I heard that it actually had come to pass somewhere." (From *Gimpel the Fool*, by Isaac Bashevis Singer.)

Did Singer write about Gimpel the Fool, or did he write about the origin of life? Our survey (a couple of paragraphs above) of biological evolution was about its stages *after* the emergence of membrane-enclosed replicators (cells) on a young Earth. This, largely Darwinian, evolution has been understood at least in outline. By contrast, the preceding stage, called prebiotic evolution, is a mystery: its potential routes appear to be insufficiently robust. We do not know the nature of the chemical paths that led to the first "really living" cells. We also do not know whether those cells were based on RNA or other macromolecules. One difficulty of working on this fiendishly intractable problem is that all traces of the first cells are, most likely, irretrievably lost. That means that even a construction of a "live" cell in a lab may not tell us exactly how life arose on Earth in the first place. Nevertheless, several heroic groups, notably Jack Szostak and colleagues, are at work to *"synthesize chemical systems capable of Darwinian evolution, based on encapsulation of self-replicating nucleic acids in self-replicating membrane vesicles."*[20] The behavior of membrane vesicles that they discovered as a part of this effort is unexpectedly rich, and suggests potentially relevant prebiotic routes to the first cells.[20,21] Over the years, the proposed life-origin scenarios ranged from a steady accumulation of complexity[22] to a colossally improbable fluctuation that may happen, e.g., once in a lifetime of our (Hubble-volume) universe and gives rise, from scratch or nearly so, to a replication-competent autotrophic cell ("autotrophic" denotes a cell capable of making its own components from simple compounds). The above fluctuation would be kind of inevitable if the universe (or, using the modern language, a multiverse; see below) contains an infinite number of stars and planets (e.g., Ref. 23). I return to the subject of multiverses near the end.

If the world is physics, what about philosophy? Richard Feynman's disdain for the subject would be obvious to anyone who read his Lectures on Physics: *"These philosophers are always with us, struggling in the periphery to try to tell us something, but they never really understand the subtleties and depths of the problem."*[24] The beginnings of philosophy, in ancient Greece and even earlier, were synonymous with the emergence of science. Later, much later, specific subsets of science that had defined themselves through experiments and mathematics were escaping, in droves and for good, from philosophy's vagueness and play with words that often covered up the absence of content. A succession of parades, throughout the 20th century, by obvious charlatans like the Marxist philosophers in the former Soviet Union, the "intellectuals" like Jacques Derrida in France, or philosophers like him in other places, did nothing to increase respect for the subject amongst scientists who bothered to take a look at the writings involved. Some philosophers knew that their highfalutin word plays were "not even wrong." Their other brethren took their own gobbledygook quite seriously, and were, therefore, not charlatans through intent. But the net result, *vis à vis* the discerning world outside, was the same for both groups. There are (relatively rare) folks who call themselves philosophers (a good example is Daniel Dennett) but actually do what I would call science, neuroscience in the case of Dennett. I mention this to make sure I do not offend people who are not "philosophers," in my book.

The escapes from philosophy continue. One of the latest is the subject of consciousness. As recently as 50 years ago, this problem was a preserve of certifiable cranks, theologians and philosophers, who proclaimed (and the claim is extant as I write) that consciousness, in addition to being a difficult problem (very true), is unlikely to be solvable in the context of "reductionist" science (how do they know?). Recent scientific work in this arena was led by Christof Koch, the late Francis Crick and other neuroscientists. The problem of consciousness was subdivided into subproblems. Some of them, including *"neural correlates of consciousness,"* have become, by now, a legitimate part of neuroscience. In less than 3,000 years (if you think it's a long time, ask a geologist), the subject of consciousness went from a poorly articulated notion at a campfire in Mesopotamia, through a messy arena that sustained, fruitlessly, generations of theologians and philosophers, to the modern, partially successful attempts to define the issues sharply and rigorously enough, and thereby to enable experimentally testable (as distinguished from unfalsifiable) predictions. This kind of evolution is in store, I think, for just about every problem that is claimed to be a philosophical one.

I conclude with a conjecture about the nature of philosophy or, more accurately, with a reformulation of its definition *vis à vis* science. I find it odd that the "deepest" questions, for example the nature of causality, the nature of time, the nature of "reality" *vis à vis* "observers," the meaning (if any) of life, are kind of partitioned, in an ill-defined way, between physics and philosophy. The former is one of humanity's greatest accomplishments: it led the way, hand in hand with mathematics, in our learning of how to think more accurately and critically, so that we do not end up fooling ourselves the moment a question is even halfway subtle. In contrast, the philosophy, once "exact" sciences separated from it, has become a depressing sight. *"The history of philosophy is, by and large, the history of failed models of a brain."*[25] This assessment, harsh as it is, leaves out some of philosophy's other shortcomings, including its reliance on inherently ambiguous human languages.

"If your horse dies, we suggest you dismount."[26] This, in a nutshell, is what I propose to do. It is time to see that "philosophy" was a transient, scaffold-like enterprise, with a beginning and the end. The latter could already be glimpsed in the 19th century. The proposed reformulation is "constructive," in that it relegates all subjects of philosophy to specific branches of science. These branches, as we already saw, can be defined

(not "proven" to be, but *defined*) as subsets of physics. The reformulation can be stated in "operational" terms: suppose that one is presented with a "philosophical" statement that claims to have a specific, verifiable-in-principle content. Suppose, furthermore, that a close examination of that statement confirms that it is, indeed, likely to have "content". If so, my conjecture (and the resulting reformulation) is that the above statement can *always* be classed as belonging to science, not "philosophy." For example, a "content-positive" philosophical statement about the nature of mathematical proofs, or about human motivation, or about space-time can be viewed, without a stretch, as a statement in mathematics, in neurosciences, and in fundamental physics, respectively. Whether such "assignments" are always possible, and whether, as a result, the philosophy departments at universities are based on a long-term, unfortunate misunderstanding remains to be seen.

"*There are two, and only two, kinds of statements: trivial or incorrect.*" These words, by Lev Landau, are said to have been uttered by him fairly often, and not entirely in jest. He meant, for example, that mathematical proofs can be viewed as being, in a sense, tautologies,[27] irrespective of the subtlety or length of a proof. (The notion of tautology in mathematics transcends the more narrow meaning of "tautology" in everyday discourse.[27]) Landau also meant, more informally, that our statements about the world, if they are accurate enough to reflect the world's design — from planets and stars to humans — may be akin to proofs in mathematics, i.e., that they are "trivial" in a narrow, *nonderogatory* sense, and that all other, less accurate statements are simply irrelevant ("incorrect"). There is a connection, here, to a view of the world discussed, over the last decade, by several authors (Refs. 1 and 2, and references therein). It involves the assumption that mathematical objects are "discovered," not "invented." An antecedent of this view, referred to as Platonic, was known to the Greeks, and is a (mostly unstated) premise of many mathematicians. A vastly general perspective, referred to as Multiverse-IV,[1,2] is that the currently known physics (which can be formulated as a set of mathematical propositions) is an infinitesimally small subset of all possible "physics," whose range encompasses, quite literally, all possible mathematics. (Multiverses I to III are models that do not go as far as Multiverse-IV.[28,29]) Viewed this way, the seemingly extreme classification that I began with, i.e., that everything is physics, was quite timid. The proposed unification of physics and mathematics[1,2] would account for the old puzzle of the *"unreasonable effectiveness of mathematics in physics,"*[30] as mathematics and physics would be, in that view, the same thing. Hence the title of the present chapter.

These and related concepts, widely discussed by cosmologists, include the possibility that *nothing* determines the choice of vacuum state for a Hubble-type universe. (See Refs. 1 and 31 for the definitions of a vacuum state and a Hubble volume.) In this view, the universe we inhabit is an infinitesimally small speck within a multiverse that contains every possible type of a vacuum state. If so, then, for example, the electron-to-proton mass ratio, a fundamental constant in the observable cosmos, would be just a (variable) parameter, akin to the distance between a star and its planet. Both numbers would be determined by stochastic events ("historical" accidents) that attended the formation of this or that big-bang universe.[1,31] The values of "fundamental" parameters would seem to be "restricted," then, solely by the fact that we, the "observers," must be present to measure the properties involved, i.e., that life, let alone "intelligent" life, requires "accommodations," e.g., the properties of a universe that are compatible with the existence of (relatively) stable atoms, stars and planets. Thus, the laws of physics (and consequently of everything else) that we register by studying physics in our part of the cosmos may be determined not necessarily by fundamental principles but instead by the contingency of having to have an "observer" in place to perceive those laws. This notion is referred to as the anthropic principle.[31-33] At first sight, these concepts, including Multiverse-IV,[1,2] are far-fetched enough to be untestable and therefore would forever belong to "metaphysics," a set of unfalsifiable speculations. Recall, however, that the distinction between physics and metaphysics is determined by whether a conjecture in question is testable at least in principle, and not by whether that conjecture is bizarre or involves currently unobservable entities. As discussed by others (Refs. 1, 2, 31 and 34, and references therein), both the anthropic principle and specific models of multiverses have explanatory powers, and may become a part of mainstream physics. Ramifications of these remarkable concepts and conjectures are so much beyond my mettle that this would be a good place to stop.

Acknowledgments

I thank Ahmed Zewail, who organized the 51st Welch Conference, for inviting me to attend it, and for encouraging the piece above. I am grateful to Christof Koch and William Dunphy for their helpful comments on the manuscript. Work in the author's laboratory is supported by grants from the National Institutes of Health.

References

1. Tegmark M. (2003) Parallel universes. In: Barrow JD, Davies PCW & Harper CL (eds). *Science and Ultimate Reality: From Quantum to Cosmos*, pp. 1–18. Cambridge University Press, Cambridge, UK.
2. Tegmark M. (2008) The mathematical universe. *Found Phys* (in press).
3. Weinberg S. (1987) Newtonianism, reductionism and the art of congressional testimony. *Nature* **330**: 433–437.
4. Anderson PW. (1972) More is different. *Science* **177**: 393–396.
5. Laughlin RB, Pines D. (2000) The theory of everything. *Proc Natl Acad Sci USA* **97**: 28–31.
6. Elsasser WM. (1958) *The Physical Foundation of Biology.* Pergamon Press, London, UK.
7. Stoltzfus A. (1999) On the possibility of constructive neutral evolution. *J Mol Evol* **49**: 169–181.
8. Lynch M. (2005) The origins of eukaryotic gene structure. *Mol Biol Evol* **23**: 450–468.
9. Lynch M. (2007) The evolution of genetic networks by non-adaptive processes. *Nat Rev Genet* **8**: 803–813.

10. Varshavsky A. (2006) The early history of the ubiquitin field. *Pro Sci* **15**: 647–654.
11. Dawkins R. (1976) *The Selfish Gene.* Oxford University Press, Oxford, UK.
12. Blackmore S. (1999) *The Meme Machine.* Oxford University Press, Oxford, UK.
13. Weinberg RA. (2006) *The Biology of Cancer.* Garland Science, New York, NY.
14. Vogelstein B, Kinzler KW. (2004) Cancer genes and the pathways they control. *Nat Med* **10**: 789–799.
15. Einhorn LH. (2002) Curing metastatic testicular cancer. *Proc Natl Acad Sci USA* **99**: 4592–4595.
16. Hardman JG, Limbird LE, Gilman AG. (2001) *The Pharmacological Basis of Therapeutics.* McGraw Hill Companies, Inc., New York, NY.
17. Folkman J. (2007) Angiogenesis: An organizing principle for drug discovery? *Nat Rev Drug Discov* **6**: 273–286.
18. O'Hare, T, Corbin AS, Druker BJ. (2006) Targeted CML therapy: Controlling drug resistance, seeking cure. *Curr Op Genet Dev* **16**: 92–99.
19. Varshavsky A. (2007) Targeting the absence: Homozygous DNA deletions as an immutable marker for cancer therapy. *Proc Natl Acad Sci USA* **104**: 14935–14940.
20. Chen IA, Salehi-Ashtiani K, Szostak JW. (2005) RNA catalysis in model protocell vesicles. *J Am Chem Soc* **127**: 13213–13219.
21. Hanczyc MM, Szostak JW. (2004) Replicating vesicles as models of primitive cell growth and division. *Curr Opin Chem Biol* **8**: 660–664.
22. Benner SA, Ricardo A, Carrigan MA. (2004) Is there a common chemical model for life in the universe? *Curr Opin Chem Biol* **8**: 672–689.
23. Ellis GFR, Brundrit GB. (1979) Life in the infinite universe. *Quat J Royal Astr Soc* **20**: 37–41.
24. Feynman RP, Leighton RB, Sands M. (1963) *The Feynman Lectures on Physics*, Vol. 1. Addison-Wesley Publ. Co., Reading, MA.
25. Wilson EO. (1994) *Naturalist.* Warner Books, Inc., New York, NY.
26. Stein, H. *"If your horse dies, we suggest you dismount".* A bon mot attributed to American economist Herbert Stein (1916–1999).

27. Byers W. (2007) *How Mathematicians Think*. Princeton University Press, Princeton, NJ.
28. Tegmark M. (2007) Many lives in many worlds. *Nature* **448:** 23–24.
29. Carroll SM. (2006) Is our universe natural? *Nature* **440:** 1132–1135.
30. Wigner EW. (1967) *Symmetries and Reflections*. MIT Press, Cambridge, MA.
31. Guth AH, Kaiser DI. (2005) Inflationary cosmology: Exploring the universe from the smallest to the largest scales. *Science* **307:** 884–890.
32. Weinberg S. (1987) Anthropic bound on the cosmological constant. *Phys Rev Lett* **59:** 2697–2610.
33. Barrow JD, Tipler FJ. (1986) *The Anthropic Cosmological Principle*. Clarendon Press, Oxford, UK.
34. Wolfram S. (2002) *A New Kind of Science*. Wolfram Media, Inc., Champaign, Il.

chapter 3

Physical Biology
4D Visualization of Complexity

*Ahmed H. Zewail**

The integration of physics and chemistry with biology — physical biology — offers new opportunities for the deciphering of its complexity. Because of the collective interactions of the many elements involved, 4D visualization of structural dynamics is essential to an understanding of the mechanism of the function. Here, we provide an overview of the principles of 4D visualization and highlight the potential of space–time imaging through some examples of applications, ranging from chemical reactions and phase transitions to molecular assemblies and biological cells. Some "big questions" are raised in the hope that the new tools and concepts will provide an understanding of what complexity, and emergence, actually mean.

1. Prologue

The aim of physical biology as a new discipline is, as the title of the book implies, the integration of physics and chemistry united to explore the complexity of biology. Understanding mechanistically how physical forces and interactions govern biological function, from the molecular to the cellular scale, uncovers the nature of microscopic processes such as protein folding/misfolding, self-assembly and order, and the unique function of life's matrix, the triatomic water in cells. On the other hand, the study of information flow and circuitry of the cell provides the possibility for painting maps of interactive elements which are important especially in

*Physical Biology Center, Arthur Amos Noyes Laboratory of Chemical Physics, California Institute of Technology, Pasadena, CA 91125, USA, e-mail: zewail@caltech.edu.

the programmatic diagnoses of diseases in medicine. In physical biology, the focus on physical methods and concepts for elucidating the complexity of structures and dynamics involved is distinct from the aim of mapping the engineering of information flow; i.e., the networks or wirings in cells, systems biology.

Because the elements of biological machines are defined on the scale of macromolecules, one is concerned with how they interact, communicate, and define a nanometer-scale function. This machinery derives its power from the control it exerts, with atomic-scale precision, which is responsible for the so-called emergence. Emergence is a new addition to the lexicon of biology, and other fields, but its precise definition is still amorphous. Given that biological machines operate in the nonequilibrium state, irrespective of viewpoint, we have to understand how the pieces are made, how their

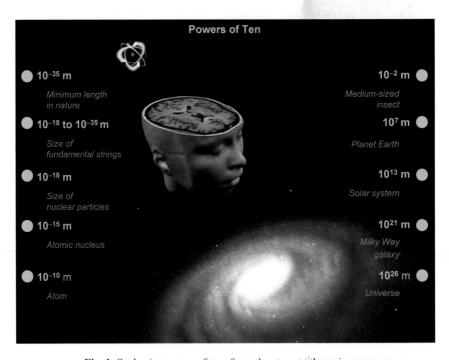

Fig. 1. Scales in powers of ten, from the atom to the universe.

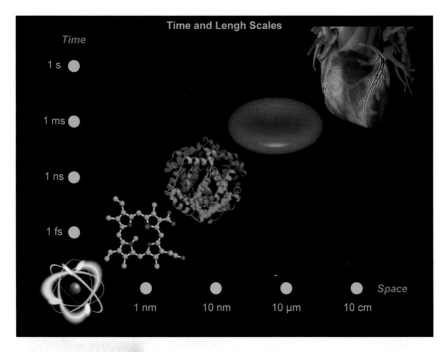

Fig. 2. Time and length scales, from atoms to organs. Shown are schematically represented atom, heme molecule, hemoglobin, red blood cell, and a heart.

physical forces exert control, and how feedback and feedforward of elements allow for function and robustness of information flow.

The emerging disciplines of physical, systems, and synthetic biology are clearly exciting frontiers in this century. For all, understanding mechanisms would remain dark and elusive, or at best speculative, unless the processes involved are directly visualized. The goal is to dissect complexity by following the behavior (function) in the four dimensions of space and time, of structure and dynamics. Such an understanding must account for the unique features of complexity — selectivity, diversity, directionality, and organization — and for the molecular forces controlling function. The mechanism may be "reduced" to elementary steps using the language of atoms (quantum mechanics), or we may need new concepts that tell us why the "whole is greater than the sum of its parts" and why coherence and order are possible under the random thermal conditions of a physiological temperature.

2. Visualization: From the Very Small to the Very Big

Realization of the importance of visualization and observation is evident in the exploration of natural phenomena from the very small to the very big. A century ago, the atom appeared complex, a "raisin or plum pie of no structure," until it was visualized on the appropriate time and length scale. Similarly, with telescopic observations a central dogma of the cosmos was changed and complexity yielded to the simplicity of the heliocentric structure and motion in the entire solar system. From the atom to the universe, the length and time scales span extremes of powers of ten (Fig. 1). The electron in the *s*-orbital of hydrogen has a "period" of sub-femtosecond, and the size of atoms is on the nanometer scale or less. Our universe's lifetime is ~ 13 billion years and, considering the light year ($\sim 10^{16}$ m), its length scale is on the order of 10^{26} m. In between these scales lies the world of life processes with scales varying from nanometers to centimeters and from femtoseconds to seconds (Fig. 2).

For the atom, not only was the language (quantum mechanics) developed but also the behavior was controlled — it has essentially been tamed. Since the first conceptual idea that *"there are only atoms and the void"* postulated by Democritus more than two millennia ago, we now know the components of the atoms and how to detect them, count each one, and cool them to sub-kelvin or trap them with light. Of major impact in molecular sciences is the ability to observe atoms at rest at angstrom resolution and atoms in motion at femtosecond resolution. However, for macromolecules, complexity arises from the collective interactions of thousands of atoms to form structures as well as from their dynamics, which determine the functions and the rates of such functions.

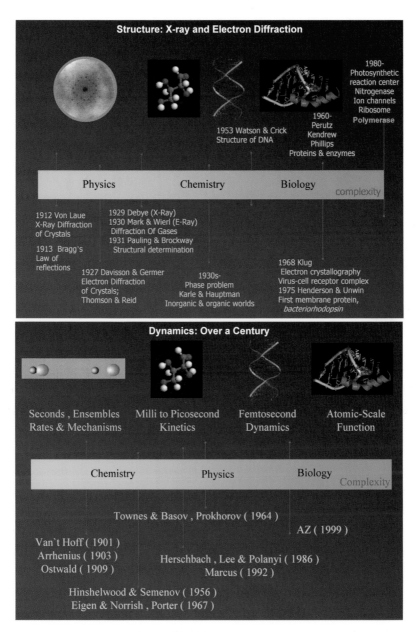

Fig. 3. (*Top*) Highlights of X-ray and electron diffraction determination of structures. (*Bottom*) Milestones, over a century of developments, in dynamics.

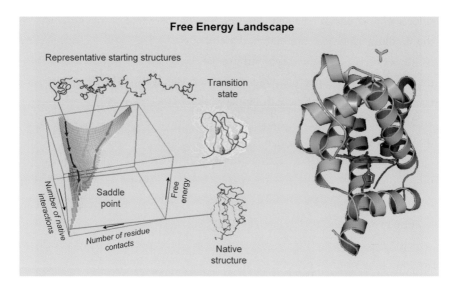

Fig. 4. (*Left*) Free energy landscape; see the chapter by Dobson. (*Right*) The structure of myoglobin.

Over a century of developments, X-ray and electron techniques have led to structural determination, beginning with the structures of two atoms (sodium chloride) and now culminating in the structural determination of more than 10^5 atoms in molecular mechanics. Similarly, for dynamics, the time scale at the beginning of the 20th century was practically in the seconds (hydrolysis of sugars); now, it has reached the atomic scale of femtosecond (Fig. 3). In the early days of DNA structural determination (1950s), it was believed, *"If you want to know the function, determine the structure,"* a statement made by Francis Crick that dominated the thinking of the time. But as we learn more about complexity, it is clear that the so-called "structure-function" correlation is insufficient to establish the mechanisms in complex systems. An example that may illustrate this point comes from the function of proteins.

Both the structures of the hemoglobin and myoglobin have been determined, but we still do not understand how they fold, how they selectively recognize oxygen, how oxygen is liberated from the cage into the medium, and how the matrix water assists folding, and the role it plays in directionality, selectivity, and recognition. Measuring the timescales involved through spectroscopic probes does not provide the structural transformations describing the complex energy landscape and involving many nuclear motions (Fig. 4). Visualization of the changing structures during the function is what is needed. The wavelength of the probing radiation or particle must be on the scale of interatomic separations (sub-nm), and for this reason, X-rays and electrons are ideal. However, if used as diffraction techniques in Fourier (reciprocal) space, then an inversion methodology is needed to deduce the structure in real space.

3. 4D Electron Microscopy

Our effort at Caltech in recent years has focused on the development of 4D visualization with ultrafast electron microscopy; development of electron diffraction has also been part of this effort. Electron microscopy provides real-space direct imaging with atomic-scale resolution. Here, I briefly highlight the concept and a few recent examples from physics, chemistry, and biology (the article by Sir John Thomas in this volume provides a lucid and comprehensive overview of the revolutionary advances in microscopy and imaging). The questions of interest, which are pertinent to physical biology, are numerous and include the following few: How do the elements of a polymer molecule of thousands of atoms, following the rules of genetic expression, interact to render a native structure capable of a specific function? Why, when they misbehave, do we contract diseases such as Alzheimer's? and Why do atomic-scale conformational changes of some macromolecular systems, as in prions, give rise to diseases?

The essential concept in 4D ultrafast electron microscopy (UEM) is based on the premise that trajectories of coherent and timed single-electron packets can provide the equivalent image obtained using N-electrons in conventional microscopes (Fig. 5). Recently, this goal of obtaining real images and diffraction patterns using timed, single-electron packets was achieved. In the new design, a femtosecond optical system was integrated with a redesigned electron microscope operating at 120 keV or 200 keV. By directly illuminating the photocathode above its work function with extremely weak femtosecond pulses, both real-space images and diffraction patterns can be obtained. A high-frequency train of pulses separated by nanosecond or longer intervals allows for the recording of the micrographs in a second or so (see below). Single-electron imaging circumvents the repulsions between electrons (space-charge) and the concomitant decrease in the ability of the microscope optics to focus electrons, as well as to provide the necessary

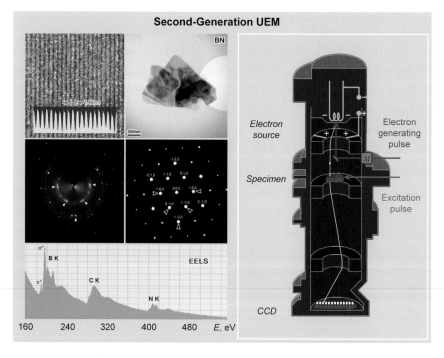

Fig. 5. (*Left*) Microscopy image (atomic scale), diffraction patterns (atomic resolution), boron image map, and electron energy loss spectrum, all taken with Caltech's 200 keV UEM. (*Right*) Single-electron trajectory schematized in UEM; see text and references.

stability of the electron flux during image processing. Depending on the required spatiotemporal resolutions, UEM can, in principle, operate in two modes: single-electron and single-pulse imaging, and the energy spread can be minimized by controlling the excess energy above the work function and/or invoking reverse-chirp methodology (see references). With frame-referencing at different times, the sensitivity of change is optimal for the isolation of transient structures.

At the atomic-scale resolution, we had to consider the length and time scales of electron imaging in UEM, namely, the energy-time and space-time relationships. Of particular importance are the fundamental limits of fermionic electrons in determining the coherence volume in imaging and diffraction. These issues have been discussed elsewhere, but suffice it to mention here that unlike bosonic photons which can occupy the same space over the duration of a coherent pulse, the Pauli exclusion principle restricts the volume in phase space for electrons. In general, the phase space can be thought of as divided into cells and electrons belonging to this volume element are indistinguishable; if more than one electron can occupy the volume, statistics is then used to deal with the problem. On the other hand, if the microscope operates in the "dilute regime," i.e., less than one electron per cell, then the "degeneracy parameter" is much less than one and the problem can be considered without yielding to statistics. The analogy with light (bosons) is useful, considering that the degeneracy factor is about 10^{-4} for a blackbody radiation (incandescent source at 3000 K and frequency of 5×10^{14} Hz), but about 10^9 for a common helium-neon laser.

The coherence volume, with a quantum size proportional to $\hbar^3$, is related to the classical value which can be calculated from a knowledge of the speed (or wavelength λ) at which the accelerated electrons pass the specimen, the spread in the electrons' energies, and the ratio of λ/α, where α is the divergence angle of the source. Using typical values for a 120 kV acceleration ($\lambda = 3.348$ pm), the volume is obtained to be about 5×10^5 nm^3. For our second microscope operating at 200 kV, it approaches 10^7 nm^3 (10^{-14} cm^3) at λ of 2.507 pm. In the single-electron pulse mode, the number of electrons per cubic centimeter is 10^{12}. Thus, the degeneracy factor is orders of magnitude less than one and each electron interferes with itself! In real space, and for a given contrast, the spatial resolution is only limited by the number of electrons.

4. Applications of 4D Electron Imaging

The space-time resolutions, and sensitivity, provide the impetus for investigating diverse dynamical phenomena of complex molecular, cellular, and material structures. Some examples from different fields illustrate the scope of applications.

4.1. Physics

Structures of nonequilibrium phases, which are formed by collective interactions, are elusive and less explored as they are inaccessible to conventional studies used for the equilibrium state. In order to understand the nature of these optically-dark phases, it is important to observe changes in structure at atomic-scale resolution. Such direct observations were recently made for the superconducting cuprates. The specific material studied is oxygen-doped LCO ($La_2CuO_{4+\delta}$); the undoped material is an antiferromagnetic Mott insulator, whereas doping confers superconductivity below 32 K and metallic properties at room temperature. Following near infrared pulse initiation, a structural phase transition was observed, defining a nonequilibrium state that is born in ~30 ps and lasts for ~300 ps with a major structural change along the *c*-axis of the material. Perhaps one of the most striking findings is the correlation found between photon and carrier doping (at the superconducting level); see references section. But a more pertinent example is that of metal-insulator transitions, which are common in the transformations of solid-solid materials and serve as an example of complex structural dynamics studied by both UEM and ultrafast electron crystollagraphy (UEC).

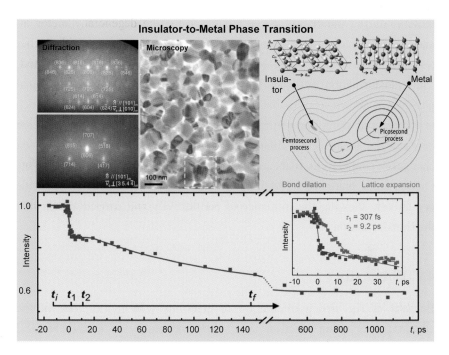

Fig. 6. Insulator-to-metal phase transition in vanadium dioxide. Equilibrium structures are shown, together with dynamics of atomic motions, as studied by UEM and UEC.

The physics of vanadium dioxide structure is a prototypical case. This room-temperature insulator has a monoclinic structure. Pairing and tilting of the vanadium pairs along one axis (Fig. 6) is at the heart of the insulating behavior. Part of the glue that keeps the paired atoms close to each other comes from the localization of two electrons on the vanadium sites, and from the gain in exchange energy when spins point in opposite directions. Excitation removes a fraction of these electrons and weakens or dilates the bonds, which is sufficient to unleash a collective relaxation of the structural distortion, a delocalization of the charge, and a loss of magnetic order.

Deciphering the complexity of motions became possible when mapping in time was made at once for different Bragg interferences (spots). The most important information comes from the observation that the fastest atomic motions during the change occur along one particular axis of the crystal, and are related to pairs of vanadium atoms moving apart from one another (Fig. 6). Only at later times do the other crystallographic planes expand. The transformation pathway between stable monoclinic and tetragonal phases was shown to pass through an unstable, tetragonal unit cell, which is compressed along one of the axes, thus establishing a direct connection between the femtosecond dilation of the V–V bond and the equally fast changes in conductivity that can be measured using conventional techniques. For the same material, UEM images (Fig. 6) were obtained for nm-scale crystallites, showing the influence of connectivity and carriers density on the transformation.

4.2. Chemistry

Because of the strong interaction of electrons with matter, one of the advantages of using electron imaging is the ability to study chemical reactions under collisionless conditions, and at nm-scale molecular interfaces. A "textbook" case for the former is that of the non-concerted elimination reaction of haloethanes (ICF_2CF_2I). The methodology of using different timed pulses in ultrafast electron diffraction (UED) allows for the isolation of the structure(s) of the reactant, intermediates-in-transition, and the product. The specific reaction studied involves the elimination of two iodine atoms from the reactant (haloethane) to give the product (haloethylene). Because of its fleeting nature, the structure of the intermediate has not been determined previously, and the challenge was in determining the structural dynamics of the entire reaction. This was achieved by referencing the observed diffraction to well-isolated frames at different times. In this way, we determined the bond distances and angles (Fig. 7). Moreover, the molecular structure of the $CF_2CF_2I^\cdot$ intermediate was established from the frame referencing at a time after the femtosecond breakage of the first bond. For other systems, reactions in the ground and excited states were similarly studied under collisionless conditions, in the absence of solvent perturbation (see references).

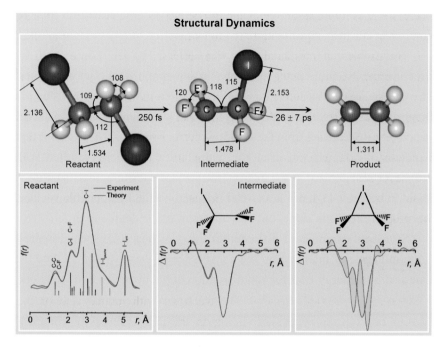

Fig. 7. Structural dynamics of the indicated chemical reaction, with the reactant, intermediate, and product structures determined under collisionless condition.

At interfaces, molecular assemblies, such as water networks, are unique in their behavior. Surface water of cells has unique properties that are distinct from those of liquid water, while amorphous ice, which constitutes the bulk of matter in comets, has very different properties. The directional molecular features of hydrogen bonding and the different structures possible, from amorphous to crystalline, make the interfacial collective assembly of water on the mesoscopic scale much less understood. Structurally, the nature of water on a substrate is determined by forces of orientation at the interface and by the net charge density, which establishes the hydrophilic or hydrophobic character of the substrate. However, the transformation from ordered to disordered structures and their coexistence must depend on the time scales for the movements of atoms locally and at long ranges.

Therefore, it is important to elucidate the nature of these structures and the time scales for their equilibration.

This problem of interfacial water was addressed by determining both the structure and dynamics using hydrophobic or hydrophilic surface substrates. The interfacial and ordered (crystalline) structure was evident from the Bragg (spots) diffraction and the layered and disordered (polycrystalline) structure was identified from the Debye-Scherrer rings (Fig. 8). The temporal evolution of both phases, after the temperature jump, was studied with monolayer sensitivity. On the hydrophilic surface substrate, the structure is found to be cubic (I_c), not hexagonal (I_h), and on the hydrophobic surface, the structure remains cubic, but very different in the degree of order.

Structural dynamics is different for the two phases. Of special interest is the influence of interface on the coexistence of these structures, and the time scales of energy transfer and disruption of the hydrogen bond network.

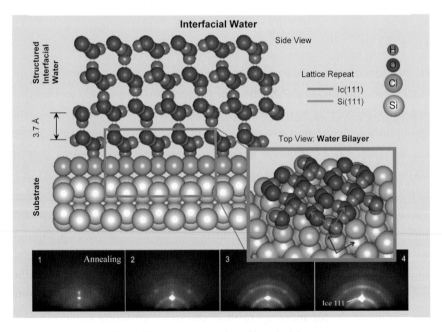

Fig. 8. Structural dynamics of interfacial water.

At the microscopic level, several conclusions were drawn. First, the reaction coordinate for breaking hydrogen bonds involves a significant contribution from the O···O distances, as evidenced in the depletion with time of the corresponding peak in the radial distribution function. Second, the time scale of energy dissipation in the layered structure must be faster than that of desorption, as no loss of water molecules was observed. Third, the time scale of the dynamics at the interface is similar to that of water at protein surfaces. Finally, the order of water molecules at the interface is of a high degree. Using molecular dynamics, in collaboration with Parrinello's group, the nature of forces (the atomic-scale description) and the degree of order were examined, elucidating features of crystallization, amorphization and the time scales involved. There are still questions pertinent to the topology of the energy landscape, spatial voids at the interface, and energy transfer from the substrate. Work is continuing on this problem which is relevant to other fields, including biology.

4.3. Biology

As mentioned earlier, biological complexity is a continuum of changing length and time scales. Among the phenomena of strongly correlated systems, self-assembly and self-organization are two that exemplify the controlled molecular and nonlinear interactions — physics and chemistry join in here! The simplest of assembled membrane-type structures is a bilayer of fatty acids. These long carbon chains self-assemble on surfaces (substrates) and can also be made as "2D crystals." To achieve crystallinity, the methodology of Langmuir–Blodgett films is invoked, providing control over pH, thickness, and pressure. In our studies, a temperature-jump of the substrate was introduced to heat up the adsorbed layers in direct contact with either a hydrophobic or a hydrophilic substrate. This initiating femtosecond infrared pulse has no resonance for absorption to the adsorbate layer(s). The studies made for monolayers, bilayers and multilayers of fatty acids and phospholipids provided an opportunity to determine the structural dynamics at the interfaces of nanometer scale and to examine changes due to the transition from 2D to 3D dimensionality.

The transient anisotropic expansion of fatty acid and phospholipid layers is vastly different from that observed in the steady state at equilibrium. On the ultrashort time scale, the expansion is along the -CH_2-CH_2- chain, but the amplitude of change far exceeds that predicted by (incoherent) thermal expansion. If heating is that of an equilibrated system, the change in the value of c_0, the unit cell length along the chains, with temperature, should be independent of the number of -CH_2-CH_2-CH_2- sub-units in the chain, as thermal expansion determined by the anharmonicity simply gives $\Delta c_0/c_0 = \alpha$, where α is the thermal expansion coefficient, typically very small, $10^{-5}\,\mathrm{K}^{-1}$. For a 10-degree rise, this expansion would be on the order of $10^{-4}\,\text{Å}$, while the observed transient change is as large as 0.01 Å. The in-

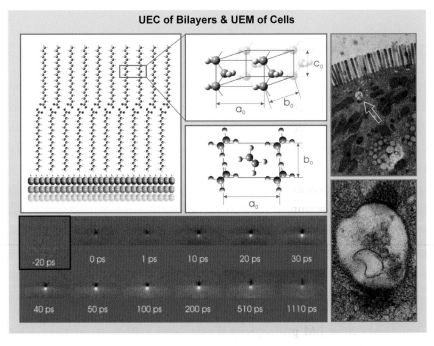

Fig. 9. Bilayers and cells studied by UEC and UEM.

chain large amplitude of expansion is understood even for harmonic chains, provided the system is in the nonequilibrium state, as detailed elsewhere (see references and also Fig. 9).

The impulsive force at short times transmits a large change in the value of c_0 as the disturbance (wave-type) accumulates to give the net effect that is dependent on the number of C atoms. In other words, as the disturbance passes through the different bonds, the diffraction amplitude builds up and exhibits a delay, ultimately giving rise to a large total amplitude for the change. This picture also explains the dependence of expansion on the total length of the chains, the increase in the initial maximum amplitude as the temperature of the substrate increases, and the effect of substrate strong (hydrophilic) vs. weak (hydrophobic) binding. The fact that the initial change in intensity (and elongation) occurs on the 10 ps time scale and that the distance traveled is ~20 Å (for a monolayer), the speed of propagation should be sub-kilometers per second, which is close to the propagation of sound waves. The speed could be of a higher value, reaching the actual speed of sound in the layers, a ballistic propagation in the chains. The model of coherent coupling among bonds in the underdamped regime of harmonic motions yields results vastly different from those of the diffusive behavior in the overdamped regime, but it has some common characteristics to the Fermi–Pasta–Ulam model of anharmonically-coupled chain dynamics. Preliminary MD results show the increase in $-CH_2-CH_2-CH_2-$ distance near the silicon surface by 0.08 Å in about 5 ps. With the same approach, studies of the self-assembly were made to elucidate the formation of interchain stacking with the void channels in between at zero pressure. Further research in this area will explore the role of hydration, and the extension to ion channels.

Biological imaging has the potential of producing structures for thousands of complexes and conformations important in cell biology. Cryoelectron microscopy (cryoEM) is an essential part of this endeavor. In the "single particle analysis" technique, purified proteins are spread into thin films across EM grids, plunged into liquid ethane to immobilize the proteins within vitreous ice, and imaged in an electron microscope. The key advantages of cryoEM-based methods are several. First, because the proteins are imaged one-by-one, in isolation rather than in a crystal, no time-consuming searches for crystallization conditions are necessary; second, the proteins can be imaged in physiologically relevant buffers and concentrations rather than those special conditions that produce crystals; and third, a mere $\sim 10^8$ particles are estimated to be needed to determine the structure.

Despite steady progress, the highest resolution achieved by single particle cryoEM is 4–8 Å, but typical structures are still determined at a resolution of 1–2 nm. The limit on resolution is not fundamental as the current resolution is determined by radiation damage among some other factors, such as particle heterogeneity and image contrast analysis. High energy electrons break covalent bonds, deposit thermal energy, and occasionally even knock out atomic nuclei. Thus, radiation damage limits the useful dose to ~ 20 e$^-$/Å^2, a flux issue. Because near-atomic resolution reconstructions require pixel sizes of approximately 1 Å^2, the quantum noise ("shot noise") present in the images makes it difficult to align the images precisely; misalignment directly degrades reconstruction resolution.

However, a major consequence of radiation damage is the movement of specimen during the exposure, which causes blurring of images. At about 80 K, proteins in vitreous ice are essentially motionless. Nevertheless, when hit by an electron beam, and covalent bonds are broken, electrons are liberated and heat is deposited in the macromolecule and surrounding ice. Pressure builds up within the ice as radiolytic fragments move at least a van der Waals radius apart. Residual positive charge accumulates as secondary electrons (electrons originally present in the sample) are emitted, causing internal repulsions. As a result, the deteriorating macromolecule might move and rotate appreciably. By reducing the exposure times by many orders of magnitude (from $\sim 10^0$ to somewhere between 10^{-6} and 10^{-12} seconds) beam-induced specimen movement becomes negligible on the time scale of the UEM experiment. This concept is currently under experimental scrutiny in a collaborative effort with my colleague Grant Jensen.

Three significant features of biological UEM are noteworthy. First, by dramatically reducing specimen movements, the resulting images should be much sharper. Sharper images will not only contribute more accurate information to 3D reconstructions, but will also allow for information to be merged correctly because the images are alignable with higher precision. This advance may make single particle analysis with near-atomic resolution a common methodology. Second, while this "freezing-in-time" concept is important to imaging of single (i.e., noncrystalline) macromolecules, UEM also opens the possibility of recording femtosecond (or longer) time scale dynamics. Thus, it is conceivable, for instance, to obtain time frames of single particles after an excitation pulse warms the sample, excites a conformational change, or releases a photocaged reactant. There are no fundamental physical reasons why multiple electron pulses could not be configured to simultaneously record whole tilt series, producing dynamically-resolved 3D tomographs of particles or cells. Finally, in biological UEM the regularity of pulsed dosing may result in the control of energy redistribution and heat dissipation and such controls are currently under experimental examination.

With UEM, images of cells derived from the small intestines of a four-day old rat were obtained. The specimen was prepared using standard thin-section methods. The cells were positively stained with uranyl acetate causing them to appear dark on a bright background. Figure 9 depicts the UEM images of the cells at two different magnifications. These images were obtained using the pulse trains of UEM with an exposure time of a few seconds; such exposure times compare well with standard EM imaging. In the figure, both the microvilli and the sub-cellular vesicles of the epithelial cells are visualized.

The images and diffraction patterns discussed above for materials and biological cells are "snapshots" or "frames" at a particular point in time. However, by delaying a second initiating optical pulse to arrive at the sample in the microscope with controlled time steps, we obtained a series of such snapshots with a well-defined frame time (movie). So far, such movies have been made for materials exhibiting a first-order phase transition (see references) using the UEM1 apparatus. More recently, we have developed UEM2 with additional capabilities for resolving the electron kinetic energy and for scanning the electron probe. In this UEM2, atomic-scale spatial and energy resolutions were achieved (Fig. 5). The latest in these studies is the success in UEM recording of protein single crystal (catalase) images, making visible in real space the lattice separations of 6.85 and 8.75 nm. We have also achieved, in UEM-2, a new limit of resolution, being able to visualize the 3.4 Å separation of a graphitized specimen, without the field emission gun (FEG) required in conventional (atomic-scale) electron microscopy. Moreover, the number of electrons has now been increased to permit bright- and dark-field scanning UEM.

4.4. Nanoscale Mechanical and Melting Phenomena

Another dimension of UEM is that of the *in situ* study of mechanical, melting or crystallization phenomena on the nm scale. Recently, we reported the serendipitous discovery of a mechanical nanoscale molecular phenomenon, a switchable channel or gate (Fig. 10) observed in a material of crystalline quasi-one-dimensional (1D) semiconductor Cu-TCNQ (TCNQ = 7,7,8,8-teracyanoquinodimethane, $C_{12}H_4N_4$). Remarkably, the switching, after a shock, not only is reversible with the near infrared pulses being on or off, but also returns the material in space to the original structure. The functional behavior is robust in the relatively low-fluence regime. At significantly higher fluences, we observed, in the microscope, the internal dilation and the reduction of the copper ions to form islands of neutral copper metal structures.

The strong electron acceptor (π-acid) TCNQ undergoes a facile redox reaction at room temperature with metals such as silver and copper. In the single crystals of the resulting Cu-TCNQ charge-transfer complex, Cu^+ and $TCNQ^-$ form discrete columnar stacks in a face-to-face configuration with strong overlap in the π-system. Further, the copper atoms are bound in a

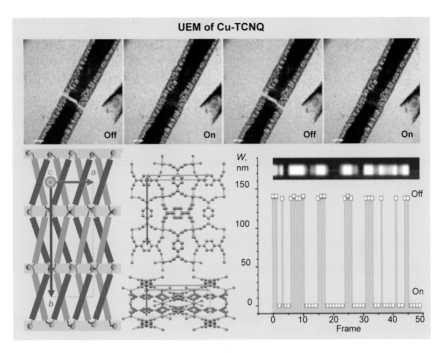

Fig. 10. Nanoscale mechanical phenomena observed by UEM in the quasi-1D Cu-TCNQ material. The nanogate width is denoted by *W*.

four-coordinate, highly distorted tetrahedral geometry to the nitrogen atoms on the cyano groups of the TCNQ molecules. The strong through-space interactions between the π-electrons and the resulting quasi-1D structure of the material in the solid state impart interesting structural and electronic properties in the field of the low-dimensional organic solids, and in the exploration of one-dimensional semiconducting nanostructures.

Not only is the controlling excitation the key to gating of the channel at lower fluences, but also it is sufficient at 1.6 eV to facilitate charge transfer and hence weakening of copper bonding and enhancement of its mobility at higher fluences. The surface energy of metal clusters composed of a few atoms is quite large, and there is a strong driving force for the smaller clusters to coalesce (Ostwald ripening). This behavior greatly enhances the formation of larger copper clusters and further reduces the ordered structure of the Cu-TCNQ material until finally, at relatively high fluence, the crystal separates into its constituent atomic and molecular components. The results discussed herein for channel formation and reductive metal clustering open the door to numerous further studies, including those of nucleation, charge/energy transport, and ultrafast dynamics of dislocations and coherent nuclear motions. The findings may be of value in applications involving molecular nanoswitches and channels, as well as optical pulse memory.

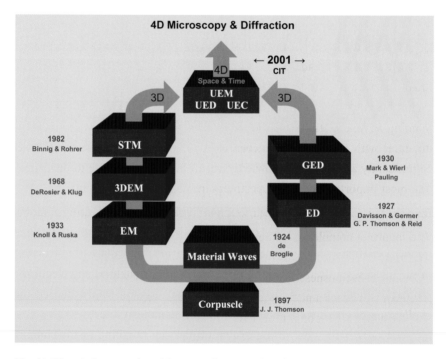

Fig. 11. Historical perspective with some milestones of developments in 2D and 3D imaging, in real and in Fourier space. The culmination in 4D microscopy and diffraction is shown in the confluence at the top.

5. Epilogue

At the end, it is the understanding of the nature of physical forces at different time and length scales, and the mechanism of function, which will provide the real meaning of emergence and complexity. In all of the above-mentioned applications, 4D electron imaging, in real or in Fourier space, is our method of choice at Caltech. As evident in many cases, e.g., those of chemical reactions, phase transitions, and nanoscale mechanical phenomena, both the spatial and the temporal resolution of imaging are essential for the trilogy construct of structure-dynamics-function. With time being the fourth dimension in 4D microscopy and diffraction, which have their historic roots in 2D and 3D methods (Fig. 11), we anticipate myriad of applications.

Uncovering the nature of biological complexity will surely benefit from the advances made in 4D visualization. Despite the enormous progress made so far, we are back to a 50-year-old physics question by Erwin Schrödinger — *What is life?* — but with a different perspective. In the coming 50 years, studies of the molecular basis of life in physical biology, together with the new approaches of systems and synthetic biology, will significantly advance our knowledge regarding what I believe to be some of the most important questions pertinent to biological complexity:

Life. How can a soup of chemical molecules form life? Does the origin of life involve a complexity of new paradigms beyond what we know now?

Humans. How does a network of organs result in a conscious being? How does our consciousness result in altruistic or destructive desires?

Organs. How does a network of ions, water, and proteins, for example, form the most powerful computing machine, the brain? How do we acquire, or lose, memory through the dense traffic of billions of neurons?

Cells. How does a "microfactory" with thousands of components attain dynamic stability? What is the origin of selectivity and fidelity in such a crowded environment, and what causes robust self-organization and self-assembly in noisy environments and nonequilibrium functional states?

Molecules. How do macromolecules fold or misfold uniquely and how do they recognize others in, for example, drug delivery and function? And, what makes a triatomic water molecule the most complex to understand and the most useful matrix of life?

These are "big questions" and the issues involved require knowledge of quantum and statistical physics, molecular bonding and synthesis, and network systems engineering and nonlinear information-flow dynamics. The hope is that the various powerful tools of molecular and systems biology, together with visualization, will make the science of biological complexity in the 21st century as revolutionary in its impact as was quantum mechanics in the 20th century.

The integration of physics and chemistry with biology is important, but it would be naive to think that the fundamentals of physics or chemistry as disciplines are at "the end of science," as claimed by some. Until today we still do not understand the real forces behind, for instance, dark energy and dark matter, nor do we understand why most of our universe is unknown, why the physical constants of nature are constant, why they take on their unique values, and what the purpose of duality, uncertainty, and chaos is. In chemistry, the nature of water in different phases, the forces behind self-assembly, and the recognition of molecules to others are frontiers that are still pregnant with major questions. Naturally, as we progress in acquiring new understanding, new questions will emerge. Lastly, because we cannot violate the second law, ordered biological phenomena, such as a tree, have to be considered in the context of universal interactions as the net entropy must increase. Thus, biological complexity, although local, it is part of our cosmic light-life realm, a concept recognized millennia ago (Fig. 12), and is appropriate for ending this piece.

Fig. 12. The significance of light-life interaction as perceived millennia ago, since Akhenaton and Nefertiti.

Acknowledgments

I wish to acknowledge the support of this research by the National Science Foundation, the Air Force Office of Scientific Research, and the Gordon and Betty Moore Foundation. Part of the biological imaging section represents the collaboration with Grant Jensen (NIH grant R01 GM081520-01) in the Physical Biology Center whose members are listed in Fig. 13. The dedicated effort by the group members who carried out the research discussed here on 4D visualization is acknowledged in the publications listed below. I would like to thank Dmitry Shorokhov for his care in formatting the text and for helpful discussion.

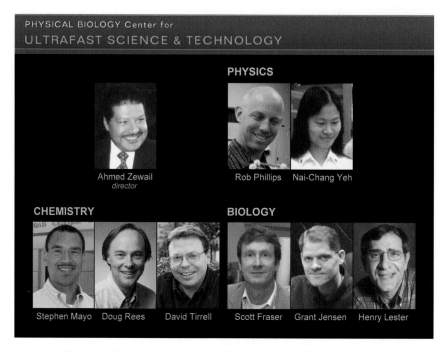

Fig. 13. Colleagues involved in the Physical Biology Center at Caltech.

Selected Publications

The following representative publications from Caltech are provided for readers interested in more details of the work on 4D visualization discussed in this chapter.

1. Zewail AH. (2008) Visualizing complexity: Development of 4D microscopy and diffraction for imaging in space and time. In: Chiao R, Phillips W, Leggett A, Cohen M, Ellis G, York D, Bishop R & Harper C (eds), *Visions of Discovery: Shedding New Light on Physics and Cosmology*, in press. Cambridge University Press, London.
2. Carbone F, Baum P, Rudolf P, Zewail AH. (2008) Structural preablation dynamics of graphite observed by ultrafast electron crystallography. *Phys Rev Lett* **100:** 035501.

3. Flannigan DJ, Lobastov VA, Zewail AH. (2007) Controlled nanoscale mechanical phenomena discovered with ultrafast electron microscopy. *Angew Chem Int Ed* **46**: 9206–9210.
4. Baum P, Zewail AH. (2007) Attosecond electron pulses for 4D diffraction and microscopy. *Proc Natl Acad Sci USA* **104**: 18409–18414.
5. Baum P, Yang D-S, Zewail AH. (2007) 4D Visualization of transitional structures in phase transformations by electron diffraction. *Science* **318**: 788–792.
6. Park HS, Baskin JS, Kwon O-H, Zewail AH. (2007) Atomic-scale imaging in real and energy space developed in ultrafast electron microscopy. *Nano Lett* **7**: 2545–2551.
7. Lobastov VA, Weissenrieder J, Tang J, Zewail AH. (2007) Ultrafast electron microscopy (UEM): Four-dimensional imaging and diffraction of nanostructures during phase transitions. *Nano Lett* **7**: 2552–2558.
8. Seidel MT, Chen S, Zewail AH. (2007) Ultrafast electron crystallography II. Surface adsorbates of crystalline fatty acids and phospholipids. *J Phys Chem C* **111**: 4920–4938.
9. Gedik N, Yang D-S, Logvenov G, Bozovic I, Zewail AH. (2007) Nonequilibrium phase transitions in cuprates observed by ultrafast electron crystallography. *Science* **316**: 425–429.
10. Zewail AH. (2006) 4D ultrafast electron diffraction, crystallography, and microscopy. *Annu Rev Phys Chem* **57**: 65–103.
11. Lin MM, Shorokhov D, Zewail AH. (2006) Helix-to-coil transitions in proteins: Helicity resonance in ultrafast electron diffraction. *Chem Phys Lett* **420**: 1–7.
12. Lobastov VA, Srinivasan R, Zewail AH. (2005) Four-dimensional ultrafast electron microscopy. *Proc Natl Acad Sci USA* **102**: 7069–7073.
13. Srinivasan R, Feenstra JS, Park ST, Xu S, Zewail AH. (2005) Dark structures in molecular radiationless transitions determined by ultrafast diffraction. *Science* **307**: 558–563.
14. Ruan C-Y, Lobastov VA, Vigliotti F, Chen S, Zewail AH. (2004) Ultrafast electron crystallography of interfacial water. *Science* **304**: 80–84.
15. Ihee H, Lobastov VA, Gomez UM, Goodson BM, Srinivasan R, Ruan C-Y, Zewail AH. (2001) Direct imaging of transient molecular structures with ultrafast diffraction. *Science* **291**: 458–462.

chapter 4

Revolutionary Developments from Atomic to Extended Structural Imaging

*John Meurig Thomas**

Of the three kinds of primary beams (neutrons, X-rays and electrons) suitable for structural imaging, the most powerful are coherent electrons. The source brightness of such electrons significantly exceeds achievable values for neutrons and X-rays: their minimum probe diameter is as small as 0.1 nm, and their elastic mean-free path is *ca* 10 nm (for carbon), much less than for neutrons and X-rays. Moreover, because electrons are focusable and may be pulsed at sub-picosecond rates, and have appreciable inelastic cross-sections, electron microscopy (EM) yields information in four distinct ways: in real space, in reciprocal space, in energy space and in the time domain. Thus, besides structural imaging, energy landscapes of macromolecules may be explored, and under optimal conditions elemental compositions, valence states and 3D information (from tomography) may be retrieved by time-resolved EM.

Advances in designs of aberration-corrected high-resolution electron microscopes have greatly enhanced the quality of structural information pertaining to nanoparticle metals, binary semiconductors, ceramics and complex oxides. Moreover, electron tomography sheds light on the shape, size and composition of bimetallic catalysts attached to nanoporous supports. With energy-filtered electron tomography, chemical compositions of sub-attogram quantities located at the interior of microscopic objects may be retrieved non-destructively.

Radiation damage is the main problem which prevents the determination of the structure of single biological macromolecules at atomic resolution using any kind of microscopy, irrespective of the primary beam. Great advances have recently been made via the EM of biological macromolecules and machines embedded in vitreous ice. Electron cryomicroscopy is now used to analyze the structures of molecules arranged as crystals or as single particles; and cryoelectron tomography reveals nuclear pore complex structure and dynamics.

*Department of Materials Science, University of Cambridge, Cambridge, CB2 3QZ, UK.

Electron energy loss spectroscopy and imaging, especially energy-filtered (S)TEM tomography is an important development in the imaging of both inorganic and biological systems; and Zewail's revolutionary single-electron (4D) ultrafast-EM and diffraction contribute richly to the structural imaging of both atomic and macromolecular species at sub-picosecond time scales.

1. Introduction

If our aim is to determine and visualize, at near-atomic or sub-nanometer resolution, the structure, architecture and general architectonic features of various macromolecular or cellular or extended inorganic materials, few instruments currently surpass, in the panoply of powerful procedures that it offers, the modern high-resolution electron microscope. This chapter, therefore, deals predominantly, but not exclusively, with the primacy of electron microscopy in its various manifestations and the several ways in which it reveals information that interests us.

Visualization alone is not enough. To achieve maximum insight from any worthwhile means of imaging, one should — for relatively simple or quite complicated inorganic systems — also learn about elemental composition, degree of structural flexibility, valence states and modes of bonding of the constituent atoms. And for more complicated biomolecular and cellular entities, the functional role of (or mechanistic details associated with) the entities investigated need, ideally, to be revealed by the particular technique of imaging that one adopts. To be specific, in the act of visualizing the double-stranded (ds) DNA genome in say, a bacteriophage, one needs to know not only the compact arrangement of the DNA but also how it is topologically organized to facilitate efficient ds DNA release during infection.

It is instructive at the outset to recall that perfectly reliable experimental methods already exist for imaging the two extreme kinds of structures that define the continuum of materials from atoms, on the one hand, to living cells, on the other. Intermittently throughout this chapter, we shall cite a number of revolutionary examples from the inorganic world, where the limitations that arise from electron-beam damage are generally small.

Highly resolved visual images are, therefore, more readily produced from most inorganic materials; and their inclusion in what follows is judged appropriate because they point the way to what may be ultimately achievable with biologically relevant materials, which are more vulnerable to damage inflicted by electrons and other interrogating beams.

2. Imaging Atoms by Field-Ion Microscopy (FIM)

With inorganic materials, attaining atomic resolution is relatively easy. A revolutionary method became available in 1951 when E.W. Müller introduced the technique of *field-ion microscopy* (FIM), which he and, later, others[1–4] extended to encompass *atom-probe microanalysis* (APM), so that it became routinely possible to visualize individual atoms at fine tips of metals and alloys, to identify their nearest neighbors, and to detect vacancies and also atomic steps and kinks at their surfaces (see Figs. 1 and 2). FIM and APM have become indispensable tools not only to the metallurgist and materials scientist, but also to surface scientists interested in the fundamentals that define their sub-disciplines. Field-evaporation combined with time-of-flight mass-spectrometry yields clear pictures of individual atoms. Look at the nine atoms on the (001) platform growing progressively smaller, one atom at a time.

In the work of Smith *et al.*[2,3] and Kruse *et al.*,[4] we see how the phenomenology of adsorption, surface migrations, corrosion and model studies of heterogeneous catalysis has been enlarged by the application of FIM and APM.

Atom-probe tomography (APT),[5] whereby the surface atoms at the tip of the metal or alloy are progressively field-evaporated, reveals structural and elemental (compostional) details of the interior of the solid. This technique yields three-dimensional information with single-atom sensitivity and subnanometer spatial resolution, albeit in a destructive manner.

A new type of atom probe known as *local electrode atom probe* (LEAP),[6,7] is capable of mapping the position of single "impurity" atoms in semiconductor nanowires made, for example, from indium arsenide, InAs, and to image the interface between Au — the catalyst used in growing the InAs nanowire — and the nanowire itself in 3D with 0.3 nm resolution. Figure 3 shows a striking example of the imaging and compositional study rendered feasible by LEAP.

Laser-assisted atom-probe tomography (LAAPT) is a technique capable of rapidly analyzing large volumes of non-electrically conducting material, thereby yielding the spatial distribution (3D mapping) of individual guest atoms within a host solid. The technique involves[8,9] the application of a

Atom Probe Microanalysis
GDW Smith (Oxford)

- Single atoms field evaporated from specimen and identified by time-of-flight mass spectroscopy.
- Local composition measured directly by counting atoms.
- Spatial resolution is approx. 2nm laterally 0.2nm in depth.

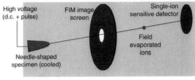

FIM Image Formation

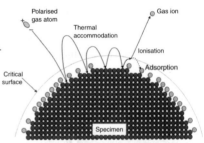

Field Ion Microscope (FIM)

- Specimen in form of needle, 100nm end radius.
- Voltage applied to specimen generates high field.
- Gas ionized at apex generates image on screen.

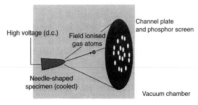

Ni_7Zr_3 Penn State University

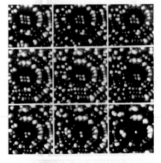

Fig. 1. So far as inorganic materials are concerned, especially metals and alloys, field ion microscopy, introduced by Müller in 1951[1] proved revolutionary. It is capable of imaging individual atoms and revealing vacancies and interstitials, as these pictures by Smith *et al.*[2,3] and that in Fig. 2 demonstrate.

Fig. 2. The arrow in this picture denotes an interstitial atom of Rh. (From Kruse *et al.*[4] Free University of Brussels (Rh tip).)

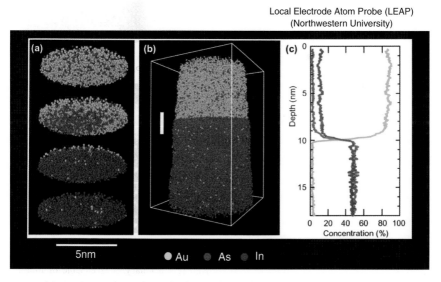

Fig. 3. The LEAP technique (see text) is capable of mapping semiconductor nanowires in atomic detail.[8]

high voltage to the base of a cryogenically cooled, needle-shaped specimen and continuous pulsing of the tip of the specimen with a focused laser. With silicon, for example, the sample is formed into a needle-shaped specimen with a focused ion beam. The applied voltage creates a high electric field, 20 to 40 Vnm^{-1}, at the apex of the needle. The laser pulse transiently heats the specimen tip, thereby providing just enough thermal energy for a single ion to escape from the end of the needle.

It remains to be seen whether LAAPT will be applicable to the tomographic-compositional study of biochemically and biophysically relevant materials. It seems possible that proteins and cells attached to an inorganic substratum, especially those that are relatively rich in heavier atoms (such as sulphur, iron, cobalt etc.), may indeed be imageable in this way.

3. Imaging Atoms and Molecules by Scanning Tunneling Microscopy (STM)

Ever since the powerful surface-imaging technique of STM was introduced in 1982 by Binnig and Rohrer,[10] it has spread widely into many branches of physical and biological science. STM images, formed as a result of recording (through the tunneling current that passes between a fine metal tip and the surface under investigation) a contour map of constant local density of the (electronic) states of the sample at the Fermi level, provide atomic resolution of adsorbed particles. Typical examples of how individual atoms of Pd or Au may be readily imaged at low temperatures (5 to 10 K) when they are bound at a thin layer (three monalayers or so) of MgO, which is in epitaxial contact with a Ag (001) substratum.[11] (It is interesting to note that while Pd atoms are arranged in a random fashion, Au forms an ordered array on the surface, a fact that is explicable in terms of charge transfer from the MgO to Au atoms.)

Low temperatures are also essential for the imaging of small molecules (like O_2 prior to or just after chemisorptive dissociation on a Pt (111) surface[12]). Besenbacher and colleagues have published a series of elegant STM studies of various metal and oxide surfaces, from which they have been able to deduce important mechanistic insights into the elementary processes of adsorption, surface migration and catalytic turnover at various solid surfaces.[13,14]

Besenbacher *et al.*[15] have also imaged guanine quartet networks stabilized by cooperative hydrogen bonds when adsorbed at Au (111) surfaces (see Fig. 4) recorded at 150 to 170 K. (Quadrupletes formed from guanine may lead to structures of DNA that are quite different from normal ones.[16]) The STM image clearly shows that each unit mesh is composed of four molecules; and, indeed, by superimposing the guanine-quartet structure determined by X-ray crystallography[17] on G-quadruplex DNA crystals and the STM image of Besenbacher *et al.*, there is a good correspondence between the two, independently determined, guanine networks.

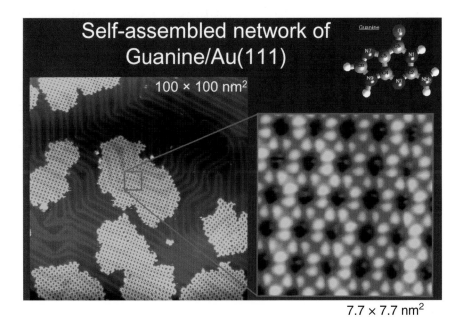

Fig. 4. Scanning tunneling microscopy (STM) reveals the atomic arrangements of self-assembled networks of guanine at a Au (111) surface.[15]

In situ STM imaging, combined with direct electrochemical measurements (such as electron transfer) of species like cytochrome *c* or *Pyrococcus furiosus ferredoxin* (PfFd) assembled on (thiolate-modified) Au(111) surfaces, has been elegantly achieved by Ulstrup and colleagues.[18–20] Well defined *in situ* images of PfFd molecules, immobilized on either mercaptopropionic or cysteine-modified Au(111) surfaces in stable monolayers or submonalayers, directly revealed (Fig. 5) that adsorbed protein molecules retain their expected lateral dimensions with uniform distribution of the protein molecules over the surface; and direct electron transfer rates, characterized by cyclic voltammetry, were found to be slow compared to similar processes in azurin or cytochrome *c*. Whilst spatial resolutions are inadequate to reveal intramolecular (atomic) features or the location of the Fe and S atoms of the surface-bound PfFd or other proteins, there is no doubt that the STM technique of Ulstrup *et al.*[18] can pinpoint the individual micromolecules themselves, as may be seen in Fig. 5, taken from their work.

Atomic force microscopy (AFM) is, potentially, better at probing processes on insulating surfaces than STM, although the degree of spatial resolution that it can attain is not high enough yet. AFM and related techniques — which involve the ingenious use of "optical tweezers," as demonstrated by Bustamante (elsewhere in this text) — are invaluable in determining force–velocity (F–V) relationships within various molecular machines, such as those relevant to the situation where RNA polymerase II (RNAPII) is in the act of transcribing messenger RNAs in eukaryote cells, during highly regulated processes.

Aberration-corrected electron microscopes (described below — see Sec. 5) offer a powerful, non-destructive method of determining atomic structure at the surfaces of metals and other solids. They have also recently revealed unexpected features in the structure-compostion relationship of bimetallic nanoparticles. In so far as industrially-important nanoparticle (*ca* 6 nm diameter) platinum is concerned, it is the combination of the improved spatial resolution (made possible by aberration-corrected lenses[21]) and the so-called exit wavefunction restoration procedure of Kirkland[22] that enables images and models such as those shown in Fig. 6 to be recorded and drawn.[23] Particularly important is the ability, now made possible, to locate and characterize atomic steps, kinks and vacancies at the thin edges of the exterior of close-packed surfaces of nanoparticle catalysts such as Pt, which is extensively used in both the auto-exhaust system and in fuel cell technology.

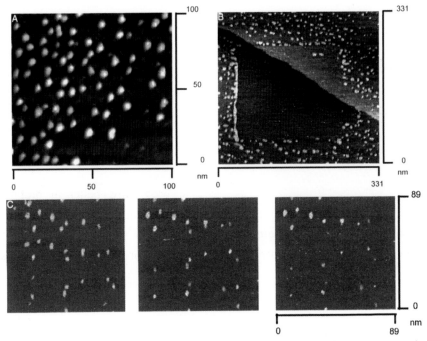

(A) In situ STM image of YCC adsorbed on Au(111). 10 mM phosphate buffer, pH 7.5. Constant current mode. Scan area 100 × 100 nm². Working electrode potential –0.16 V (SCE). Bias voltage –0.2 V. Tunneling current 0.5 nA. (B) In situ STM image of YCC adsorbed on Au(111), 10 mM phosphate, pH 7, after tip-induced desorption of YCC in a 190 × 190 nm² area at –0.1 V bias voltage. Scan area 330 × 330 nm². Working electrode potential –0.04 V (SCE), –0.2 V bias voltage. Tunneling current 0.9 nA. (C) In situ STM images of YCC adsorbed on Au(111). 10 mM phosphate, pH 7.5. Constant current mode. Tip current 1.0 nA. Tip potential –0.31 V (SCE). Soaking in 50 µM YCC, 10 mM phosphate, pH 7.5 for 5 h at room temperature. Substrate potentials: left, –0.36 V; middle, –0.16V; right, +0.04 V (SCE). The images in the middle and to the right were recorded 2.4 and 4.9 min, respectively, after the image to the left.

Fig. 5. Molecules of cytochrome-C_y, formed as monlayers at Au (111) surfaces are also readily imaged by STM, and also amenable to studies of electron transfer by *in situ* methods devised by Ulstrup *et al.*[18–20]

Nanocrystalline Pt Catalyst Particles

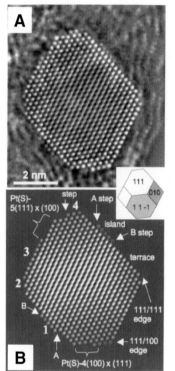

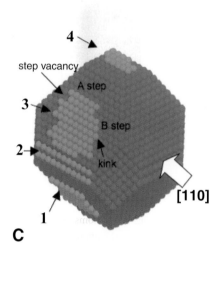

Fig. 6. Exit-wave restoration methods in electron microscopy (worked out by Kirkland *et al.*[22,24]) are able to image atomic distributions at the exterior surface of nanocrystalline particles of Pt.[24]

Dissociation of small, diatomic species (CO, O_2, N_2) occurs much more rapidly at monatomic steps than at terrace sites.[24]

Aberration-corrected electron microscopy has also uncovered an unexpected compostional feature in Pd–Au bimetallic nanoparticles, where a shell of three layers of Pd surrounds another shell of three layers of Au which, in turn, envelopes a four-layer core of Pd. This seemingly bizarre compositional arrangement[25,26] is quite unexpected on the basis of the bulk properties of the two metals. (The surface energy of Au is significantly

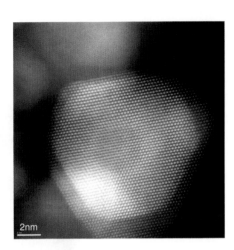

Fig. 7. Columns of Au atoms, visible as white spots, in this nanoparticle imaged by Yacaman *et al.*[25] are separated from one another by *ca* 0.2 nm.

smaller than that of Pd, so one would expect Au to be at the outermost shell.) In Fig. 7, taken from Yacaman's work, we clearly see columns of Au atoms, separated from one another by distances of *ca* 0.2 nm.

4. Imaging Biological Cells by X-Ray Tomography[27,28]

Soft X-ray microscopy, pioneered by Lawrence Bragg,[29] Cosslett and Nixon[30] in Cambridge and by Kirkpatrick and Baez[31] in Stanford has been resuscitated with the advent of accessible sources of synchrotron radiation.

It has features that combine those associated with both light and electron microscopy. It is fast and relatively easy to accomplish (like light microscopy) and it produces high-resolution, absorption-based images (like electron microscopy). One of its supreme advantages is that it can examine whole, hydrated cells (at better than 35 nm resolution, currently) and reveal details of sub-cellular structures. In the energy range of the photons used, typically between the K-shell absorption edges of carbon (284 eV) and oxygen (542 eV), organic material absorbs approximately nine times as much as does water, thus producing a quantifiable natural contrast and dispensing with the need for contrast enhancement procedures (such as staining).

Computed tomography, which involves reconstructing projections of an object viewed from different directions (see Fig. 8),[32] derives from the mathematical principles first described by Radon.[33] A projection of an object at a given angle in real space is a central section through the Fourier transform of that object, the so-called "central slice" theorem.[34] The method of back projection, used in X-ray tomography (and in electron-tomography; see Sec. 6 below), is based on inverting the set of recorded images, projecting each image back into an object space at the angle at which the original image

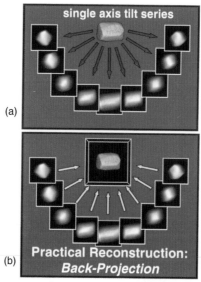

A schematic diagram of tomographic reconstruction using the back-projection method. In **(a)** a series of images are recorded at successive tilts. These images are back projected in **(b)** along their original tilt directions into a three-dimensional object space. The overlap of all of the back-projections will define the reconstructed object.

Fig. 8. Illustration of the principles employed in nanotomography, either by X-rays or electrons.[32]

was recorded. Using a sufficient number of back projections, from different angles, the superposition of all the back projected "rays" will reconstruct the original object (Fig. 8).[34,35]

The recent work of Larabell and LeGross[27,28] using X-ray tomography (with a set-up ascdepicted in Fig. 9, where a Fresnel zone plate lens is used

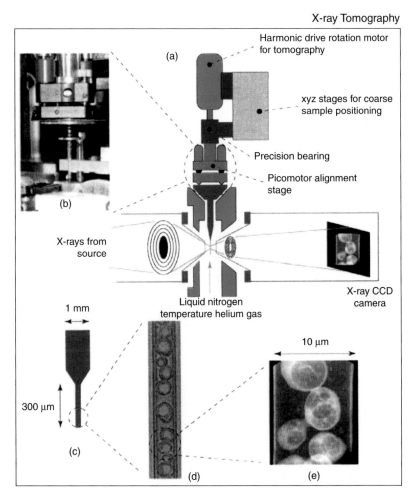

Fig. 9. Principles used at the (US) National Center for X-ray Tomography (NCXT) for the three-dimensional imaging of cells[27,28] (see also Fig. 10 and Ref. 38).

to focus the (monochromatic) X-rays from a synchrotron), has yielded detailed images of whole yeast cells (*Saccharomyces cerevisial*) such as those shown in Fig. 10. (For whole-cell imaging, the destructive effects of radiation-damage must be attenuated by cryo-cooling, as first demonstrated by Glaeser.[36]) These workers have also shown[28] how X-ray tomography has uncovered complex structural details of whole bacterial cells (*Escherichia coli*). Other workers using soft X-rays have recently achieved a spatial resolution of 15 nm.[37] In the hard X-ray region (typically 8 to 11 keV), because of the difficulty in fabricating the requisite zone-plate (Fresnel) lenses, such spatial resolution is not yet attainable, and values of around 60 nm are the best that can be achieved at present.[38]

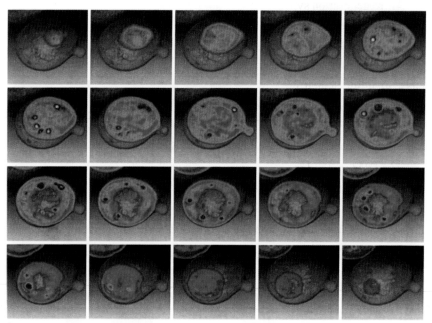

Computer-generated sections through the tomographic reconstruction of an early budding yeast. Structures have been assigned different colors, which indicate degree of X-ray absorption. Dense lipid droplets appear white and other cell structures are colored shades of blue, green, and orange with decreasing density. Yeast cell 5μm diameter.

Fig. 10. Low-resolution X-ray tomograms of early budding yeast cells.[27]

It is thought that, in the near future, work at places such as the (US) National Center for X-Ray Tomography (NCXT), will contribute greatly to cell biology because the achievable spatial resolution should be extended to between 10 and 15 nm; and the development and deployment of multiple wavelength (multicolor) X-ray probes will enable the simultaneous localization of multiple proteins to be undertaken.[28]

Further mechanisms of image contrast remain to be fully exploited in X-ray tomography. For example, phase-contrast imaging, in addition to (or in place of) absorption may be used.[38] Notwithstanding the technical difficulties encountered in fashioning more powerful zone-plate lenses, much is expected from work at NCXT and similar facilities (capable of cryo-fixed biotomography) in Japan, Germany and Taiwan.[39]

Third generation synchrotrons equipped with undulators open up further advances in X-ray diffraction microscopy (also termed *diffraction imaging microscopes*, DIMs). Because these facilities deliver intense sources of coherent electromagnetic radiation, they make it possible to achieve coherent X-ray imaging, which uses in-phase X-rays, thereby offering a "lensless" alternative to ordinary X-ray microscopy (that require zone-plates). This form of X-ray imaging was first suggested by Sayre[40] and first demonstrated at the Brookhaven Synchrotron in 1999 by Miao *et al.*[41]

The DIM method is, in principle, an optimal method for X-ray imaging, as there is no intensity loss of high spatial frequency information owing to a so-called lens contrast transfer function, designated CTF. (In effect, CTF tells us how much the real image is corrupted by an imperfect lens — see the discussion on "*Electron Microscopes*" below.) The theoretical resolution limit is close to the wavelength used to create the diffraction pattern. However, limitations concerning sample configuration (isolated particles are required) and long data acquisition times (many hours) render the technique one of rather low throughput as compared with what may be achieved by X-ray microscopy based on zone-plate imaging. As with the latter, cryogenic sample stages must be used to mitigate the deleterious effects of radiation damage inflicted during data collection.

The procedure (to view cells and sub-cellular entities) using DIM entails three steps:

o   A tilt series of diffraction patterns are generated using coherent X-rays, providing the amplitude of the diffracted wave field;

o   The phases of the wave field are obtained using variants of phase-retrieval algorithms[42] developed in other branches of optics; and

o   The series of tilt images are recorded by means of Fourier transformation of each individual (fixed angle) data set.

A major development was recently reported by Chapman *et al.*[43] who showed how to accomplish *ab initio* 3D X-ray diffraction microscopy using coherent X-rays. They achieved quite a high resolution in all three dimensions: 10 nm in *x* and *y* directions and 50 nm in *z*. Further improvements of this procedure are awaited with the advent of free-electron X-ray lasers.

4.1. A Possible New Means of Imaging Cells by Mass Spectrometry

The ability of mass spectrometry to generate intact biomolecular ions efficiently in the gas phase has led to widespread applications in proteomics[44] and, *inter alia,*[45] biological imaging.[46] The so-called MALDI approach (*matrix assisted laser desorption/ionization*) has been at the forefront of these developments. Very recently, Northern *et al.*[45] have introduced a new technique, known as NIMS (*nanostructure-initiator mass spectrometry*), which is a tool for spatially defined mass analysis. There is high hope that NIMS can be developed for biomedical applications such as fundamental studies on single cells. Currently, the lateral resolution of NIMS is about 150 nm. It is felt that ultimately, it will permit mass analysis of single cells.

5. Electron Microscopy: A Veritable Cornucopia of Techniques for Atomic and Extended Structural Imaging

In the physical sciences, encompassing most aspects of engineering, nanotechnology and many facets of earth science, the transmission electron microscope (TEM) and the scanning transmission electron microscope (STEM) are indispensable tools, especially because most of the specimens investigated in these fields are rather beam-insensitive: they withstand the radiation dose of the electron beam without undergoing irrevocable destruction. But nowadays the most advanced methods of visualizing bio-macromolecular species, viruses and cells, even though they are extremely beam sensitive, also rely heavily on both TEM and STEM. In recent years, enormous advances have been made on three fronts for both TEM and STEM studies. There have been major technical developments in:

o   The instruments themselves, which now achieve greater resolution than hitherto;

o   Specimen preparation, particularly of biological materials which has made it routinely possible to study unstained samples, thanks to cryoelctron microscopy; and

o   Image analysis and associated computational procedures have been transformed, thus facilitating the retrieval of structural information down to less than 0.4 nm resolution.

Also, very recently there have been revolutionary advances in ultrafast (time-resolved) electron microscopy and electron crystallography, which are described later in Sec. 8.

In Fig. 11 are shown two schematic sketches of the essence of STEM flanking the center-piece, which summarizes the essence of TEM. Even a cursory glance at the signals identified at the extreme left of this figure shows the extent of the wealth of information that may be retrieved using scanning transmission electron microscopy. And this does not include the extra, and vitally important, information concerning visualization that comes from introducing tomographic analysis (that entails systematically tilting the specimen in small increments about an axis perpendicular to the direction of the incoming electron beam, just as with X-ray tomography, described above).

Also shown in the lower half of Fig. 11 are the contrast transfer functions (CTFs) of two distinct conditions of spherical aberration constant, C_s. It is seen that when C_s values are relatively high, the interpretable spatial frequency detectable (and imageable) with such a microscope goes down to no lower than 0.19 nm, whereas, with an aberration-corrected electron microscope ($C_s \leq 0.1$ mm), the interpretable spacing in the imaged object reaches as far down as 0.08 nm. (The oscillations in the 1D curve denoting the CTF signifies the change of phase with decreasing separation distance.)

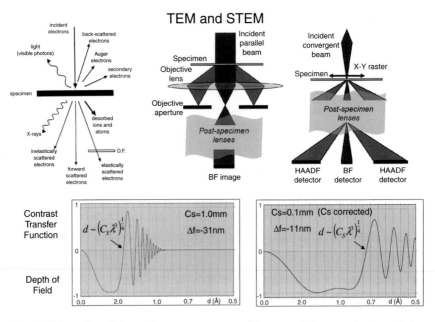

Fig. 11. Schematics of the various arrangements of lenses and detectors in transmission electron microscopy (TEM) and scanning transmission electron microscopy (STEM) (*top*); and the nature of the contrast transfer function (CTF) for lenses of two different values of the coefficient of spherical aberration (Cs) (*bottom*). See text for more details.

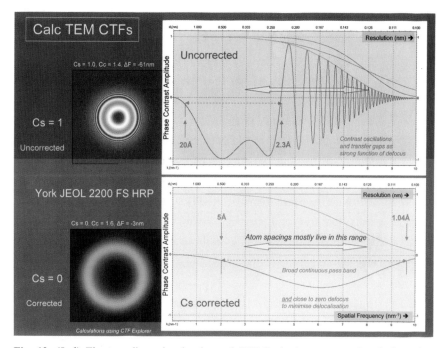

Fig. 12. (*Left*) The two-dimensional values of CTF for both uncorrected and aberration corrected objective lens and (*right*) the respective values of the one-dimensional CTFs.

Figure 12 shows (on left) 2D representations of CTFs for the York University[47] aberration-corrected electron microscope; and the accompanying 1D CFT plots shows how the "corrected" lens leads to feasible imaging of spacings in the object down to 0.104 nm.

Electron optical lenses are far from perfect; consequently, the recorded image is corrupted by the spherical aberration of the lens; and that is why as stated earlier, it is necessary to talk of a contrast transfer function (CTF). In a TEM, information is recorded according to an approximately sinusoidal CTF, which modulates the relative contributions of different spatial frequencies to the image. The CTFs shown in Fig. 11 are damped as a result of the contribution of chromatic aberration. The positions of the first cross-over of these curves yield the values of the interpretable resolution, D, which is given by:

$$D \sim (C_s \lambda^3)^{1/4}$$

This immediately tells us that, as the electron wavelength λ decreases (i.e., as the accelerating voltage increases), the resolution is improved, as is also the case if the coefficient of spherical aberration is reduced. Because of the large depth of the field of the electron microscope as compared with the size of many of the objects under investigation (notably viruses), all the features at different heights in the object are simultaneously in focus and the recorded 2D image is a projection of the 3D scattering matter in the object. In other words, features at the top and bottom and at all levels in between become superimposed on the image and are therefore difficult to unravel. To get to the 3D structure of the object, one has to look at the specimen from different directions; and this can be done either by tilting the object inside the microscope, or (especially for viruses) by using particles setting in different orientations on the grid. The whole problem of unravelling the projected data from the object was solved by DeRosier and Klug[48] nearly 40 years ago, who exploited the mathematical device of Fourier transform, which splits the density in the image into all its 2D spatial frequency components. The precise way in which this is done is described in Crowther's recent Leeuwenhoek Lecture[49] of 2006, to which we return below.

Thorough methods of coping with the problems arising from such complications as CTF and variation of defocus are continually being explored, especially by exponents of biological TEM. Table 1 summarizes the essence of two recent important papers.

We should note some other key features of Fig. 11. Bright field images (BF) are recorded via the forward scattered beam. High-angle annular dark field (HAADF) ones from those beams that experience large angles of scatter, such that they obey Rutherford's law, not Bragg's. In other words, the intensity of the scattered electrons is proportional to the square (very nearly) of the atomic number of the atoms that cause the scattering. This means that heavy-atom species distributed on a light-atom support

Table 1. Two Recent Publications Highlighting the Importance of the Contrast Transfer Function (CTF) in Macromolecular Imaging

1	Jensen GJ, Kornberg RD. (2000) Defocus-gradient corrected back projection *Ultramicroscopy* **84**: 57. 3D reconstructions of icosahedral viruses from cryo-electron microscope images have reached resolutions where the microscope depth of field is a significant resolution-limiting factor.
2	Fernandez JJ, Li S, Crowther RA. (2006) CTF determination and correction in electron cryo-tomography. *Ultramicroscopy* **106**: 587. This work addresses the determination and correction of the CTF of the electron microscope in cryo electron tomography.

(e.g., Au in biological matrices or Pt on SiO_2) stand out very clearly in HAADF images.

It is also to be noted that the purpose of the objective lens in TEM is to serve as a Fourier transformer as shown in the ray diagram (Fig. 13).

Another important difference between TEM and STEM imaging concerns the simple question of minimizing beam-damage because, with STEM, unlike TEM, the energy that the beam endows to the specimen is more readily dissipated, as schematized in Fig. 14.

Figure 12 also shows that electron-energy-loss spectroscopy (and imaging) may also be carried out with both STEM and TEM instruments. And electron tomography, involving the recording and processing of data as symbolized in Fig. 8, may also be carried out. Because of the restrictions imposed by the dimensions of the pole-pieces of the microscope, a "missing wedge" of data inevitably occurs.[32]

5.1. Imaging of Extended Inorganic Structures

TEM has been used successfully to determine, at atomic resolution, a large number of inorganic materials, especially metals and alloys, semiconductors and minerals, which, in general, are not beam-sensitive. A striking example is the powerful, shape selective siliceous catalyst, ZSM-5[50,51] (see Fig. 15).

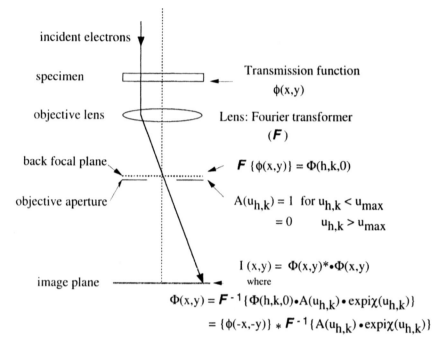

The objective lens of a microscope serves as a Fourier transformer. The diffraction pattern formed at the back focal plane is further Fourier transformed to yield the image. There are phase changes in the electron waves for (h,k) reflections (h and k being the indices in reciprocal space) as a function of wave vectors u(h,k). The phase contrast imaging performance of an HREM is governed by the contrast transfer function χ/u(h,k), which constains the spherical aberration coefficient of the lens and the extent of its defocus. A(u(h,k)) is an objective lens transfer function: the intensityI (x,y) at the image plane is proportional to the projected electrostatic potential density in the "weak-phase" approximation. Φ(x,y) yields the structure in real space (the asterisk signifies convolution).

Fig. 13. Ray diagram for a TEM set up that shows the objective lens to be a Fourier transformer. The diffraction pattern formed at the back focal plane of the lens is further Fourier transformed to yield the magnified image of the object. (From Thomas, Terasaki, Gai, Zhou, Gonzalez-Calbet. (2001) *Acc Chem Res* **34**: 583.)

Computational procedures worked out by Moodie and Cowley (known as the multi-slice method)[52] enable matching of images observed for given specimen thickness, under well-defined settings of defocus of the lens, its spherical aberration and the accelerating voltage and that calculated on the basis of a given structural model. Many inorganic solids have had their

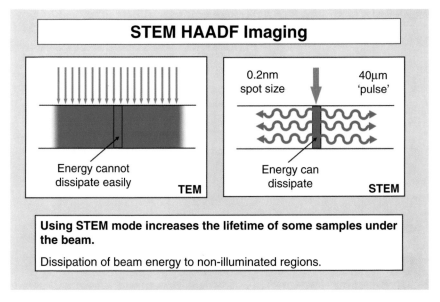

Fig. 14. Because of its mode of operation, a STEM instrument inflicts less beam damage than its TEM counterpart.

structures determined in this way.[52–54] That of the Bronsted acid catalyst ZSM-5 $(H_n^+(H_2O))_{16}[Al_nSi_{96-n}O_{192}]$ $(n < 27)$ is one such example (Fig. 15). The ray diagram shown in Fig. 13 helps us to understand the principles involved; the key point to note is that the objective lens of a TEM is effectively a Fourier transformer, which produces the diffraction pattern (in reciprocal space) at its back focal plane. This, in turn, is further Fourier transformed to yield the magnified (real space) image of the specimen.

Also shown in Fig. 15 is the same SiO_2 open-structure framework (bottom right) where a few Si^{4+} tetrahedral sites have been tenanted by Ti^{4+} ions to yield the powerful redox(oxidation) catalyst known as TS-1.[55] Long before single crystal X-ray structural analysis of ZSM-5 became possible (owing to lack of single crystals of adequate size), a combination of HREM[50,51] and ^{29}Si magic-angle-spinning NMR,[56] which can resolve crystallographically distinct tetrahedral sites in silicalite (the pure SiO_2 end-

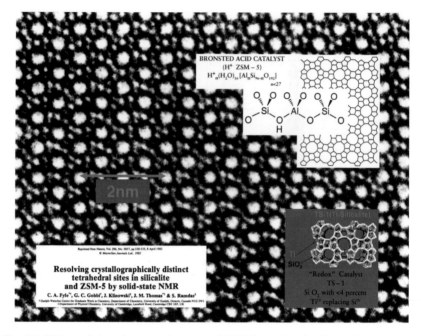

Fig. 15. High-resolution-electron microscopy (HREM) combined with magic-angle-solid-state NMR, yielded the structure of the microporous silica known as silicalite-I before it was determinded by X-ray single crystal diffractometry. When some Si^{4+} ions in silicalite are isomorphously substituted by[34] Al^{3+} ions, a Bronsted solid acid catalyst, known as ZSM-5 is produced. When a few Si^{4+} ions are replaced by Ti^{4+} ions, a redox catalyst, known as TS-1 is produced[56] (see Fyfe CA, Gobbi GC, Klinowski J, Thomas JM, Ramdas S. (1982) *Nature* **296**: 530.)

member of ZSM-5 and TS-1), had yielded the structure of this important zeolitic catalyst.

Another example of an extended inorganic structure, where "local" crystallography (via HREM) elucidates the nature of polymorphic inter-growths and imperfections, is contained in the recent work of Wright and Zhou *et al.*,[57,58] where the change in stacking order is readily seen (Fig. 16). It is to be emphasized that, whereas single-crystal X-ray crystallography generally requires samples that contain some 10^{15} unit cells or more — over which the evaluated structure is spatially averaged — the merit of structure

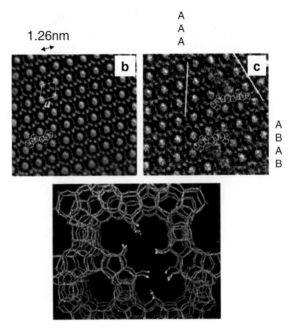

Fig. 16. HREM can readily identify local structure, and the co-existence of distinct crystallographic phases as in this case, with the microporous catalyst known as zeolite beta.[57,58]

determination by HREM is that only some 10^5 (or even fewer) unit cells are involved in the structure determination.

In Fig. 17 is shown the structure determined by Terasaki (see Ref. 54) of the mesoporous silica known as MCM-48, where it transpires that there are branching "cylinders" of SiO_2 of diameter 2 to 4 nm.[58] The method employed for doing so entailed using the amplitudes from the intensities of the electron-diffraction spots and the extraction of the phases of the corresponding reflections from the high-resolution image. As is well known, phases are more important for solving the structure, whereas amplitudes are more important for refining it. In the MCM-48 structure, solved in this fashion, the pores and channels possess crystallographic order, but the SiO_2 framework itself is amorphous, as revealed by ^{29}Si magic-angle-spinning

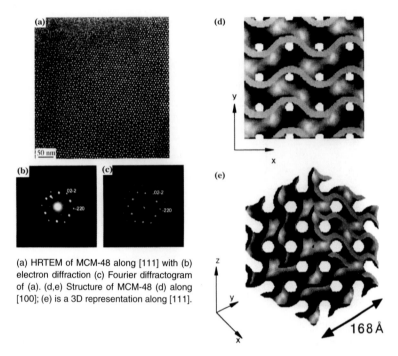

(a) HRTEM of MCM-48 along [111] with (b) electron diffraction (c) Fourier diffractogram of (a). (d,e) Structure of MCM-48 (d) along [100]; (e) is a 3D representation along [111].

Fig. 17. Electron-crystallography (see text) was used by Terasaki to solve the structure of a mesoporous silica known as MCM-48, where the channels are well ordered, but like the silica framework separating them, is amorphous.[54]

NMR.[59] The technique of electron crystallography used by Terasaki *et al.*, has its origins in the classic work of Klug[60] Henderson and Unwin,[61] Kornberg[62] and others.

The chain silicates known as pyroxenoids (general formula $MSiO_3$ ($M \cong Ca, Mg, Mn, Fe$)) are composed of corner-sharing SiO_4 tetrahedra and differ from one another only in the repeat distances along the chain length, within which there are different degrees of tilts of the SiO_4. The archetypal member is wollastonite[63,64] ($CaSiO_3$), and other well-known members are rhodonite ($MgSiO_3$) and pyroxmangite ($FeSiO_3$);[65] all these are very stable in an electron beam. In 1973, when I started my HREM studies, only 100 KV electron microscopes were commercially available, and their

Imaging Wollastonite (CaSiO₃)

1973   1979   2007

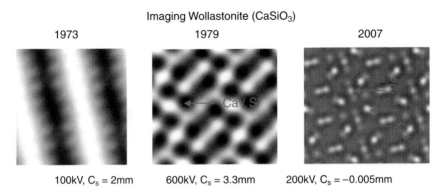

100kV, C_s = 2mm   600kV, C_s = 3.3mm   200kV, C_s = −0.005mm

Fig. 18. (Computed) comparisons, made by AI Kirkland, of how the pyroxenoid mineral, wollastonite, would appear in TEM with a large spherical aberration (2 mm) in a 100 KV instrument, vintage 1973, (*left*, and with a 3.3 mm coefficient in a 600 KV instrument (*center*)). The merits of having aberration correction (*right*) such that, with the appropriate orientation, individual SiO_4 tetrahedra and discrete ions (Ca^{2+}) should become readily visible.[66]

spherical aberration was quite substantial, typically 2 mm. In subsequent years there was a strong trend to improve microscope resolution by utilizing high voltages, the 600 KV instrument in Cambridge (UK) being a case in point. There was indeed an appreciable improvement in resolution; and it was just about possible to discern discrete SiO_4 tetrahedra (in projection) with such instruments. The first two images in Fig. 18 show those calculated for wollastonite (with the assumption of absence of Shott noise) from which the position of the superimposed Ca^{2+} and SiO_4^{2-} ions are clearly detectable. The third image is that computed (also using the multi-slice procedure[52]) and it shows that, with an aberration-corrected electron microscope operated at 200 KV, not only do the positions of the Ca, and Si atoms stand out clearly, so also do those of the oxygens in the SiO_4 tetrahedra. (Work is in progress[66] to demonstrate the validity of these computations and to open up more detailed studies into stratigraphic variations of inorganic structures at unit-cell, and sub-unit cell level).

5.2. Imaging of Biological Molecules

Here we are concerned with visualization from cells to molecules (see Fig. 19). With cells, both optical microscopy and X-ray microscopy, as outlined above, are well-recognized techniques. But to proceed to higher resolution, cryoelectronmicroscopy is the method of choice, following the pioneering work of several investigators, especially Dubochet *et al.*,[67] who showed the merits of "preserving" the structures of virus and related materials in rapidly frozen amorphous ice. The potential of pursuing electron cryomicroscopy has been adequately described by others — see, for example, Henderson,[68] Wah Chiu *et al.*,[69] and Crowther[70] — and it is exploited effectively at the National Center for Macromolecular Imaging in the Baylor Medical College, under the direction of Wah Chiu. There are many other centers worldwide, not least that located at the Laboratory of

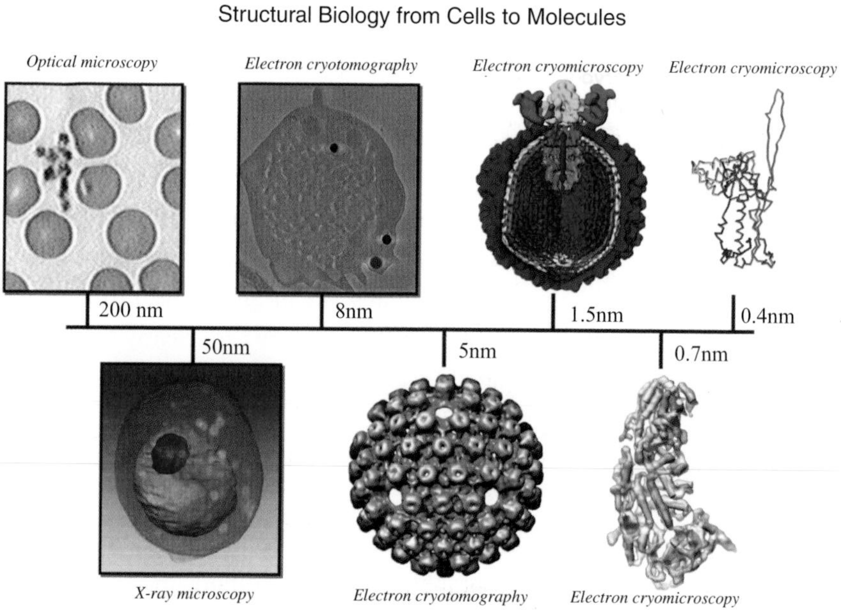

Structural Biology from Cells to Molecules

Fig. 19. Landmarks in the Quest to visualize cells and their macromolecular constituents. (After Wah Chi, with acknowledgement.)

Molecular Biology at Cambridge, UK; and certain aspects of the work have been admirably summarized recently by Crowther.[70] The "pipeline" used in biological cryoelectron microscopy is shown in Fig. 20, which is to be compared with the procedure described by Crowther,[70] as shown in Fig. 21.

Advances in electron cryoelectronmicroscopy (cryoEM) have made possible the structural determination of large biological machines at a resolution range of 0.6 to 0.9 nm. The detailed computations and visualization methodologies necessary for the so-called "structural mining" of the computed cryoEM maps of these sub-cellular machines have been recently reviewed by Wah Chiu *et al.*[71] As a consequence of the detailed analysis that have been worked out, some of the secondary structure elements can be accurately identified and, in some instances, structural mechanisms may be formulated that relate to their biological function. In Fig. 20, the presence of 12 proteins in the GroEL (that was imaged) stands out clearly as do the

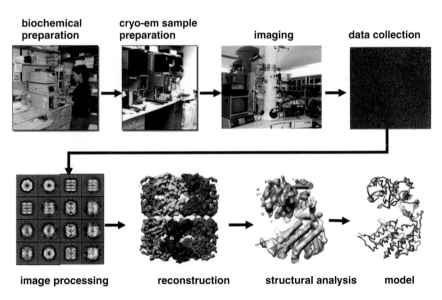

biochemical preparation **cryo-em sample preparation** **imaging** **data collection**

image processing **reconstruction** **structural analysis** **model**

Fig. 20. The "pipeline" in biological cryoEM — see text. (After Wah Chiu, National Center for Macromolecular Imaging.)

α-helices and β-sheets of the protein. It is noteworthy that the combination of cryoEM to study large biological assemblies at low resolution with X-ray crystallography to determine near atomic structures of assembly fragments is now fast expanding the horizon of structural biology.[72]

RA Crowther, whose work was responsible for the first structural elucidation of a virus (the bushy stunt),[73] in his recent Leeuwenhoek lecture,[70] has elaborated further the manner in which cryoEM is nowadays

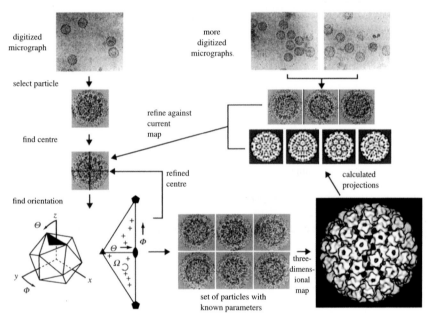

Outline of the scheme for processing the images of virus particles. The micrograph has first to be digitized on a film scanner to convert the blackness of the film into an array of numbers suitable for processing in a computer. Particles are selected and boxed out, and approximate centers and orientations relative to the icosahedral symmetry axes are found by self-common lines (Crowther 1971). A preliminary three-dimensional map is calculated from an initial set of images. From the map, a set of projections can be calculated, effectively simulating particle images but with a greatly reduced noise level. This set of projections provides a basis for determining by cross-common lines the centers and orientations of additional particles selected from further micrographs (Crowther et al. 1994). The whole process is iterated, increasing the number of particles included, refining their parameters more accurately and increasing the resolution of three-dimensional map.

Fig. 21. The "single particle" approach to the solution of virus structures by electron microscopy. (After Crowther.[70])

used to image virus particles. One does not start with a crystal lattice of virus particles condensed into an ordered solid; instead, one proceeds by the so-called "single-particle" method, now almost universally adopted in this field.[74–76] Figure 21 explains the essence of the procedure; and Fig. 22 shows a map of the recently determined hepatitis B core protein, which was completed from *ca* 6300 particle images from 34 electron micrographs. Figure 23, taken from the laboratory of Wah Chiu in Houston, shows both cryoEM single-particle images of the infectious P22 virus and a map of its structure.[77] Table 2, also taken from the work of Wah Chiu, summarizes the information pertaining to viruses that is retrievable from the appropriate cryoEM images.

Table 2. Summary (After Chiu W[71,77]) of the Kind of Information that Cryoelectron Microscopy Yields

It can:
Reveal all molecular components of a spherical virus;
Trace the backbone of a virus capsid protein;
Reveal conformational changes of capsid particles in a maturation process;
It transpires that all double-stranded DNA viruses have similar but distinct structural details and are probably critical for respective host infection.

CryoEM is an indispensable method for capturing the nature of the molecular components of phages, a fact well represented by Fig. 24 (of the epsilon 15 bacteriophage).[69]

Only relatively few 2D crystals of biological molecules have so far yielded to electron microscope structural elucidation. Four molecular structures for which atomic models were determined, at resolutions of 0.4 nm or better are enumerated in Table 3, taken from a review[68] by Henderson (entitled "Realizing the Potential of CryoEM"). The methods used were

J.M. Thomas

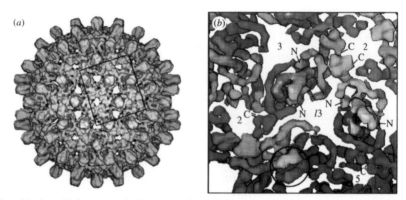

Map of the hepatitis B core protein (Böttcher et al. 1997). (a) View of the whole shell. Map computed from approximately 6300 particle images from 34 micrographs. (b) Enlarged view of the boxed area indicated in (a). This form of the shell is made from 240 copies of the core protein, so after icosahedral averaging there are four computationally independent subunits, which are here colored blue, red, yellow and green. The red–blue and yellow–green molecules form closely associated dimers. The positions of the icosaheral twofold, threefold, fivefold and local threefold axes are indicated (2, 3, 5, I3) as are the positions of the N- and C-termini of the protein. The fold of the four subunits is identical, and the thicker tubular regions correspond to α-helices. Bundles of four helices (an example is encircled) form the spikes protruding from the surface.

Fig. 22. The hepatitis B core protein. (After Crowther.[70])

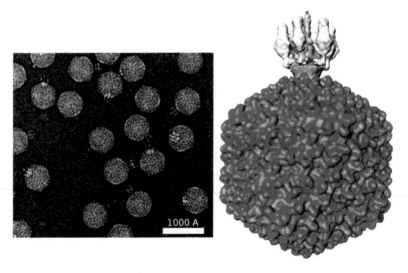

Fig. 23. Image of the infectious P22 virus derived by cryo-electronmicroscopy (single particle method) from Chang *et al.*[77] (Courtesy, Wah Chiu.)

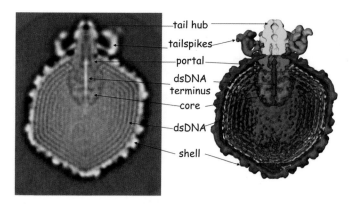

Fig. 24. The molecular components of the ε15 bacteriophage structure derived from cryoelectronmicroscopy. (After Wah Chiu *et al.*[69])

first worked out on bacteriorhododopsin.[78] Several others, including helical structures have also been determined at lower resolutions (*ca* 0.5 to 0.8 nm). Apart from those biological materials listed in Table 3, it is relevant to note that Kühlbrandt and Neg Wang determined[79] the 3D structure of a plant light-harvesting complex by electron crystallography at 0.6 nm resolution; and that even earlier Kornberg and Uzgires[80] showed the ease of forming 2D crystals of antibodies and their suitability for image analysis by electron microscopy on lipid monolayers. Their technique is applicable to other large biological molecules that are, like antibodies, difficult to crystallize in three dimensions. Kornberg *et al.*[81] grew 2D crystals of yeast RNA polymerase preserved in vitreous ice and determined the structure by both electron crystallography and single-particle techniques.

In closing this section, it is relevant to emphasize that the key factor governing the structural information retrievable in any electron micrograph of biological macromolecules is radiation damage. Whereas the degree of electron-beam damage is almost infinitesimal in the case of metals, semiconductors, most oxidic and certain other inorganic structures, biological macromolecules are extremely vulnerable to radiation damage. The (desirable) elastic scattering of electrons by the atoms making up a biological

Table 3. Structural Analysis by Electron Microscopy of a Number of Biological Macromolecules or Molecular Assembles (after Henderson[68])

Structure	Resolution (nm)
2D Crystals (High resolution, better than 0.4 nm)	
Bacteriorhodopsin	0.35, 0.30
Plant LHC II	0.34
Tubulin dimer	0.37
Aquaporin	0.38, 0.40
2D Crystals (Low resolution; better than 0.9 nm)	
Deoxycholate bacteriorhodopsin	0.60
Halorhodopsin	0.50
Porin PhOE	0.60
Plant photosystem II RC	0.80
Glutathione transferase	0.60
Glycerol channel GlpF	0.69
EmrM multidrug transporter	0.70
Helical Structures	
Acetychlorine receptor	0.40
Bacterial flagellum	0.40
Microtubule	0.80
Calcium A TPase	0.80
Tobacco mosaic virus	1.00

structure is accompanied by a multiplicity of inelastic interactions — of which we may take considerable advantage (see Sec. 7) — which deposit energy in the covalent bonds which hold the macromolecules together. Each C–C single bond in a protein, for example, upon rupture, suffers a change of bond distance that increases from 0.15 to 0.34 nm (the latter being the van der Waals distance) with concomitant generation of free radicals which, in turn, set off a cascade of other reactions that produce a variety of compounds very different from the original specimen under investigation. These radiolytic products, which are very similar whether the incident radiation comes from electrons, X-rays or neutrons,[68] include small, mobile fragments as well as a few with additional cross-links. The smaller fragments can be restrained from moving from their initial location by cooling the specimen to low temperatures; and the lower the temperature employed, the more effectively are these small species immobilized. It follows that keeping the specimen at the lowest temperature possible, in principle, ensures that the structure of the radiation-damaged specimen will retain its similarity to the pristine starting sample to the greatest degree. As Henderson[68] and others[82] have pointed out, the choice is between specimens cooled to near liquid N_2 or those cooled to near liquid He temperatures. Use of liquid He-cooled specimens requires greater care owing to the decreased conductivity of the carbon support films (and this causes charging) and the lower latent heat of evaporation of He results in boiling, vibration of the specimen support stage and hence a lower resolution in the image. Notwithstanding these non-trivial practical considerations, by paying great attention to these problems Baumeister and Unwin and their respective colleagues, in particular, have demonstrated the great merit of operating the imaging conditions at liquid He temperatures. This has been especially important in the pioneering electron tomography of biological machines (see Sec. 6).

6. Electron Tomography (ET)

Although ET was first demonstrated nearly 40 years ago, only recently has it begun to take center-stage, in the study of both inorganic and biological materials. For a short review covering both sub-disciplines, see Ref. 32.

6.1. Electron Tomography in the Chemical and Materials Sciences

The general principles outlined above for X-ray tomography are essentially the same (see Fig. 8) as those that apply in ET. Both bright field (BF) and dark field (DF) as well as HAADF conditions have all been used in ET, the

BF conditions being utilized predominantly in the biological sciences. In the materials and chemical sciences, HAADF–STEM tomography has proven to be particularly informative. Because high-angle scattering is largely an incoherent process, unlike conventional BF and DF scattering where coherence reigns, images formed under HAADF–STEM conditions do not suffer from confusing contrast changes like diffraction contrast that are associated with coherent scattering. The sensitivity to composition (because of the obedience to Rutherford scattering and its near Z^2 dependence), together with the incoherence of the scattering process, enables small heavy metal clusters or nanoparticles to be made highly visible within a matrix or support such as the situation that prevails when individual nanocluster catalysts,[83] like $Ru_{10}Pt_2$, are dispersed within the interior walls of a mesoporous silica support. Here (see Fig. 25), a TEM–BF image fails to detect the extensive presence of minute clusters of the $Ru_{10}Pt_{22}$ nanocatalysts (for selective hydrogenerations[84]), whereas their pressure is readily revealed in STEM–HAADF. The precise distribution, shape and composition (determined by electron-stimulated X-ray emission within the electron microscope) of $Ru_{10}Pt_2$ cluster catalysts in the mesoporous silica, known as MCM-41, is shown in Fig. 26.[85,86] ET is an invaluable technique for catalyst characterization in that it pinpoints the exact location of indivdual clusters of nanocatalysts.[87]

Given that there is now considerable interest in using both colloidal Au (consisting of some 2- to 4 nm diameter particles) as well as well-defined Au clusters, such as Au_{11} and Au_{67}, for locating the sulphur-rich regions of proteins in biological systems — see the recent work of Leapman *et al.*[88] — the importance of using HAADF–STEM conditions for ET work cannot be overemphasized. And the recent use[89] of Au nanoparticles as vectors to deliver tumor necrosis factors (TNFs) in the administration of drugs in cancer therapy also indicates that, rather than using TEM–BF imaging in ET, traditionally favored by biologists, HAADF–STEM conditions should be employed.

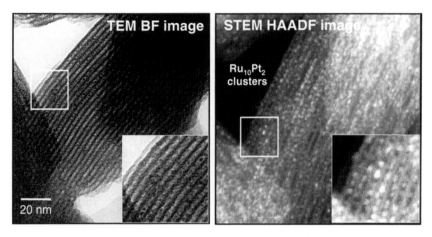

Fig. 25. The superiority of STEM HAADF over TEMBF for detecting nanoparticles and clusters of metals or bimetals is illustrated here. (From John Meurig Thomas, PA Midgley *et al.* (2004) The chemical application of high-resolution electron tomography: bright-field or dark field? *Angew Chemie Intl Ed* **43:** 6745.) (See also text.[86])

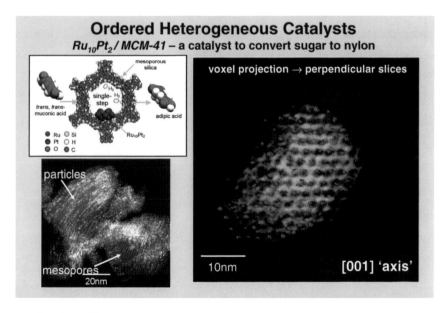

Fig. 26. Bimetallic clusters of $Ru_{10}Pt_2$ (hydrogenation catalysts) are readily visible in dark-field STEM tomograms.[85]

6.2. Electron Tomography in the Biological Sciences

The main merit of ET in the biological sciences is that it bridges the gap between the information attainable at an atomic level by X-ray crystallography and that provided by the dynamic observations of light microscopy (using mainly fluorescence excitation) with a spatial resolution of hundreds of nanometers. ET does much better than X-ray tomography as far as resolution is concerned; and the work of the leading practitioners in this field, notably Baumeister,[82] McEwan and Marks,[90] and Marsh and McIntosh et al.,[91] underscores the great value of ET in providing local information from "disordered" systems, especially from a variety of biological machines that operate in the various regions within the living cell. ET is able to supply 3D information with a resolution of 4 to 6 nm, once the difficulties with sample preparation, and beam damage have been overcome. Baumeister, in particular, has expatiated on all the requirements of cryo-ET and ways of operating liquid He-cooled microscopes, in such a manner as to minimize radiation damage and so as to maximize the efficiency of the computational processing required to achieve the desired ends. This is usually done with biological specimens maintained in a frozen-hydrated state with the rest of the constituents immobilized in a native configuration, and contributing also to increased stability under the electron beam.

Baumeister's approach (in using the frozen-hydrated specimens) has enabled him to characterize not only whole cells but also different organelles and their components,[82,92] which include the architecture of eukaryotic cells and nuclear pore complexes (NPCs)[93] from *Dictyostelium discoideum* (see Fig. 27). The approach of Marsh *et al.* to sample preparation is different from that of the Baumeister group. They apply a high pressure (*ca* 2100 bar) during a 10- to 20 ms period before cooling the biological material in liquid N_2. Figure 28 shows the organellar organization, and the compactness of the packing of the various sub-cellular species surrounding the Golgi region.[90]

Compilations of tomographic 3D reconstructions of cells and cellular structures are given in Refs. 90 to 96, the last three papers focussing on the architectural features of retrovirus envelope protein, vaccinia virus and flagella, respectively. Chromatin fibers, bacterial surface layers, membrane pores and mitochondria are further examples that have had their structures elucidated by cryo-ET.

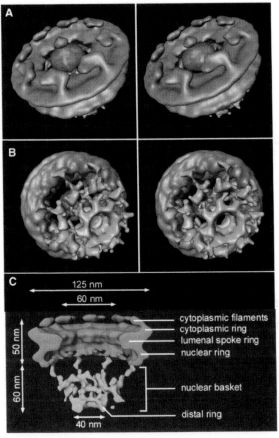

Structure of the Dictyostelium NPC. **(A)** Cytoplasmic face of the NPC in stereo view. The cytoplasmic filaments are arranged around the central channel; they are kinked and point toward the CP/T. **(B)** Nuclear face of the NPC in stereo view. The distal ring of the basket is connected to the nuclear ring by the nuclear filaments. **(C)** Cutaway view of the NPC with the CP/T removed. The dimensions of the main features are indicated. All views are surface-rendered (nuclear basket in brown).

Fig. 27. Structure of *Dictyostelium* nuclear core protein determined by electron tomography (ET) by Baumeister *et al.*[82]

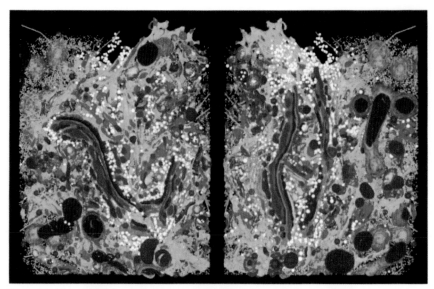

3D model of the organellar organization surrounding the Golgi region of pancreatic β cells shown in two views rotated wity respect to each other around a vertical axis. The model was derived from a tomogram of a thin section of plastic-embedded freeze-substituted cells. The Golgi complex with seven cisterna (C1–C7) is at the center. The color coding is as follows: C1 in light blue; C2 in pink; C3 in cherry red; C4 in green; C5 in dark blue; C6 in gold; C7 in bright red. The other organelles are color coded as follows: ER in yellow; membrane-bound ribosomes in blue; free ribosomes in orange; microtubules in bright green; dense core vesicles in bright blue; clathrin-negative vesicles in white; clathrin-positive compartments and vesicles in purple; mitochondria in dark green. The width across each panel is 3.25 μm. Reproduced from [15••] with permission.

Fig. 28. Virus of the organellar organization, from ET, surrounding the Golgi region of pancreatic β shells. (Ater Marsh and McIntosh *et al.*[91])

7. Electron-Energy-Loss Spectroscopy (EELS) and Imaging

The great merit of EELS and electron-loss imaging and energy-filtered TEM (EFTEM), is the wealth of compositional and structural information it yields, as is apparent in Figs. 29 and 30. Valence states, bond distances, and composition may all, under ideal conditions, be retrieved from the kind of spectra shown in Fig. 29.[53,97–99] Even plasmon-loss spectroscopy[100] and the Compton scattering envelopes[101] can access information that is well-

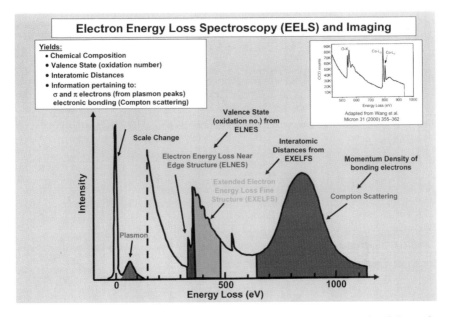

Fig. 29. Electron energy loss spectroscopy (EELS) yields a wealth of structural and electronic information, summarized here[97–99] (see RD Leapman, *et al.* (2004) *Curr Opin Neurobiol* **14:** 591; RF Egerton (2001) *Electron Energy Loss Spectroscopy in the Transmission Electron Microscope*, Plenum Press.)

nigh inaccessible by other means, such as, respectively, the composition of alkali-metal alloys and the electronic bonding in amorphous carbon.

One of the most eloquent demonstrations of the analytical power of EFTEM is seen in Fig. 31, where a sequence of EELS spectra is recorded along a line (shown dashed at top left), where the zero-energy loss intensity was also measured. This example involves a minute cluster that has nucleated inside a specimen of "304" stainless steel. From the resulting energy landscape depicted in Fig. 31, where six distinct elemental components are identified, the precise composition of (and the spatial variation within) the minute precipitate may be determined.[102]

By extending this methodology further, my colleague Paul Midgley has devised[103] an ingenious technique that may be labeled "volume

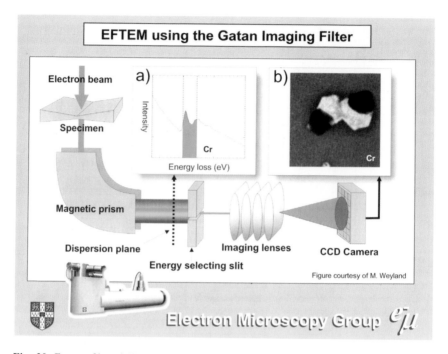

Fig. 30. Energy filtered (EF)TEM is carried out using a so-called Gatan Imaging Filter, schematized here.

spectroscopy," whereby the composition of a minute volume in the interior of a specimen may be non-destructively and non-invasively determined down to the sub-attogram (10^{-18} g) level. This new method of quantitative analysis combines EFTEM measurements at every tilt angle that is involved in the concomitant energy-loss tomography (see inset at top left of Fig. 32). Midgley *et al.*[103] have taken advantage of this elegant approach in the nanochemical characterization of nylon-multiple-wall carbon nanostructures.

Although the electron-beam susceptibility of biological macromolecules will not allow the degree of spatial resolution achievable with inorganic systems,[104] EELS and EFTEM have an important role to play in cellular biology, as may be gleaned from, *inter alia*, the work of Leapman *et al.*[105] who mapped phosphorous-containing ribosomes within a cell (see Fig. 33).

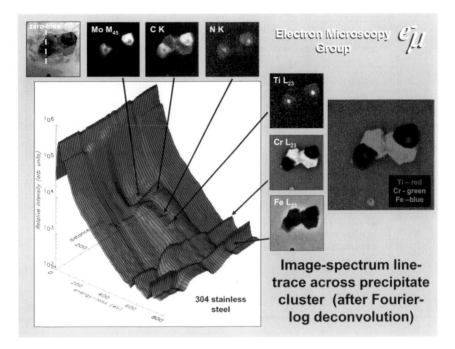

Fig. 31. A line trace (*top left*) through a minute precipitate in a stainless steel specimen is the starting point in constructing this "landscape" which is composed of EELS spectra at discrete points along the (zero-loss) line. From the EELS spectra, traces of Mo, C, N, Ti, Cr and Fe are detected. (After Midgley and Thomas.[102])

There is no discernable $PL_{2;3}$ EELS signal for the cytoplasm: but there is for the ribosome. Even though the attainable resolutions are modest, the location of individual ribosomes at different heights within sections of *C. elegans* cell may be unmistakeably identified.

A combination of EFTEM and TEM electron tomography could yield further insights into cellular structures and compositions. For example:

o The total distribution of biomolecules, including proteins and nucleic acids, would emerge from *nitrogen* EELS signals;

o The proteins that contain high levels of the amino acids cysteine and methionine from *sulfur* signals; and

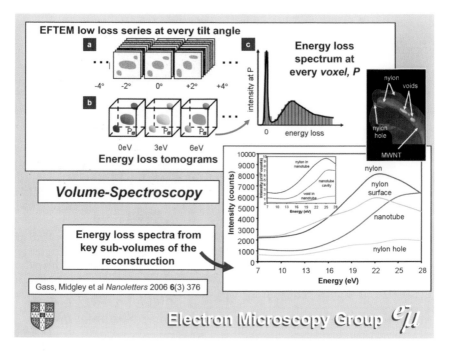

Fig. 32. By recording both a series of frames for tomographic analysis, and an EELS spectrum at each angle of tilt, Midgley *et al.*[102] has introduced a method of chemical analysis known as volume spectroscopy. This enables the composition of sub-attogram quantities of the specimen to be determined by non-destructive methods.

○ The distribution of nucleic acids, phosphorylated proteins and phospholipids from *phosphorus* signals.

This methodology could be further extended for the detailed 3D mapping of certain key inorganic elements within the cells. The importance of potassium can hardly be overemphasized; and the growing realization that RNA folding is strongly dependent on the concentration of Mg^{2+} ions, makes this an important determinant. Bustamante and Tinoco *et al.*[108] have shown, for example, that a stable ribozyme does not form in low Mg^{2+} concentrations.

In the cell nucleus, the relative concentrations of phospholipids and phosphoproteins are low enough for the phosphorous distribution to provide

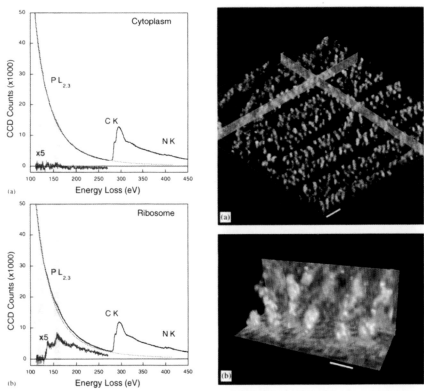

(a) EELS from 10nm diameter region of cytoplasm from cell in C. elegans, obtained from STEM spectrum-image. (b) EELS from 10nm diameter region of single ribosome. No significant phosphorus is detected in the cytoplasm, whereas a clear phosphorus signal is evident in the spectrum from the ribosome with a P $L_{2,3}$ signal-to-background fraction of approximate 0.1 at an energy loss of 160 eV. The background is fitted from 110 to 130 eV and extrapolated above the phosphorus edge (dashed curve); subtraction of the fitted background produces the characteristic P $L_{2,3}$ shape in spectrum (b). Energy windows used for EFTEM ratio mapping are indicated on each spectrum.

Volume-rendered, tomographic reconstruction of phosphorus in a section of C. elegans cell: (a) Rows of ribosomes are evident along stacks of endoplasmic reticulum membranes. Slices through the reconstruction of the x-z and y-z planes are also shown. Bar = 100 nm. (b) Higher magnification of volume-rendered phosphorus distribution showing individual ribosomes located at different heights within the section. Bar = 20 nm.

Fig. 33. Leapman *et al.*[105] have combined EELS and volume-rendered tomography to reveal rows of ribosomes in endoplasmic reticulum membranes.

information about the packing density of DNA within chromatin (see Bazett-Jones *et al.*[106] and Leapman *et al.*[107])

8. Time-Resolved Electron Microscopy

Time-resolved studies by optical microscopy are much easier to perform (on cells and other semi-micro materials) than those by electron microscopy, owing to the necessity to operate with the specimens in an environment that permits the passage of electrons so as to yield the required images (or diffraction patterns).

My own work, done several decades ago, on the catalytic oxidation of single crystals of graphite by nanoparticles of a range of metals, began with a time-lapse, cinematographic study of the process using a hot-stage (optical) microscope that enabled the oxidation to be followed at 1 bar pressure of air or O_2. Greater insights were obtained when we investigated this system electron-microscopically with a hot-stage that "tolerated" an ambient pressure of O_2 up to 3 m bar.[109,110] The stage in which the specimen was mounted and the dynamics of the catalysis investigated, involved a differential pumping system which was a primitive form of the methods pioneered and perfected by PL Gai.[111–114]

(A movie clip, being an excerpt of the televised broadcast (see Ref. 110) is attached to this article.)

Gai[114] has perfected her *in situ* environmental TEM in such a fashion as to achieve atomic resolution in the image of nanoparticles participating in catalytic turnover. The set up devised by her and Boyes and adapted by others[115] is schematized in Fig. 34; and a typical result, illustrating how time-resolved and temperature-resolved heterogeneously catalyzed reactions may be followed by electron microscopic imaging, is shown in Fig. 35. The highest degree of time-resolution so far achieved by Gai *et al.*[116] falls in the range of milli- to microseconds. (This is to be compared with my own work, shown above, that recorded catalytic events on the seconds to minutes time

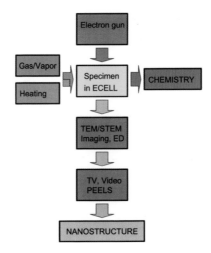

Fig. 34a. Block diagram illustrating the Gai-Boyes set-up used for "environmental" electron microscopy (ETEM).

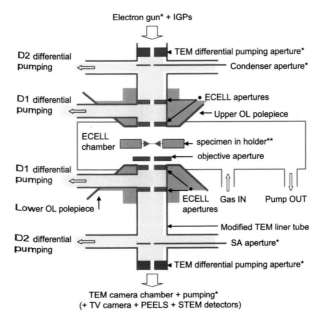

Fig. 34b. Details of the ETEM procedure pioneered by PL Gai PEELS stands for parallel recording of EELS spectra. (After PL Gai. In: AI Kirkland, JL Hutchison (eds), *Nanocharaterisation*, Roy Soc Chem (2007) p. 268.)

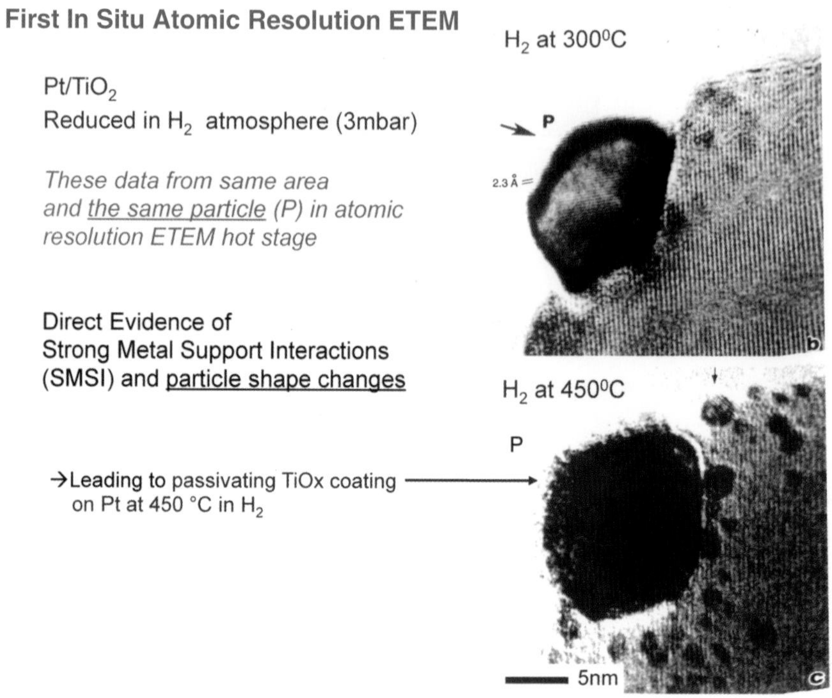

First In Situ Atomic Resolution ETEM

H$_2$ at 300°C

Pt/TiO$_2$
Reduced in H$_2$ atmosphere (3mbar)

*These data from same area
and the same particle (P) in atomic
resolution ETEM hot stage*

P

2.3Å

Direct Evidence of
Strong Metal Support Interactions
(SMSI) and particle shape changes

H$_2$ at 450°C

P

→Leading to passivating TiOx coating
on Pt at 450 °C in H$_2$

5nm

Fig. 35. The first *in situ*, atomically resolved images of a Pt nanoparticle catalyst supported on TiO$_2$ in an atmosphere (3m bar) of H$_2$ at high temperature is reproduced here from Gai's work[111–116] (see ED Boyes and PL Gai. (1997) *Ultramicroscopy* **67**: 219; PL Gai. (1998) *Advanced Materials* **10**: 1259.)

scale). Gai *et al.*[116] have also followed, *in situ*, the twinning transformation of tungsten carbide at elevated temperatures (Fig. 36).

Again the time-resolution fell in the milli- to microsecond range.

8.1. Ultrafast Electron Microscopy (UEM)

Some 16 years ago, in an article entitled "Femtosecond Diffraction," I speculated[117] as to whether the work that Zewail and colleagues had just accomplished using species to be structurally investigated in molecular

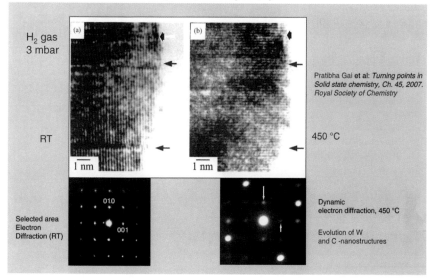

In situ ETEM: dynamic atomic level twinning transformations in tungsten carbide
(Images recorded from the same area during reactions.)
(PL Gai, University of York)

Fig. 36. Atomically resolved *in situ* ETEM image of WC exposed to H_2/He gas at: **(a)** room temperature and at 450°C. **(b)** The elimination of the twin boundaries could be followed in real-time (at microsecond rates).[116]

beams could ever be carried out on species in the condensed (solid or surface) state. If it were ever achieved, so I opined, it would mark the dawn of a new era. Since that time, enormous progress has been made by Zewail and his colleagues[118–128] in the now burgeoning area of ultrafast structural determinations (see Table 4). Effectively, these workers showed how to unite the time domain with the spatial domain in such a manner as to achieve unprecedented real-time and real-space capabilities. Adding the fourth dimension, time, to the three spatial ones that are retrievable by high-resolution electron microscopy, opens up new vistas and applications by using timed, single-electron packets.

All this is achieved through coherent electron packets, which are liberated from a photocathode with femtosecond laser pulses, and are accelerated

Table 4. Ultrafast Structural Determinations: A Selection of Relevant Papers

Title and Author(s)	Ref.
Femtosecond Diffraction, Thomas	117
Ultrafast Electron Crystallography: The Dawn of a New Era, Thomas	127
A Revolution in Electron Microscopy, Thomas	128
The Prospects of Exploiting 4D Ultrafast Electron Microscopy in Solid-State and Biological Chemistry, Harris and Thomas	129
4D Ultrafast Electron Diffraction; Crystallography and Microscopy, Zewail	125
Visualizing Complexity: Development of 4D Microscopy and Diffraction, Zewail	126
Direct Imaging of Transient Molecular Structures with Ultrafast Diffraction, Ihee *et al.*	118
Dark Structures in Molecular Radiationless Transistors Determined by Ultrafast Diffraction, Srinivasan *et al.*	119
Ultrafast Electron Crystallography of Interfacial Water, Ruan *et al.*	120
Nonequilibrium Phase Transition in Cuprates Observed by Ultrafast Electron Crystallography, Gedih *et al.*	121
4-Dimensional Visualization of Transitional Structures in Phase Transformations by ED, Baum *et al.*	122
Four-dimensional Ultrafast Electron Microscopy of Phase Transitions, Grinolds *et al.*	123
Atomic-scale Imaging in Real and Energy Spare Developed in Ultrafast Electron Microscopy, Park *et al.*	124

at 120 keV, and contain as little as one electron per pulse. Such design ensures that no space-charge broadening of the electron beam occurs and that electron trajectories can be brought into sharp focus. The photocathode is illuminated with very weak (E ≈ 500 pJ) fermtosecond pulses — which constitute a high-frequency train (12.5 ns or longer separation) — such that the ultrafast electron microscope (UEM) provides, in principle, the ability to benefit from multiple pulsing with single-shot time resolution. For example, with a periodic sample, each electron pulse could, by dextrous manipulation, be applied to a different but equivalent region. Moreover, by lengthening the interval between pulses, any reversible damage inflicted by an electron pulse on samples such as metals or other inorganic objects (e.g., Si, Ge, GaAs) may be allowed to self-heal prior to further repeated investigations. By obtaining diffraction patterns, Zewail and co-workers have reached atomic resolution, and by microscopy they have already obtained images that are resolved on the nanometer scale.

The conceptual design of the UEM, in which a femtosecond optical system is integrated with a high-resolution electron microscope, is depicted in Fig. 37. The laser system consists of a diode-pumped mode-locked Ti: sapphire laser oscillator which generates sub-100-fs pulses at 800 nm with a repetition rate of 80 Mhz and an average power of 1W. Part of the beam may be used to heat or excite the sample and to define the zero of time, while the remainder is frequency-doubled in a nonlinear crystal to yield 400-nm femtosecond pulses for generating the train of electron pulses. Single electron detection is achieved by using a phosphor scintillator in conjunction with a charge-coupled device camera.

This approach by Zewail differs fundamentally from that of Gai and Boyes outlined above — they used continuous beams of electrons. With his set-up (Fig. 37), Zewail has demonstrated the versatility of this so-called 4D ultrafast electron microscopy (4D UEM) — see Fig. 38 and Table 4, and, in particular, his comprehensive review in Ref. 125 and in this volume. With 4D UEM, dynamical evolution of structure and the concomitant charting of energy landscapes[130,131] may be directly mapped so that there is reason to

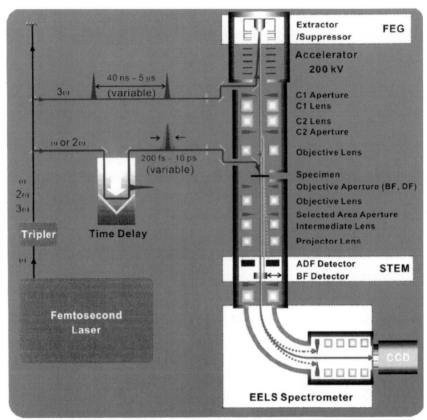

Conceptual design of Caltech's UEM2. A schematic representation of optical, electric, and magnetic components are displayed. The optical pulse train generated from the laser, in this case having a variable pulse width of 200 fs to 10 ps and a variable repetition rate of 200 kHz to 25 MHz, is divided into two parts, after harmonic generation, and guided toward the entries of the designed hybrid electron microscope. The frequency-tripled optical pulses are converted to the corresponding probe electron pulses at the photocathode in the hybrid FEG, whereas the other optical pump beam excites (T-jump or electronic excitation) the specimen with a well-defined time delay with respect to the probe electron beam. The probe electron beam through the specimen can be recorded as an image (normal or filtered, EFTEM), a diffraction pattern, or an EEL spectrum. The STEM bright-field detector is retractable when it is not in use. See the text for details.

Fig. 37. Schematic illustrating the conceptual design of the ultrafast electron microscopy (UEM) inaugurated by Zewail *et al.* (See Table 4 for further references.)

UEM: Crystals and Biological Cells

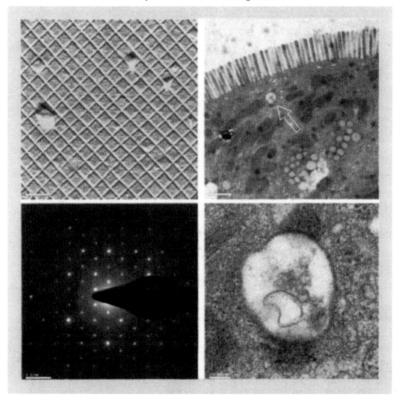

UEM images obtained for materials (left) and biological cells (right). The pattern obtained in the diffraction mode for crystalline gold is also shown (bottom, left).

Fig. 38. Both crystals and cells are amenable to investigation by the UEM as schematized in Fig. 37. (From AH Zewail. (2007) Visualising complexity: development of 4D microscopy and diffraction for imaging in space and time. In: *Visions of Discovery: Shedding New Light on Physics and Cosmology*, Cambridge University Press, UK.)

believe that greater appreciation of biological function may be gained by its application.

Radiation damage, as discussed above, is a critical determinant in all studies (involving electrons, X-rays and neutrons). Quite apart from the fact that X-ray crystallography cannot cope with certain local structural problems, where there may be few (or no) unit cell repeats, it must not be forgotten that X-rays are intrinsically inferior to electrons as probing beams. As pointed out by Henderson,[68] electron-scattering cross-sections are higher than those of X-ray scattering from molecules by some five to six orders of magnitude. Moreover, notwithstanding the notorious bugbear of electron-beam damage, electrons are less damaging to specimens per useful scattering event than are X-rays. If, and this is quintessential, the suppression of electron-beam damage to be expected from the multiple and controlled pulsing, the coherent (single-shot) approach inherent in the technique encapsulated in Fig. 37 permits the collection of diffraction data at higher dose, one cannot but speculate that Zewail's 4D UEM will not only reveal new insights into the dynamics of biophysically relevant processes, but also open up the possibility of arriving at structures of even higher resolutions than those recently obtained (by Crowther,[70] Chiu,[72] and others[72]) by utilizing cryogenic procedures.

When beam-damage is not a critical issue, as in the study of numerous

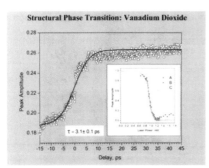

UEM of the metal-insulator phase transition in vanadium dioxide. Shown are the temporal evolution and, in the inset, the hysteresis observed when the fluence was varied.

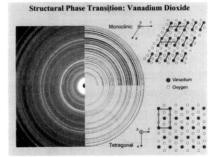

Patterns obtained in the diffraction mode of UEM showing the two phases, monoclinic (top) and tetragonal (bottom), of vanadium dioxide. The two relevant structures are displayed on the right.

Fig. 39. The 4D-approach, which adds the time-dimension to the three that are obtained from direct imaging and diffraction, was recently[122] put to good effect by Zewail *et al.* to follow the phase transition of VO_2 from its insulator to metallic state. The temporal evolution (*left*) can be followed in an unprecedented manner at time-scales of picoseconds.

inorganic solids, the 4D UEM technique has already generated information about the dynamics of change in the solid-state that is quite spectacular. Time-scales with half-lives smaller than picoseconds have been obtained in, for example, the charting of the insulator–metal phase transition in VO_2,[122] a phenomenon that has long been an enigma (see Fig. 39). By following intensity changes of the numerous (*hhl*) reflections (see Fig. 40), Zewail *et al.*[122] have been able to probe the spatial changes within the structure, during the phase-transition, in hitherto unequalled details — nearly ten orders of magnitude better than ever before — and, it transpired, somewhat surprisingly that, in VO_2, there is a stepwise mechanism for atomic motions.[122] First, light absorption causes the bond between pairs of V atoms to dilate on the femtosecond time scale. Then, only after a longer time (picoseconds) does the system relax along the orthogonal direction, thereby transforming a compressed tetragonal structure to a stable metal product.

8.2. New Developments in Time-Resolved Structural Studies

Electron diffraction microscopy[133] or *lensless diffractive imaging*[134] has elicited much interest of late in view of: (a) the increasing availability of coherent electron-wave sources (as in a field-emission-gun-equipped electron microscope or in the set-up of Zewail as shown in Fig. 37); and (b) the progressively improving prospects of having coherent soft X-rays generated by a tabletop soft X-ray source. Miao *et al.*[133] have shown that, by combining coherent electron diffraction with the oversampling phasing method, the 3D structure of a zeolite solid (Linde type A) can be determined *ab initio* at a resolution of 0.1 nm from 29 simulated noisy diffraction patterns. In principle, this form of microscopy can be used to determine the 3D structure of nanocrystals and nanocrystalline samples at ultra high resolution, even beyond the capabilities of lens-based electron microscopy. If these hopes can be fulfilled, significant progress can be made by combining the necessary experimental arrangements to capitalize on the kind of rapid pulsing methodology of Zewail.

Sandberg *et al.*'s[134] tabletop (soft X-ray) approach, which, at present, has reached only 214 nm resolution, is also a promising new approach that could, in due course, be adapted for time-resolved studies. One immediate obstacle to be surmounted is to work at shorter wavelengths by high harmonic generation, which it is believed should ultimately improve the

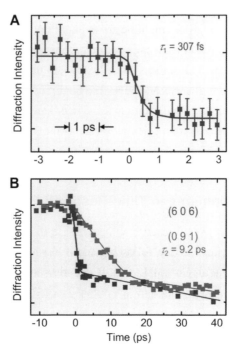

Ultrafast dynamics of structural phase transition in vanadium dioxide. (A) Intensity change of the (6 0 6) Bragg spot with time. A decay with a time constant τ_1 of 307 fs was obtained on the total time scale of ± 3 ps (40) (B) Intensity change of (6 0 6) (blue) and (0 9 1) (red) spots with time. For all investigated Bragg spots, two different types of dynamics were observed: a femtosecond decay similar to the blue trace is measured for (8 0 6), (8 2 6), (8 $\bar{2}$ 6), (8 4 6), (8 $\bar{4}$ 6), (6 0 6) (7 1 4) (4 $\bar{1}$ 7), (10 2 4), and (6 $\bar{2}$ 8); a decay with a time constant τ_2 of 9.2 ps, similar to the red trace, is measured for (0 9 1), (0 8 4) and (0 8 $\bar{2}$) on the total time scale of 40 ps. The temporal range values (Δt_1 and Δt_2) over which the decay is pronounced are 760 fs and 15 ps (40). This difference indicates a stepwise mechanism for atomic motions (see text).

Fig. 40. The ultrafast-dynamics of phase change now possible using UEM-2 (see Fig. 37), enables the intensities of numerous Bragg peaks to be followed (effectively by "dark-field" intensity measurements) down to the femtosecond scale in the insulator metal transition of VO_2.

resolution to some tens of nm. The authors of this paper[134] state: "As the laser repetition rates are increased from 3 kHz to tens of kHz, the soft X-ray flux will be simultaneously increased and acquisition time will be dramatically reduced from hours to minutes. Also, as computing power increases and reconstruction times shrink, a tabletop soft X-ray lensless microscope will become increasingly practical for routine use in biological imaging, nanoscience, and metallurgy in support of next-generation lithography. Moreover, the femtosecond time resolution afforded by high harmonics will enable tabletop time-resolved imaging on femtosecond time scales." Whilst the optimism encapsulated in this statement might seem a little premature, no one can deny that, less than a decade ago, it seemed inconceivable that the ultrafast structural determinations pioneered by Zewail *et al.* would be achievable.

Acknowledgments

I am deeply grateful to A.H. Zewail for the numerous stimuli that he provided during the preparation of this article; to my colleague P.A. Midgley for his ready cooperation; and to the following for provision of illustrations and many useful ideas: Wah Chiu, P.L. Gai, R. Henderson, C.J. Kiely, N. Kruse, R.D. Leapman, J.C.H. Spence, W. Baumeister, G.D.W. Smith, J. Ulstrup, C.A. Larabell, F. Besenbacher, M.J. Yacaman and H.-J. Freund.

References

1. Muller EW. (1951) *Z Physik* **31:** 136.
2. Miller MK, Smith GDW. (1989) *Atom Probe Microanalysis: Principles and Applications to Materials Problems*, *Mater Res Soc*, Pittsburgh.
3. Smith GDW, Bagot PAJ, see website of GDW Smith, Department of Materials, University of Oxford.
4. Medvedev VK, Suchorski Yu, Voss C, Visant de Bocarme T, Bär T, Kruse N. (1998) *Langmuir* **14:** 6151.

5. Miller MK. (2000) *Atom Probe Tomography: Analysis at the Atomic Level.* Kluwer/Plenum, New York.
6. Imago Scientific Instruments, Madison, Wisconsin.
7. Kelly TF, Gibb TT, Olson JD, Marteus RL, Shephard JD, Weiner SE, Kunichi TC, Strait DR. (2004) *Microscopy Microanal* **10**: 373.
8. Perea DE, Allen JE, May SJ, Wessels BW, Seidman DN, Lauhon LJ. (2006) *Nano Letters* **6**: 181.
9. Thompson J. (2007) *Science* **317**: 1370.
10. Binnig G, Rohrer H, Gerber C. (1982) *Phys Rev Lett* **49**: 57.
11. Sterner M, Risse T, Martinez U, Giordano L, Heyde M, Rust HP, Pacchioni G, Freund HJ. (2007) *Phys Rev Lett* **98**: 096107.
12. Winterllin J. (2000) *Adv Catal* **45**: 131.
13. Lauritsen JV, Besenbacher F. (2006) *Adv Catal* **50**: 97.
14. Besenbacher F. (1996) *Rep Progr Phys* **59**: 1737.
15. Otero R, Schöck M, Molina LM, Laegaard E, Stengaard I, Hammer B, Besenbacher F. (2005) *Angew Chemie Intl Ed* **44**: 2270.
16. Lundquist WI, Klug A. (1989) *Nature* **342**: 825.
17. Laughlin G, Murchie AIH, Norman DG, Moore MH, Moody PCE, Lilley DMJ, Luisi B. (1994) *Science* **265**: 520.
18. Hausen AG, Boisen A, Nielsen JU, Wackerbarth H, Chorkendorff Ib, Andersen JET, Zhang J, Ulstrap J. (2003) *Langmuir* **19**: 3419.
19. Chi Q, Farver O, Ulstrup J. (2005) *PNAS* **102**: 16203.
20. Zhang J, Chi Q, Albrecht T, Kuzretsov AM, Gruble M, Hausen AG, Wackerworth H, Ultstrup J. (2005) *Electrochimica Acta* **50**: 3143.
21. Haider M, Rise H, Uhlemann S, Schwan E, Kabius B, Urban K. (1998) *Ultramicroscopy* **75**: 53.
22. Kirkland AI, Meyer R. (2004) *Microscopy Microanal* **10**: 401.
23. Honkala K, Hellman A, Remediakis IN, Logadottio A, Carlsson A, Dahl S, Christensen CH, Norskov JK. (2005) *Science* **307**: 555.
24. Gontard LC, Chang LY, Hetherington CJ, Kirkland AI, Ozkaya D, Dunin-Borkowski RE. (2007) *Angew Chemie Intl Ed* **46**: 3683.
25. Yacaman MJ, Ferrer D, Jorres-Castro A, Gao X, Ortiz-Mendez U. (2007) *Nanoletters* **6**: 1701; see also Kiely CJ *et al.* (2005) *J Mater Sci* **15**: 1755.
26. Herzing A, Watanabe M, Edwards JK, Hutchings GJ, Kiely CJ. (2007) *Farad Disc* **6**: 1701.

27. Larabell CA, Le Gros MA. (2004) *Molecul Biol Cell* **15:** 957.
28. Le Gros MA, McDermott G, Larabell CA. (2005) *Curr Opinion in Structural Biology* **15:** 593. See also Gu W, Ethin LD, Le Gros MA, Laralull CA. (2007) *Differentiation* **75:** 529.
29. Phillips DC. (1979) In: WL Bragg (ed), *Biographical Memoirs of Fellows of the Royal Society* (1979) **25:** 1.
30. Cosslett VE, Nixon WC. (1960) *X-ray Microscopy.* Cambridge University Press.
31. Kirkpatrick J, Baez J, see "Google" entry on X-ray Microscopy.
32. Midgley PA, Ward EPW, Hungria AB, Thomas JM. (2007) *Chem Soc Rev* **36:** 1477.
33. Radon J, Werh B, Salchs K, Wiss G. (1917) *Leipzig Math-Phys Kl* **69:** 262.
34. Deans SR. (1983) *The Radon Transform and Some of its Applications.* Wiley, New York.
35. Herman GT. (1980) *Image Reconstruction from Projections: The Fundamentals of Computerised Tomography.* Academic Press, New York.
36. Glaeser RM. (1971) *J Microscopy* **12:** 133.
37. Chao WL, Hartneck BD, Liddle JA, Anderson EH, Attwood DT. (2005) *Nature* **435:** 1210.
38. Atwood DL. (2006) *Nature* **442:** 642.
39. Yin G-C, Tang M-T, Song YF, Chen FR, Liang KS, Ko CH, Shieh HPD. (2006) *Appl Phys Lett* **442:** 642.
40. Sayre D. (1980) In: M Schlinker (ed), *Imaging Processes and Coherence in Physics*, p. 229. Springer, Berlin.
41. Miao J, Charalambous P, Kinz J, Sayre D. (1999) *Nature* **400:** 342.
42. Machesini S, He H, Chapman HN, Hau-Riege SP, Nay A, Howells MR, Weierstall U, Spence JHC. (2003) *Phys Rev B* **68:** 140101.
43. Chapman H, Barty A, Noy A, Marchesini S, Hau-Riege SP, Cui C, Howells MR, Rosen R, He H, Speme JHC, Beetz T, Shapiro D. (2006) *J Opt Soc Am A* **23:** 1179.
44. Aebersold R, Mann M. (2003) *Nature* **422:** 198.
45. Northen TR, Yanes O, Northen MT, Marrinuscci D, Uritboonthai W, Apon J, Golledge SL, Nordström A, Suizdak G. (2007) *Nature* **449:** 1033.

46. Chaurand P, Stoeckli M, Capriolo RM. (1999) *Anal Chem* **71:** 5263.
47. Gai PL, Boyes ED. York University, N.K, private communication, Sept. 2007.
48. De Rosier DJ, Klug A. (1968) *Nature* **217:** 130.
49. Crowther RA, The Leeuwenhoek Lecture, *Phil Trans Ray-Soc B* (2007) doi:10.1098/rstle. 2007. 2150 (in press).
50. Thomas JM, Millward GR, Bursell LA. (1981) *Phil Trans Roy Soc* **A300:** 43.
51. Thomas JM, Millward GR. (1984) *J Chem Soc Chem Comm* 77.
52. Spence JCH. (2003) *High-Resolution Electron Microscopy*, 3rd ed., Oxford Science Publications.
53. Thomas JM. (1983) *Inorganic Chemistry: Towards the 21st Century.* ACS Symposium Series 211, p. 445.
54. Thomas JM, Terasaki O, Gai PL, Zhou W, Gonzalez-Calbet JM. (2001) *Accnts Chem Res* **34:** 583.
55. Notari B. (1996) *Adv Catalysis* **41:** 253.
56. Fyfe CA, Gobbi GC, Klinowski J, Thomas JM. (1982) *Nature* **296:** 530.
57. Wright PA, Zhou W, Perez-Pariente J, Arrantz M. (2005) *J Am Chem Soc* **127:** 494.
58. Wright PA. (2007) *Microporous Framework Solids, Roy Soc Chem Publ*, p. 64. London.
59. Thomas JM, Klinowski J, Ramdas S, Tennokoon DTB, Hunter BK. (1983) *Chem Phys Lett* **102:** 158.
60. Klug A, Nobel Lecture. (1983) *Agnew Chemie Int Ed* **22:** 565.
61. Henderson R, Unwin PNT. (1975) *J Mol Biol* **94:** 425.
62. Kornberg RD. (1977) Structure of chromatins, *Ann Rev Biochem* **46:** 931.
63. Jefferson DA, Thomas JM, Smith DJ, Camps RA, Cato CJB, Cleaver JRA. (1979) *Nature* **281:** 5.
64. Catlow CRA, Thomas JM, Jefferson DA, Parker SC. (1982) *Nature* **295:** 658.
65. Thomas JM, Jefferson DA, Mollinson LG, Smith DJ, Crawford ES. (1978/1979) *Chemica Scripta* **14:** 167.
66. Kirkland AI, Thomas JM *et al.* (work in progress).

67. Adrian M, Dubochet J, Lepault J, McDowall AW. (1984) *Nature* **308:** 32.
68. Henderson R. (2004) *Quart Rev Biophys* **37:** 3.
69. Jiang W, Chang J, Jakana J, Weigele P, King J, Wah Chui. (2006) *Nature* **439:** 612. (See also Jiang W, Baker ML, Jakana J, Weigele P, King J, Wah Chiu. *Nature* **451:** 1130.)
70. Crowther RA, The Leewenhoek Lecture, *Phil Trans Ray-Soc B* (2007), in press.
71. Wah Chiu, Baker ML, Jiang W, Dougherty M, Schmid MF. (2005) *Structure* **13:** 363.
72. Rossmann MG, Morais MC, Leiman PG, Zhang W. (2005) *ibid.* **13:** 355.
73. Crowther RA, Klug A. (1975) *Ann Rev Biochem* **44:** 161.
74. Crowther RA. (1971) *Phil Trans Roy-Soc* **B261:** 221.
75. Crowther RA, Kiselev NA, Böttcher B, Berriman JA, Borisova GB, Ose V, Pampers P. (1994) *Cell* **77:** 943.
76. Böttcher B, Wynne SA, Crowthers RA. (1997) *Nature* **386:** 88.
77. Chang J, Weigele P, King J, Chiu W, Jiang W. (2006) *Structure* **14:** 1073.
78. Henderson R, Baldwin JM, Ceska TA, Zemlin F, Beckmann E, Downing KH. (1990) *J Mol Biol* **213:** 899.
79. Kuhlbrandt W, Neng Wang D. (1991) *Nature* **350:** 130.
80. Uzgiris EE, Kornberg RD. (1983) *ibid.* **301:** 125.
81. Asturias FJ, Chang W, Li Y, Kornberg RD. (1998) *Ultramicroscopy* **70:** 133.
82. Baumeister W. (2002) *Curr Opin Struct Biol* **12:** 679.
83. Thomas JM, Midgley PA, Yates TJV, Barnard JS, Raja R, Arslan I, Weyland M. (2004) *Agnew Chemie Intl Ed* **43:** 6745.
84. Thomas JM, Johnson BFG, Raja R, Sankar G, Midgley PA. (2003) *Accnts Chem Res* **36:** 20.
85. Midgley PA, Weyland M, Thomas JM, Johnson BFG. (2001) *Chem Commun* 907.
86. Midgley PA, Thomas JM, Laffont L, Weyland M, Raja R, Johnson BFG, Khimyak T. (2004) *J Phys Chem B* **108:** 4590.
87. Thomas JM, Midgley PA. (2004) *Chem Commun* **1253.**

88. Sousa AA, Arnova MA, Kim YC, Dorward LM, Zhang G, Leapman RD. (2007) *J Struct Biol* **159**: 507.

89. Paciotti GF, Mayer L, Weinreich D, Goia DV, Pavel N, McLaughlin RE, Tamarkin L. (2004) *Drug Delivery* **11**: 169.

90. McEwen BF, Marks M. (2001) *J Histochem Cytochem* **49**: 553.

91. Marsh BJ, Mastronade DN, Buttle KF, Howells KE, McIntosh JR. (2001) *PNAS* **98**: 2399.

92. Medalia O, Weber I, Franakis AS, Nicastro D, Genish G, Baumeister W. (2002) *Science* **298**: 1209.

93. Beck M, Förster F, Eche M, Plitzko JM, Melchior F, Gerish G, Baumeister W, Medalia O. (2004) *Science* **306**: 1387.

94. Förster M, Medalia O, Zauberman N, Baumeister W, Fass D. (2005) *PNAS* **102**: 4729.

95. Crrklapp M, Risco C, Fernadez JJ, Jimerez MV, Baumeister W, Carrascosa JL. (2005) *ibid.* **102**: 2772.

96. Nicastro D, McIntosh JR, Baumeister W. (2005) *ibid.* **102**: 15889.

97. Thomas JM, Williams BG, Sparrow TG. (1985) *Accnts Chem Res* **18**: 324.

98. Egerton RF. (2001) *Electron Energy Loss Spectroscopy in the Transmission Electron Microscope.* Plenum Press, New York.

99. Leapman RD. (2004) *Curr Opin Neurobiology* **14**: 591.

100. Sparrow TG, Williams BG, Thomas JM, Jones W, Herley PJ, Jefferson DA. (1983) *J Chem Soc Chem Comm* 1432.

101. Williams BJ, Sparrow TG, Thomas JM. (1983) *ibid.* 1434.

102. Thomas PJ, Midgley PA. (2001) *Ultramicroscopy* **88**: 187.

103. Grass MH, Kozoil KK, Windle AH, Midgley PA. (2006) *Nanoletters* **6**: 376.

104. Brydson RD. (2007) In: *Nanocharacterisation*, p. 94. RSC Publ., London.

105. Leapman RD, Fioni CE, Gorlen KE, Gibson CC, Swift CR. (2004) *Ultramicroscopy* **100**: 115.

106. Bazett-Jones DP, Hendzel MJ, Kruhlak MJ. (1999) *Micron* **30**: 151.

107. Aronova MA, Kim YC, Harmon R, Sousa AA, Zhang G, Leapman RD. (2007) *J Struct Biol* **160**: 35.

108. Wen J-D, Manosas M, Li PTX, Smith SB, Bustamante C, Ritort F, Tinoco I. (2007) *Biophysical J* **92**: 2996.

109. Thomas JM. In: PL Walker (ed), *Chemistry and Physics of Carbon*, Vol. 1. Marcel Dekker, New York.

110. *In situ* cinematographic study, by electron microscopy, of the catalytic oxidation of graphite, shown at the Royal Institution Christmas Lectures, Dec. 1987 and broadcast by BBC, January 1988.

111. Gai PL. (1992) *Catal Rev Sci Eng* **34:** 1.

112. Gai PL, Kourtakis K. (1995) *Science* **267:** 661.

113. Thomas JM, Gai PL. (2004) *Adv Catalysis* **48:** 171.

114. Gai PL, Ref. 104, p. 268.

115. Gai PL, Boyes ED, Helveg S, Hansen PL, Giorgio S, Henry CR. (Dec 2007) *Mater Res Bull*, in press.

116. Gai PL, Boyes ED. (2007) In: KDM Harris, PP Edwards (eds), *Turning Points in Solid-State, Materials and Surface Science* in press. RSC Publishing, London.

117. Thomas JM. (1991) *Nature* **351:** 694.

118. Ihee H, Lobastov VA, Gomez UM, Goodson BM, Srinivasan R, Ruan C-Y, Zewail AH. (2001) *Science* **291:** 458.

119. Srinivasan R, Feenstra JS, Park ST, Xu SJ, Zewail AH. (2005) *Science* **307:** 558.

120. Ruan C-Y, Lobastov VA, Vigliotti F, Chen S, Zewail AH. (2004) *ibid.* **304:** 80.

121. Gedik N, Yang D-S, Logvenov G, Bozovic I, Zewail AH. (2007) *ibid.* **316:** 425.

122. Baum P, Yang D-S, Zewai AH. (2007) *ibid.* **318:** 788. See also Cavalieri P. (2007) *ibid.* **318:** 755.

123. Grinolds MS, Lobastov VA, Weissenrieder J, Zewail AH. (2006) *PNAS* **103:** 18427.

124. Park HS, Baskin JS, Kwon O-H, Zewail AH. (2007) *Nano Lett* **7:** 2545.

125. Zewail AH. (2006) *Am Rev Phys Chem* **57:** 65.

126. Zewail AH. (2008) *Visions of Discovery: Shedding New Light of Physics and Chemistry* in press. Cambridge University Press.

127. Thomas JM. (2006) *Agnew Chem Ind Ed* **43:** 2606.

128. Thomas JM. (2005) *ibid.* **44:** 5563.

129. Harris KDM, Thomas JM. (2005) *Crystal Growth Design* **5:** 2124.

130. Zewail AH, this volume.

131. Zewail AH. (2005) *Phil Trans Roy Soc* **A363:** 315.
132. Marezio A, McWhan DB, Remeika JP, Dernier PD. (1972) *Phys Rev B* **5:** 2541.
133. Miao J, Oshuna T, Teraski O, Hodgson KO, O'Keefe MA. (2002) *Phys Rev Lett* **89:** 155502.
134. Sandberg RL, Paul A, Raymondson DA, Hadrich S, Gaudiosi DM, Holtsnider J, Tobey RI, Bohen O, Murname M, Kapteyn HC, Song C, Miao J, Liu Y, Salmassi F. (2007) *Phys Rev Lett* **99:** 098/103. (See also Chapman HN. (2007) *Nature* **448:** 676 and Cavalleri A. (2007) *Nature* **448:** 651.)

chapter 5

Physical Biology at the Crossroads

Carlos J. Bustamante[*]

Sixty years after the beginning of the molecular biology revolution, physical scientists are becoming attracted to biology again. This time, unlike the situation six decades ago, we know a great deal about the parts and components of the cell. Moreover, the complexity that drove away many physical scientists in the past is now yielding to the reductionist approach, and complexity in itself has become a central research theme in physics. The stage is thus set for the development of renewed efforts at quantification in biological research and at modeling integrated cellular functions. This is the new playground where physical biology and their practitioners will make their impact. Some of the research challenges lying ahead are discussed.

1. Introduction

In the decade following the end of the Second World War, the biological sciences underwent a revolution of unprecedented proportion and significance, as scientists realized that understanding the molecular basis for the storage, retrieval and utilization of the genetic information was the new and most exciting research frontier. In the dawning of this new era that would come to be known as "Molecular Biology," trained physicists and physical scientists played important, and in many cases, key roles. Physical scientists were the first to understand that the description of biological systems had to be reducible to the same physical laws that govern the inanimate world;

[*]Howard Hughes Medical Institute, Department of Molecular and Cell Biology, Department of Physics, Department of Chemistry, University of California, Berkeley, CA 94720, USA, e-mail: carlos@alice.berkeley.edu.

that if our understanding of biology was to catch up with our understanding of physics and chemistry, biology as a discipline had to become quantitative and the biological phenomena reducible to their *ultima ratio*. That many of these physics-trained scientists often showed the way that led to the Molecular Biology revolution, but did not quite enter into it or did not themselves remain active in biological research, is no longer a controversial subject. It is probably safe to say that despite their initial interest, many physical scientists lost interest upon their first encounter with a subject that, despite all their efforts, only seemed to grow more complex as research progressed. To the physically trained mind, biological processes appeared to be of such complexity that they seemed to defy a rigorous approach. The vast complexity that was revealed to these pioneers argued against attempts to search for quantitative descriptions of biological processes and in favor, instead, of the tasks of classifying, organizing, recording and cataloging this complexity. To a physicist, research in biology appeared as a task of infinite detail, lacking the universality of physical laws and resembling more stamp collecting or the actions of protagonists of a Jorge Luis Borges' story bent on performing a futile inventory of all the grains of sand in a beach.

While 50 years ago it may have appeared as if there were no boundaries to this complexity, today these boundaries are in sight and becoming clearer every day. In the intervening years, biology underwent a gigantic effort at inventorizing the components of the cell — the minimal unit of living matter — a process culminating recently with the sequencing of several genomes, including the first human one. Throughout this often tedious process of classifying and cataloguing, it slowly became apparent that amid the fearful complexity of biological systems certain patterns continuously recur and constantly re-emerge, repeating themselves as minor variations of a central cannon. One of the first organizational principles to emerge as this analytical effort got underway was that the enormous diversity of biological systems reflects first and foremost the selective advantage of successful designs, designs that tend to be preserved and utilized over and over in all possible combinations, before a new design is attempted. The result of this design principle is that biological systems are not streamlined, as one would expect from an "intelligent," *de-novo* engineering effort. Instead, a particular function is often accomplished by co-opting and by combining functions that emerged in completely different contexts. Thus, more and more frequently, biologists are finding that the isolation of a new gene or the solution of a new protein crystal structure reveals motifs already existing in the data banks and that have been used over and over again in related and sometimes even unrelated tasks. Many protein families and super-families have been characterized structurally, with new members being added daily, their existence bearing witness to an organizational principle based ultimately on economy of new designs.

These types of insights, arrived at as a direct result of the effort of inventorizing the parts and their functions, have made it possible to conceive of the boundaries of the subject for the first time. Today, the frightful complexity that drove away the early physical scientists at the start of molecular biology is beginning to yield its secrets. The nature of biological complexity appears to be indeed modular and hierarchical, and it seems that two tasks will become central in future biological research: a renewed push for a quantitative description of biological systems at the molecular level, and efforts towards a new "synthesis" in biological research through integrative models of the control and coordination of the cellular functions. Both of these efforts will require the convergence of the physical and biological sciences; this is the crossroads in which physical biology will play the role that those physical scientists first envisioned almost 50 years ago. Here I describe what I believe are some of the most exciting challenges ahead.

2. The Challenges Ahead

2.1. In Search of a New Thermodynamics

Macroscopic thermodynamics was developed in the second half of the 19th century in part as a response to the need for describing the efficiency of macroscopic machines in the industrial revolution.[1] Its greatest achievement, the rationalization of the limited efficiency of steam engines and other thermal machines through the discovery of entropy and disorder, stands as one of the great triumphs of physics. As originally stated by Clausius and Kelvin, thermodynamics emerged as a science describing the energy transactions between macroscopic systems occupying states at equilibrium. Moreover, the second law, and its powerful prediction that entropy must increase in all natural spontaneous processes, is valid only for closed systems.

One of the main challenges that biology presents to the physical scientist is the unique nature of the living state. To be alive, a cell must constantly remove itself from the state of equilibrium to which it naturally tends, and keep itself, instead, in non-equilibrium steady states: a task that requires the constant expenditure of energy. What is the minimal number of external parameters that determines a unique dissipative steady state? Are there functions of steady-state systems that can play a similar role to the thermodynamic potentials of equilibrium thermodynamics? Moreover, as cells adapt to the changing conditions of their surroundings, they must undergo transitions between non-equilibrium steady states. What are the rules that dictate which steady states are accessible to a steady-state system that evolves under a new set of control parameters? We do not know the answers to these questions yet, but it seems that uncovering the principles that underlie the attainment of steady states and the transitions between them in dissipative systems will become increasingly important as we attempt to describe the non-equilibrium state of the cell and how it changes under varying external conditions.

However, the need to extend thermodynamics to describe non-equilibrium states in biology is also echoed by the need to arrive at a description of the thermodynamics of small systems. Thermodynamics attains its powerful predictive power in the so-called *thermodynamic limit*. In this limit, the thermodynamic variables satisfy the numerous thermodynamic relationships that depend on the intensiveness and extensiveness of thermodynamic parameters.[2] For example, in a macroscopic system, the thermodynamic

Gibbs free energy, G, is a first-order homogeneous function of the mole number, n, and therefore satisfies the Euler relation, $G(T;P;n) = \mu(T;P;n)n$. Similarly, because the thermodynamic internal energy is first-order homogeneous in S, V, and n, the thermodynamic parameters satisfy the Gibbs–Duhem relation, $S\,dT - V\,dP + n\,d\mu = 0$. Neither of these relationships holds for microscopic systems in general. Moreover, away from the thermodynamic limit we no longer can assume the equivalence of alternative thermodynamics formulations (entropy description, free energy description, enthalpy description) that have their counterpart in the equivalence of the statistical mechanical ensembles.

2.1.1. *The cell is not a macroscopic system*

And yet, the cell is not a macroscopic reaction vessel. A few back-of-the-envelope calculations can illustrate this statement. Imagine for this argument a simple *E. coli* cell.

We can estimate the volume of this cell as roughly $1\,\mu m \times 1\,\mu m \times 1\,\mu m$. Using this figure the reader can easily show a rather surprising fact: *a single molecule of any kind inside an E. coli cell is at a concentration of 1.6 nM!* In other words, because of the small volume of the cell, most molecules

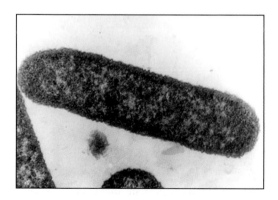

Fig. 1. Photomicrograph of *E. coli* cell.[3]

in *E. coli* do not need to be present in large numbers to insure efficient second order reaction rates and their association to their molecular targets. Indeed, most nucleic acid-binding proteins possess dissociation constants $K_d \sim 10^{-10}$–10^{-14} M. Thus, as an example, a single DNA binding protein molecule of any kind in *E. coli*, with a dissociation constant of 10^{-10} M or smaller, would be 94% or more of the time bound to its DNA binding site. So, it is not surprising that many proteins in a bacterial cell are present only in single or in very few copies. When some proteins are present in some tens or hundreds of copies, it is often because they are being used quite effectively by the cell to favor certain processes over others, or because the cell must simultaneously control may processes at once, as happens in transcription regulation of certain critical genes during the exponential growth of the cell.

This simple analysis may be somewhat surprising, for we are accustomed to thinking of chemical reactions in macroscopic systems involving Avogadro's number of molecules. However, many central functions of the cell, such as the copying of the leading strand of the DNA molecule during replication, or the segregation of the newly replicated chromosome before the completion of bacterial septation, are performed either by a single or by very few molecules. As a result, cell processes do not always display the smooth varying nature of the observables that we have come to associate with the reactions of large numbers of molecules where fluctuations are all but averaged out. Rather, inside the cell, fluctuations dominate and any real attempt to get an insight into its organization must take into account the discrete nature of the chemical transformations taking place among few reacting species. To insist somewhat more on this point, a similar calculation shows that at pH = 7 (i.e. at H$^+$ concentrations of $\sim 1 \times 10^{-7}$ M), there are only ~ 60 free protons in the whole cell. This example illustrates how macroscopic definitions and concepts from macroscopic chemistry, such as pH, are insufficient to describe the interior of the cell correctly and, in general, the chemical transformations taking place inside small vessels. These macroscopic concepts provide only an average value of the pH inside the cell, that although strictly correct, can neither properly describe the acid–base balance in the cell interior, nor the local fluctuations of proton concentration away from the average that are necessary for the myriads of cell functions and that must involve the unique donor–acceptor properties of H_2O.

2.1.2. *Molecular motors*

The limitations of the traditional (macroscopic) formalism of thermodynamics are perhaps best illustrated in the analysis of molecular motors. Most essential cellular functions such as chromosomal segregation, transport of

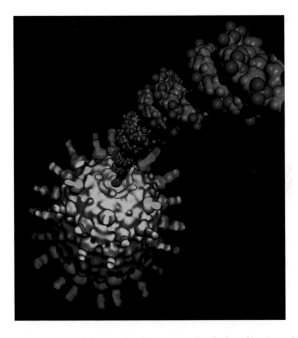

Fig. 2. Schematic drawing of the packaging motor (in dark yellow) at the base of the capsid of baceriophage phi29. During phage assembly, this motor grabs the DNA of the virus and packages it tightly inside the viral capsid at final pressures over 60 atmospheres using ATP as energy source. Single molecule manipulation methods are being used to characterize the mechanochemical transduction of these tiny engines.[4]

organelles from one part of the cell to another, or the maintenance of a voltage across the membrane, all involve directional movement of chemical species. Essential processes such as replication, transcription, and translation require the information encoded in the sequence of linear polymers to be read and copied in a directional manner, and cells must often move and orient in response to external chemical gradients and other signals. None of these processes can be carried out by simple diffusion and random collision of the reacting species. The cell needs to insure that these processes are carried out directionally. To overcome the randomizing effect of Brownian motion and perform these directional processes, cells possess molecular entities that have evolved to behave as tiny machine-like devices. These devices operate as molecular motors, converting chemical energy into mechanical work.

However, they are unlike macroscopic engines in that, because of their dimensions, the many small parts that make up these molecular motors must operate at energies only marginally higher than that of the thermal bath and hence are themselves subjected to large fluctuations. Sitting astride the line that separates stochastic from deterministic phenomena, the function of these tiny engines is then to "tame" the randomness of molecular events and to generate directional processes in the cell.

Imagine one such motor in the interior of the cell. It is not a closed macroscopic system, as the energy at which it operates is comparable to the energy of the thermal bath, and therefore, it constantly exchanges energy and matter (through the binding of ligands, fuel molecules, etc.) with its surroundings. The essential point to grasp here is that energy can flow freely in both directions, as heat dissipated from the operation of the motor to the bath and as energy from the bath to the motor. Moreover, the energy absorbed by the motor from the bath can lead to conformational changes in the molecule that can, in principle, be used to do work. This is a new phenomenon for which we have no equivalent among our daily experience with macroscopic machines. Although a thermal engine can absorb heat from a hot reservoir and pump it into a cold reservoir using part of it to do work, strictly speaking it is not absorbing energy from the thermal bath that surrounds the engine. Thus, because of their macroscopic dimensions, large machines are unable to couple with the thermal bath and undergo fluctuations. Macroscopic engines behave deterministically, because the energies at which they operate are much higher than the energies of their surroundings.

We do not understand yet how this energy exchange between a motor and its surroundings affects the thermodynamic efficiency of the motor. The motor does consume energy in the form of fuel molecules, but unlike its macroscopic counterpart, a molecular motor can use the energy of the bath to carry out some of the work and still return that energy in the form of heat upon breaking the fuel molecule's bonds. What are the theoretical limits of efficiency in such open machines? How is the efficiency affected by the fact that the heat dissipated by the motor can sometimes, through fluctuations, be negative? What are the microscopic meanings of friction and dissipation when the same molecular processes that give rise to these phenomena in macroscopic machines can actively "drive" the microscopic engines?

Clearly, a large ensemble of motors should behave as a macroscopic system and in that limit (the thermodynamic limit) we can consider the ensemble of motors and their surroundings as a closed system, and therefore, on average, the ensemble should conform to the macroscopic predictions of thermodynamics. Yet, microscopically, at the level of the individual motor, these predictions are no longer valid. Molecular motors can use, at least part of the time, the energy of the bath to perform work. We need a formalism that can appropriately describe the microscopic nature of these systems and the exchanges of energy between the systems and their thermal bath; and it must be a formalism that converges towards the macroscopic results in the thermodynamic limit.

The newly derived *fluctuation theorems*,[5–7] recently tested in microscopic experiments,[8,9] quantitatively describe the fluctuations of microscopic systems that result from the energy exchange between a system and a thermal bath. These theorems connect the realms of equilibrium and non-equilibrium statistical thermodynamics, effectively showing that it is possible to exploit the energy exchange between the system and its bath to extract equilibrium thermodynamics quantities, such as free energy, from processes in which the system is removed arbitrarily far from equilibrium. Moreover, these results also make quantitative predictions about the probability of negative dissipation between the system and its surroundings in any given process. Thus, it is possible that some of these theorems can be extended to describe the energy transactions and the efficiency of microscopic engines immersed in a thermal bath at constant temperature. In fact, some attempts at establishing these connections have already appeared.[10] This is a fertile area of growth in statistical physics that is likely to have important consequences for the connection between statistical mechanics of equilibrium and non-equilibrium, and that may play an important role in the development of a non-equilibrium thermodynamics of small systems.

2.2. The Study of Self-Assembly

One of the most amazing features of a living cell is its complex organization that extends across multiple spatial and temporal scales. Proteins fold into unique tertiary structures; membranes, with various chemical compositions organize to define spatial domains within the cell and to define the limits of the cell itself; microtubules and actin grow and shrink during a number of important dynamical processes during mitosis, cell adhesion and locomotion. Bacterial locomotion depend on self-assembly of extremely complex structures.

All of these processes depend on the ability of simpler parts to self-organize and come together through specific interactions to form well-defined complex structures and shapes. This process is called self-assembly. Self-assembly is so pervasive a feature of the living cell that we often take it for granted. And yet, the emergence of the cell as a self-sustained, self-replicating organism depends on the phenomenon of self-assembly at many different scales in space and time. Self-assembly can either lead to the formation of static structures that occupy a local or the global thermodynamic

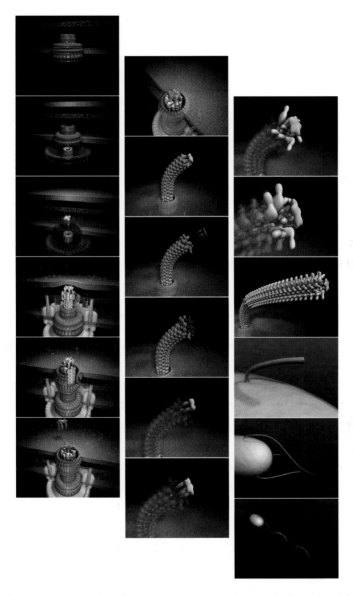

Fig. 3. The assembly of the flagellar motor demonstrates the complexity and sophistication of the self-assembly processes in the cell. This diagram indicates the sequential steps of the assembly of this rotary motor in an *E. coli* cell. (Courtesy of Professor Keiichi Namba, Graduate School of Frontier Biosciences, Osaka University.)

minimum and do not require constant energy dissipation, or of dynamic structures that require constant energy input and dissipation.[11] Both of these are observed in the cell.

Although the study of self-assembly is still in its infancy, the occurrence of self-assembly as a phenomenon in living and non-living systems has attracted the attention of scientists from many different areas of expertise. At the core of the ability of certain components to self-assemble is the existence of some interaction or interactions that are encoded in the structure of those same components either on their surface, in their shape, their charge, their polarizability, etc.[11] These interactions are typically reversible, involving a number of weak, non-covalent associations. In biological systems, the parts themselves are often chemically modified to activate or de-activate these interactions, thus allowing the cell control and regulation of the self-assembly processes.

Perhaps the more surprising feature of biological self-assembly is the robustness of the process itself, i.e., the ability of the various parts to organize themselves, with minimal frustration and minimal trapping into dead-end intermediates. It is thus apparent that self-assembly has evolved in biological systems in such a way that the assembled elements encode not just the strength, directionality and spatial specificity of the assembling interactions, but also the pathways that insure the maximum yield and minimal kinetic frustration. But what is the code? In some cases we know that the code is extremely complex as it is in the case of protein folding. In others, such as in the organization of supramolecular assemblies (assembly of viral capsids, ribosomes, nuclear pores, etc.) the processes involve a smaller number of variables and interactions and are, therefore, more accessible to analysis.

Understanding the code that controls the processes of biological self-assembly and what the physical determinants (stereo-specificity, strength, directionality, reversibility, cooperativity of assembly, etc.) of these interactions are, is an important task for physical biology. Discovering the interaction codes behind self-assembly and their variability will not only reveal one of the essential features of the living state, but also will make it possible for scientists to apply the lessons learned from biology to non-biological building blocks and endow them with some of the same properties to obtain ever more complex supramolecular architectures.

2.3. Towards an Integration of Cell Function

The culmination of the great analytical effort at fractionating, isolating, and purifying the various parts that make up the cell will not immediately lead to a qualitative jump in our understanding of how a cell works. The reason seems apparent: the complexity of biological systems is determined not so much by the number of parts they use to carry out their functions, as by the number of interactions involved in the regulation of these functions. Thus, although eukaryotes have generally larger genomes than prokaryotes, genome sizes are not correlated with the complexity of the organism.[12] Unicelullar eukaryotes have genome sizes that vary 200,000-fold, and the genome of the amoeba is about 200 times greater than that of humans.[13] Levine and Tjian (2003)[14] have argued that the complexity of biological systems correlates not with the size of the genome, but with the ratio and the number of transcription factors per gene. These quantities, the authors argue, are a measure of the amount and the complexity of regulation in gene expression. There are about 300 transcription factors in the yeast genome, 1000 in Drosophila, and possibly 3000 in humans. This analysis suggests that a different approach is needed if we wish to understand how a cell works. Such an approach must seek to study the various functions of the cell, emphasizing not so much the details of their operation but their interrelationships and regulation.

2.3.1. *Systems biology and synthetic biology*

The study of how the various cellular functions are integrated and regulated to sustain the living state is called systems biology. This area of research is becoming more important in biology as a growing number of scientists come to recognize the importance of understanding not just one or another particular function, but how these functions are coordinated and regulated to maintain the cell homeostasis. This is an area of research where physical scientists are beginning to play a significant role, as their quantitative and modeling skills are essential for describing the complexity of these functional interactions. The study of systems biology represents, thus, the opposite of the reductionist, analytical approach that has dominated research in biology for the last 100 years. The essential idea is that to understand a complex mechanism we must not only know what the parts of the mechanism are by taking them apart and seeing how they function, but we need to attempt to put them together into a functioning unit. It is in this latter process that we can hope to complete our understanding of the complex mechanism of the cell.

Building up things to understand them is not a new idea. Early in its development, Chemistry was a discipline based on taking apart, isolating and purifying the components of complex mixtures. In 1828, Wöhler proved that it was possible to synthesize urea *de novo*, and this single event changed the face of chemical science forever. Besides dealing a lethal blow to the vitalism theories of the times, Wöhler's demonstration pulled down the psychological barriers in the minds of other investigators and helped convert Chemistry, then a mainly descriptive branch of science, into a synthetic discipline; research efforts were no longer concentrated only on isolating and purifying compounds already existing but also on creating new unedited chemical species. The result of this "synthesis" phase of chemical research was the discovery of the laws and mechanisms of chemical reactions and eventually, with the advent of quantum mechanics, the development of the theory of the chemical bond. These achievements were clearly the result of the confluence of the analytical or reductionist and the synthetic approaches in chemical research.

I believe that an equivalent phase will occur in biological research. If we really wish to understand how a cell works, scientists must attempt to build a cell, possibly a cell of minimal complexity in the laboratory. The development of a minimal cell could be the turning point that converts Biology into a true synthetic science, helping to establish the minimal complexity required to sustain life and the bases for the rational design of highly specialized, streamlined organisms programmed to perform specific tasks. As it worked for the chemical sciences, we hope that the process of building such a minimal cell will reveal the fundamental principles underlying the organization of living matter.

2.4. Theory and Computation

The increasing trend towards measuring and quantifying complex biological phenomena is paralleled by the desire to model and simulate these phenomena. These efforts are being catalyzed by the fast increase of computational power as embodied in "Moore's law." Although at present the processing speed and memory capacity of computers allow only explicit coordinate simulations over time regimes of the order of nanoseconds, it seems that this limitation will be overcome as technological progress continues. A more concerning limitation of computations is their ability to "predict" the behavior of the system they describe, as opposed to simply the ability to "emulate" the known behavior of the system, that is, "postdict." The ability to predict, which is an essential part of theory, is ultimately related to the quality of the potentials used by these computations. To date, the potentials used by most codes are not good enough to use for making predictions and this is a fundamental limitation. Computational biology will fulfill its promise only when the potentials are good enough to allow scientists to make predictions about the behavior of complex systems. In fact, it can be argued that there are detailed processes whose time or length scale, complexity, or accessibility to observation are such that it is likely that we will never be able to design good enough experiments to study them in detail. For these processes, our only hope is to be able to simulate them and complete our picture through predictions of their behavior. We are not there yet. Obtaining better potentials should be an important goal of computational and physical biologists, as their development would play a decisive role in our ability to carry out the program of quantification in biology.

2.5. New Methods

Periods of fast growth in biological research, and the ability of scientists to observe, detect and quantify complex biological processes, have always followed the development of key advances in technology. Biology is perhaps more than any other discipline a technology-limited science. Because of the "combinatorial" complexity of biological systems, the development of a new method or technique and its application to a particular system inevitably leads to its use in a myriad of similar or related systems. In this manner, the impact of the introduction of a new method in biological research often multiplies many times over. It seems clear then that some of the new challenges for physical biology will require the development of new methods and techniques of physical characterization.

2.5.1. *Single-molecule methods*

The advent of single molecule manipulation and detection approaches have made it possible for scientists to investigate the fluctuating nature of individual molecules and their reactions.[15,16] The stochastic nature of chemical transactions implies that it is not possible to synchronize molecules undergoing a chemical process except for very short initial times. As a result, bulk experiments provide only averages over all the phases of the processes carried out by the molecules in the population. This is what in chemistry has been traditionally called "kinetics," i.e., the description of the dynamics of the mean of the population. In direct contrast, single molecule methods are ideal for investigating the dynamics of complex reactions, for, unlike their bulk counterparts, they make it possible to follow the *trajectories* of the individual molecules as they undergo their reactions in real time, thus avoiding the ensemble average over the unsynchronized signals of the molecules in a population. The study of certain areas, such as molecular motors, has become possible only recently through the use of methods of single molecule manipulation. Using these techniques, quantities that are most natural to these molecular entities and their reactions, i.e., forces, displacements, torques and twist angles can now be directly monitored at the single molecule level and in real time.

These methods are still in an early phase of development, but they are becoming an increasingly powerful tool for investigating the fluctuation dynamics of individual molecules.

Besides the mechanistically rich information contained in these experiments, they provide for the first time direct access to the stochastic

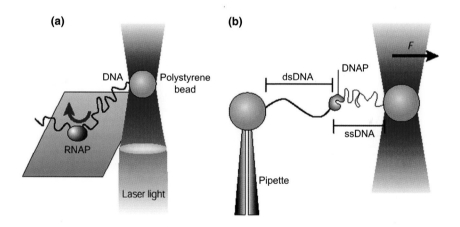

Fig. 4. Two different experimental designs that have been used to investigate the activity of single molecules of **(a)** RNA polymerase; and **(b)** DNA polymerase using optical tweezers. In the first case, the polymerase is bound to the glass slide. Transcription can be followed in real time because as it transcribes the DNA, the molecule must thread the DNA through itself, thus pulling the DNA and the bead away from the optical trap. In the second case, the replication is followed at a prefixed constant force. Since ssDNA and dsDNA have different extension at that force, conversion of ssDNA into dsDNA during replication is accompanied by a change in extension and a change in the distance between the two beads.[18]

dynamics of microscopic systems involving few molecules, a view directly relevant to our understanding of cellular processes, as discussed above. Moreover, it is almost certain that technical developments in single molecules will make it possible in the not-too-distant future to carry out experiments inside the cell.

2.5.2. *Ultra-high resolution optical microscopy*

Another technology that would have a big impact on biological research and on our understanding of the cell is the development of an ultra-high resolution optical microscope capable of providing vital images of the cell with a resolution of between 15 and 30 nm. In the last ten years, various ingenious schemes have been developed to increase the resolution of the optical microscope beyond the diffraction limit. These include the structured illumination approach,[18] stimulated emission optical microscopy,[19] and photo-activated light microscopy (PALM).[20] These methods represent

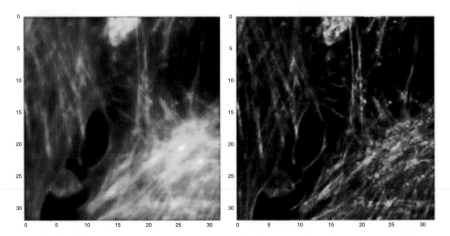

Fig. 5. Fluorescently stained bovine endothelia cells: (*left* panel) imaged conventionally and (*right* panel) with the use of structured-illumination. (Courtesy of Dr. Stephan Stranick, Chemical Science and Technology Laboratory, National Institute of Standards and Technology, USA.)

important advances towards the goals stated above, but they are not quite there yet in terms of spatial and temporal resolution. Also, recent advances involving the use of negative index of refraction lenses based on the generation of plasmon–polariton waves on the surface of metallic lenses[21] have demonstrated spectacular resolution down to ~ 30 nm resolution.[22]

There is a great deal of biology to be learned from the imaging of a cell at this super-resolution under live conditions. It may not be possible yet to accomplish this feat, but it is clear that these developments have broken the spell cast more than 120 years ago by the concept of the diffraction limit in Abbe's linear theory of imaging. The hope is that biologists will be able to count on such super-resolution instruments within the next decade.

3. Conclusions

The list of challenges described in this essay contains some of the areas in which we can expect to see important developments in the following decades. It is by no means exhaustive. The convergence of the top-down reductionist approaches and the bottom-up synthetic approach will likely lead to completely new insights into our understanding of how cells regulate and maintain their homeostasis. Moreover, the synthetic effort will allow us to develop for the first time mechanistic models of cell operation. From these studies, scientists will begin to understand in quantitative terms how the various functions of the cell are spatially and temporally coordinated and how to describe the energetics and stability of the steady states accessible to the cell under a set of external control parameters. This subset of possible or accessible states will define a domain in the parameter space consistent with organized living matter.

Barely six decades after the launching of the molecular biology revolution, concepts such as forces and torques, motors and engines, energy conversion and thermodynamic efficiency, among others, are becoming part of the *lingua franca* of biology. Biological research is thus a fertile ground once again for the convergence of a multidisciplinary group of scientists trained in areas as diverse as physics, chemistry, and engineering. The charge towards quantification in biology is now being led by scientists trained as much in physics and mathematics as in biology. But this time their contribution will not likely be limited to pointing the discipline in the right direction; rather, they are entering the field to stay, and to change the playing ground forever.

References

1. Haw M. (2007) What is the state of thermodynamics on the 100th anniversary of the death of Lord Kelvin? *Am Sci* **95(6):** 472–474.
2. Keller D, Swigon D, Bustamante C. (2003) Relating single molecule measurements to thermodynamics. *Biophys J* **84:** 733–738.
3. Office of NIH History: Exhibit on Marshall Nirenberg. http://history.nih.gov/exhibits/nirenberg/HS6_univ.htm
4. Smith DE, Tans SI, Smith SB, Grimes S, Anderson DE, Bustamante C. (2001) The bacteriphage phi29 portal motor can package DNA against a large internal force. *Nature* **413:** 748–752.
5. Jarzynski C. (1997) Nonequilibrium equality for free energy differences. *Phys Rev Lett* **78:** 2690.
6. Crooks GE. (1999) Entropy production fluctuation theorem and the nonequilibrium work relation for free energy differences. *Phys Rev* **E60:** 2721.
7. Bustamante C, Liphardt J, Ritort, F. (2005) The nonequilibrium thermo-dynamics of small systems *Phys Today* **58:** 43–48.
8. Liphardt J, Dumont S, Smith SB, Tinoco (Jr) I, Bustamante C. (2002) Equilibrium information from nonequilibrium measurements in an experimental test of Jarzynski's equality. *Science* **296:** 1832.
9. Collin D, Ritort F, Jarzynski C, Smith SB, Tinoco (Jr) I, Bustamante C. (2005) Verification of the Crooks fluctuation theorem and recovery of RNA folding free energies. *Nature* **437:** 231–234.
10. Andrieu D, Gaspard P. (2006) Fluctuation theorems and the non-equilibrium thermodynamics of molecular motors. *Phys Rev* **E74:** 011906-1–011906-15.

11. Whitesides GM, Grzybowski B. (2002) Self-assembly at all scales. *Science* **295:** 2418–2421.

12. Phillips RB. (2004) Genome complexity: Adaptive evolution or genetic drift? Does genome complexity produce organismal complexity? *Heredity* **93:** 122–123.

13. Gregory TR. (2001) Coincidence, coevolution, or causation? DNA content, cell size, and the C-value enigma. *Biol Rev* **76:** 65–101.

14. Levine M, Tjian R. (2003) Transcription regulation and animal diversity. *Nature* **424:** 147–151.

15. Smith SB, Finzi L, Bustamante C. (1992) Direct mechanical measurements of the elasticity of single DNA molecules using magnetic beads. *Science* **258:** 1122–1126.

16. Betzig E, Trautman JK. (1992) Near-field optics: Microscopy, spectroscopy, and surface modification beyond the diffraction limit. *Science* **257(5067):** 189–195.

17. Bustamante C, Macosko, JC, Wuite GJL. (2000) Grabbing the cat by the tail: Manipulating molecules one by one. *Nature Rev Molec Cell Biol* **1:** 130–136.

18. Gustafsson MGL. (2000) Surpassing the lateral resolution limit by a factor of two using structured illumination microscopy. *J Microscopy* **198(2):** 82–87.

19. Klar TA, Jakobs S, Dyba M, Egner A, Hell SW. (2000) Fluorescence microscopy with diffraction resolution barrier broken by stimulated emission. *PNAS* **97:** 8206–8210.

20. Betzig E, Patterson GH, Sougrat R, Lindwasser OW, Olenych S, Bonifacino JS, Davidson MW, Lippincott-Schwartz J, Hess HF. (2006) Imaging intracellular fluorescent proteins at nanometer resolution *Science* **313:** 1642–1645.

21. Fang N, Lee H, Sun C, Zhang X. (2005) Sub-diffraction-limited optical imaging with a silver superlens. *Science* **308:** 534–537.

22. Smolyaninov I, Hung Y, Davis CC. (2007) Magnifying superlens in the visible frequency range. *Science* **315:** 1699.

chapter 6

The Challenge of Quasi-Regular Structures in Biology

Roger D. Kornberg[*]

Most biological macromolecules function in the form of molecular assemblies. Although often conformationally heterogeneous, these assemblies contain rigid subcomplexes whose structures may be determined by cryoelectron microscopy and image processing. We have proposed the use of heavy atom clusters to facilitate the analysis, and have recently synthesized clusters of the size and reactivity towards proteins required for the purpose.

Structure determination of proteins by X-ray crystallography and nuclear magnetic resonance (NMR) spectroscopy, especially rigid domains derived by fragmentation, has become routine. Even very large, rigid multiprotein complexes, such as RNA polymerase II (12 subunits, 514 kDal) and ribosomal subunits (1.55 MDal, 793 kDal), are now amenable to X-ray analysis. Where current methodology is limited or fails entirely is in cases of structural heterogeneity and conformational flexibility, which are common in biology. Individual proteins may not crystallize for one or other of these reasons. The largest objects of biological interest fail on both counts, and are believed to lie beyond contemporary reach. Chromatin fibers, for example, are both heterogeneous, varying in composition along their length, and as flexible as the DNA that runs their length. Yet they possess an underlying regularity, as they are based on a linear array of nucleosomes. They are "quasi-regular," and the determination of their structure and those of many

[*]Department of Structural Biology, Stanford School of Medicine, Stanford, CA 94305, e-mail: kornberg@stanford.edu.

other such objects of biological interest will require improved methods of investigation.

One approach is automated screening of the crystallization conditions, which may help in cases of conformational flexibility of otherwise uniform objects. Stephen Quake and coworkers (this volume) have devised a method of screening based on microfluidics, which enables the testing of miniscule amounts of protein in a large number of conditions. Another approach is electron microscopy (EM), which does not require crystals at all, as described by John Meurig Thomas (this volume). Yet another strategy is the investigation of one molecule at a time, exemplified by the work of Carlos Bustamente and collaborators (this volume).

Here we discuss a particular EM methodology that may prove applicable to the quasi-regular structure problem. EM is capable of atomic resolution. The limitation is that high energy electrons destroy biological macromolecules, so the dose required for imaging must be spread over many identical particles, whose images are subsequently averaged to improve the signal-to-noise ratio. Averaging requires aligning the images of randomly oriented particles, which has been done in two ways. The first is by forming thin, ordered arrays (2-D crystals) so that all the particles have the same orientation and can be immediately averaged through Fourier analysis. Several proteins have been solved by this approach,[1–5] demonstrating the feasibility of near atomic resolution protein EM. Alternatively, images of single (noncrystalline) particles can be oriented by correlation searches against models. This approach is commonly referred to as "single particle analysis."[6]

The advantages of single particle analysis in comparison to other techniques of structure determination are that miniscule amounts of specimen are needed (only millions of particle images are required), the technique can be applied to any rigid, soluble particle (no crystallization is necessary), the whole process from sample preparation to model calculation is automatable, particles can be imaged in their native concentrations and buffer conditions, images of deformed or sub-stoichiometric particles can be selectively discarded, and multiple conformations can be distinguished and solved independently.

Despite these advantages, the highest resolution yet achieved using asymmetric, single particles is approximately 20 Å. Resolution is limited in two ways: imperfect images contain only part of the required information, and imperfect image alignment causes information loss during particle averaging. Coherent particle averaging requires precise knowledge of eight "alignment" parameters for each particle: two coordinates to locate the particle center, three rotational angles to describe orientation, a measure of homogeneity to allow rejection of deformed particles, the height of the particle with respect to the object plane of the objective lens (defocus),[7] and magnification.[8] The same lack of high resolution detail in a low dose image that necessitates averaging unavoidably limits the precision of alignment.

We have proposed a heavy atom EM, or "HevEM," strategy which may reduce alignment errors sufficiently to allow routine near atomic resolution structure determination by single particle analysis.[9] The strategy is to rigidly attach multiple heavy atom clusters to specific sites on the particle before imaging, and then align the images based on the projected positions of the clusters. Because heavy atom clusters are so much denser than protein or water, they can be individually seen in images, even with low electron doses and close to focus, conditions required to record high resolution protein structures.[10] Labeling at four sites should be sufficient to determine all the eight alignment parameters describing location, orientation, homogeneity, defocus, and magnification[9] to high precision.

The practical success of the HevEM approach requires uniform clusters about 1 nm in diameter and a general method for their rigid attachment to any particle of interest. We are using antibody fragments as "adaptor" molecules to bind both a heavy atom cluster and the particle to be imaged. Fv antibody fragments are the best choice because they comprise a single, rigid, soluble domain. Fvs can be engineered as single polypeptide chains (scFvs), readily manipulated genetically, and easily expressed in bacteria.[11] Fvs consist of a variable antigen-binding surface supported by a constant framework region. We have engineered a cluster binding site into the framework region of a scFv by introducing a surface cysteine residue and other, empirically determined, neighboring residues.

The fundamental advance is that rigidly fixed heavy atom clusters act as alignment markers and reveal precise orientation parameters for each particle. Fortunately, the clusters are most distinct close to focus, which is beneficial for two reasons: (1) far-from-focus images, which are currently employed to enhance particle visibility, suffer severe losses of high resolution detail due to beam incoherence; and (2) the contrast transfer functions inherent in the far-from-focus images currently used oscillate so rapidly at high resolution that correction would be all but impossible. Furthermore, the rate-limiting step of current single particle analysis is not the acquisition of good images, but rather the computation required to iteratively align tens of thousands of images. Because heavy atom clusters reveal particle alignment through a simple set of three equations, reconstructions using tens of millions of particles can be accomplished with minimal computation.

Of perhaps greatest novelty and ultimate significance, HevEM should be applicable to quasi-regular structures and impure proteins. The use of scFvs directed against regular features, such as the nucleosomes in chromatin, will reveal their distribution, and thus the underlying structure of the chromatin fiber. HevEM may even be applied to proteins *in vivo*, as scFvs can bind directly to a cell surface or may be introduced in cells by uptake or injection. The cells can then be frozen and sectioned for imaging. The structures of all the proteins rigidly attached in a large complex containing the scFv target should be revealed.

The technical challenges for implementation of HevEM are three-fold. First, a perfectly uniform cluster in the nanometer size range is required. Second, one-to-one conjugates of the cluster with scFvs of interest must be prepared. Finally, the scFvs must be modified to rigidify the cluster–protein interaction. In work done to date, the first two objectives have been accomplished. A molecularly defined, thiol monolayer-protected gold cluster, $Au_{102}(p\text{-mercaptobenzoic acid})_{44}$, has been synthesized and solved by X-ray

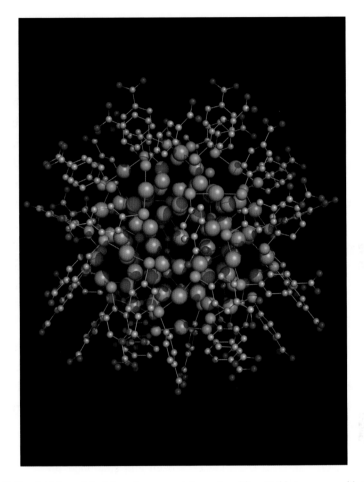

Fig. 1. Ball and stick model of $Au_{102}(p$-mercaptobenzoic acid$)_{44}$. Gold atoms are gold, except for those on the central axis of the cluster core, which are red. Sulfur atoms are cyan, oxygen atoms are red, and carbon atoms are gray. (Courtesy of Pablo Jadzinsky.)

crystallography to 1 Å resolution, revealing a 1.5 nm cluster core (Fig. 1).[12] Two procedures have been developed for the direct coupling of a cluster core to surface cysteine residues of scFvs[13]: one procedure entails activation of the cluster core by oxidation, reaction with the scFv, and deactivation by reduction; the other procedure starts with a reactive cluster, which is conjugated with an scFv and then deactivated by exchange of the surface thiol residues.

It remains to assess the rigidity of the cluster–protein interaction and to optimize rigidity by amino acid changes to the scFv. The mobility of the cluster must be less than about 2 Å for near atomic resolution of protein structure by alignment and image averaging based on cluster positions. The resolution of p-mercaptobenzoic acid moieties in the Au_{102} cluster structure to 1 Å gives reason to anticipate this limitation on cluster mobility will be achieved. The hope of HevEM for solution of the quasi-regular structure problem should then become a reality.

Acknowledgments

I am gratefully indebted to my colleagues at Stanford, Drs. Grant Jensen, Chris Ackerson, Pablo Jadzinsky, David Bushnell, Maia Azubel, and Guillermo Calero. Our research has been supported by the NIH (GM63025, AI21144) and by the NSF (CHE-0617050).

References

1. Nogales E, Wolf SG, Downing KH. (1998) Structure of the alpha beta tubulin dimer by electron crystallography. *Nature* **391:** 199–203.
2. Kimura Y *et al.* (1997) Surface of bacteriorhodopsin revealed by high-resolution electron crystallography. *Nature* **389:** 206–211.
3. Kuhlbrandt W, Wang DN, Fujiyoshi Y. (1994) Atomic model of plant light-harvesting complex by electron crystallography. *Nature* **367:** 614–621.
4. Miyazawa A, Fujiyoshi Y, Unwin N. (2003) Structure and gating mechanism of the acetylcholine receptor pore. *Nature* **423:** 949–955.
5. Gonen T *et al.* (2005) Lipid-protein interactions in double-layered two-dimensional AQP0 crystals. *Nature* **438:** 633–638.
6. Frank J *et al.* (1996) SPIDER and WEB: Processing and visualization of images in 3D electron microscopy and related fields. *J Struct Biol* **116:** 190–199.

7. Jensen GJ. (2001) Alignment error envelopes for single particle analysis. *J Struct Biol* **133:** 143–155.

8. Aldroubi A, Trus BL, Unser M, Booy FP, Steven AC. (1992) Magnification mismatches between micrographs: Corrective procedures and implications for structural analysis. *Ultramicroscopy* **46:** 175–188.

9. Jensen GJ, Kornberg RD. (1998) Single-particle selection and alignment with heavy atom cluster-antibody conjugates. *Proc Natl Acad Sci USA* **95:** 9262–9267.

10. Wagenknecht T, Berkowitz J, Grassucci R, Timerman AP, Fleischer S. (1994) Localization of calmodulin binding sites on the ryanodine receptor from skeletal muscle by electron microscopy. *Biophysical J* **67:** 2286–2295.

11. Winter G, Griffiths AD, Hawkins RE, Hoogenboom HR. (1994) Making antibodies by phage display technology. *Annu Rev Immunol* **12:** 433–455.

12. Jadzinsky PD, Calero G, Ackerson CJ, Bushnell DA, Kornberg RD. (2007) Structure of a thiol monolayer-protected gold nanoparticle at 1.1 A resolution. *Science* **318:** 430–433.

13. Ackerson CJ, Jadzinsky PD, Jensen GJ, Kornberg RD. (2006) Rigid, specific, and discrete gold nanoparticle/antibody conjugates. *J Am Chem Soc* **128:** 2635–2640.

chapter 7

The Future of Biological X-Ray Analysis

Douglas C. Rees[*]

In considering the future of X-ray analysis in 1945, J.D. Bernal predicted that "pure" X-ray analysis would solve more and more complex structures, while "applied" X-ray analysis would become indispensable in a diverse range of fields. This vision has been largely realized, particularly in biology, where X-ray crystallography has contributed substantially to our present understanding of macromolecular structure and function. What remains for the next 50 years? It seems inevitable that advances in X-ray analysis, propelled by developments in biological methodologies, X-ray technology, computation and the tools of physical biology, will solve larger and more complex biological structures, at ever higher resolutions and on faster time scales. A long-term goal will be to use these insights to develop molecularly explicit models with predictive capabilities for the structures and dynamics of cells and higher order assemblages, corresponding to the biological counterpart of weather forecasting.

1. Introduction

Following the first observation in 1934 of X-ray diffraction from a crystalline protein,[1] Dorothy Crowfoot Hodgkin recalled that J.D. Bernal "full of excitement, wandered about the streets of Cambridge, thinking of the future and of how much it might be possible to know about the structure of proteins if the photographs he had just taken could be interpreted in every detail."[2] This euphoric response to macromolecular diffraction patterns is not unusual; Hodgkin herself described the occasion when she

[*]Division of Chemistry and Chemical Engineering 114-96, Howard Hughes Medical Institute, California Institute of Technology, Pasadena, CA 91125, USA, e-mail: dcrees@caltech.edu.

initially observed diffraction from crystals of insulin in X-ray photographs as "probably the most exciting in my life"; she then "wandered the deserted streets of Oxford in her elation."[3] Driving these reactions was the conviction that X-ray crystallography would reveal the molecular structure of proteins and other macromolecules in atomic detail; this conviction is evident, for example, in the remarkable 1939 article by Bernal on "Structure of Proteins."[4] While such optimism could be fairly characterized as sheer fantasy at that time, within two decades, Bernal's dream was realized with Kendrew's determination of the crystal structure of myoglobin.[5] In the intervening period, over 40,000 structures of proteins and other macromolecules have been solved and deposited in the Protein Data Bank,[6] with the number doubling approximately every three years (Fig. 1).

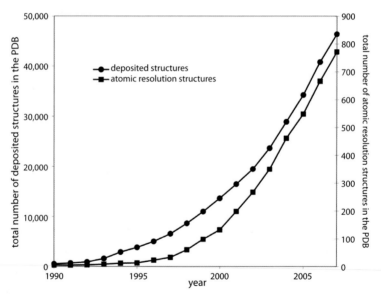

Fig. 1. Cumulative growth in the number of macromolecular structures deposited between 1990 and 2007 in the Protein Data Bank (http://www.rcsb.org[6]). The annual growth in the total number of deposited structures (left-hand axis) and in the total number of X-ray structures at atomic resolution (right-hand axis) are depicted by the circles and squares, respectively. Approximately 2% of the deposited structures are presently at atomic resolution.

In 1945, Bernal outlined his vision for the future of X-ray analysis in a lecture at the Royal Institution.[7] To provide a foundation for this vision, Bernal identified three major stages in the development of X-ray analysis up to that point. During the first stage from 1912 to 1915 (also referred to as the Bragg period), the fundamental crystallographic methodology was pioneered and "all the basic structures were worked out" for the different types of crystal packing. During the second period, crystallographic approaches were rapidly applied to a broad array of areas; this "heroic age of X-ray analysis" lasted till 1933. At that point, "It was then realized that the new structures would be essentially like the old ones," and "interest was directed towards new and more complex phenomena," including order–disorder transformations and biological structure, in a third phase that extended to the start of World War II. In Bernal's view, future developments following resumption of academic research after the war would reflect a balance between "pure" X-ray analysis that would solve more and more complex structures and "applied" X-ray analysis that "will become as essential a tool for other fields of research as the microscope and the spectroscope have been in the past." Needless to say, over the past half century, Bernal's vision has been largely realized.

As a starting point to consider the future of biological X-ray analysis, it would be appropriate to follow Bernal and trace the historical developments in macromolecular crystallography. There is a clear correspondence, for example, between these developments and the three stages identified by Bernal. The first stage, which could be referred to as the Perutz period, witnessed the development of the basic methodology, particularly the solution to the phase problem (by the method of isomorphous replacement[8]) that had thwarted earlier investigations. This first stage culminated in the structure determination of myoglobin.[5] The second stage saw the application of these basic approaches to the crystallographic analyses of increasing numbers of macromolecular structures; the end of this era may be defined in various ways, but I would suggest it is marked by the structure determination of the ribosomal subunits,[9–11] as they represent the direct descendant of the methodological lineage pioneered by Perutz — preparation and crystallization of native material and structure determination through the use of heavy atom derivatives — but on a much larger scale. During this period, Bernal's vision for the detailed interpretation of the X-ray diffraction patterns of proteins and their implications for protein (and more generally macromolecular) structure was essentially realized; today, we increasingly find that "the new structures would be essentially like the old ones," at least in broad brush, if not in detail.

Advances in X-ray synchrotron sources, computation and biological methodologies allow us to contemplate new and challenging aspects of X-ray analysis (and structural biology more generally) that were fantasy even a few years ago. With these rapidly unfolding opportunities, we need to identify issues that cause us to wander the streets, full of excitement about the future. As Bernal noted, "the predictable future is the rational projection of the past and the present"; while these extrapolations may be inspired, the same data can be interpreted in different ways leading to very different conclusions. As an example, Linus Pauling also wrote a paper in 1939 on "The Structure of Proteins."[12] Parts of this paper are fresh and relevant today, such as the importance to protein structure of hydrogen bonds and other non-covalent interactions. Pauling opined, however, that "the great complexity of proteins makes it unlikely that a complete structure determination for a protein will ever be made by X-ray methods alone," since "A protein molecule, containing hundreds of amino acid residues, is immensely more complicated than a molecule of an amino acid or of diketopiperazine. Yet despite attacks by numerous investigators, no complete structure determination for any amino acid had been made until within the last year." Even Bernal in 1945 suggested[7] that "The structure of the proteins, fundamental to the understanding of living processes, will not yield to X-ray analysis alone"; this view is further clarified by the consideration that "the attack must be combined with biochemical, centrifical, electrophoretic, electron microscope and other methods, but there is no doubt that X-rays will play a crucial part in verifying all hypothetical structures put forward on any other basis." A contemporary statement of this view would include nuclear magnetic resonance, single molecule methodologies and computation among the tools of physical biology that are brought to bear on biological systems.

Keeping in mind that an extrapolation from the past through the present by necessity provides a conservative view of the future, it seems inevitable that continuing advances in the application of X-ray analysis to the study of biological systems are to be anticipated in three general areas: the structural analysis of complex biological systems, structure determination at ever higher resolutions, and characterization of the structural dynamics of macromolecules at increasingly faster time scales. Overviews of these three areas will be discussed in turn. As this short personal perspective focuses on X-ray methods, no attempt is made to provide a balanced analysis of developments in the broader area of structural biology.

2. Structural Analysis of Complex Biological Systems

At least in the immediate future, the most significant biological insights provided by X-ray methods will undoubtedly arise from structural analyses of larger and more complex systems and assemblages. As described in S.C. Harrison's thoughtful analysis "Whither structural biology?"[13] to remain an intellectually vibrant discipline in the biological sciences, structural molecular biology will necessarily merge with structural cell biology to provide molecularly detailed representations of cellular processes in space and time. As Bernal appreciated, this goal cannot be exclusively realized through X-ray analysis alone. However, X-ray analysis is well suited to providing the atomic-level framework for this understanding. In a curious way, the future role of X-ray crystallography in providing a molecular view of cellular function is evolving from the "top down" approaches of the past to a more "bottom up" approach. In the early days, two schools of protein structure could be identified, the British and the Pasadena schools, that were distinguished by their approaches to this problem. The Pasadena school essentially pursued a "bottom up" approach that provided fundamental insights into the structure and properties of the peptide bond, including the implications for hydrogen bonding. These efforts culminated in Pauling's celebrated discoveries of the α-helix and β-sheet.[14,15] This approach failed to achieve an understanding of the tertiary structure of proteins, however, which is at the heart of the "protein structure problem." In contrast, the British school, epitomized by Bernal, Bragg and Perutz, tackled the problem by directly targeting the structure of entire proteins; this "top down" approach ultimately achieved the first three-dimensional structure of a protein.[5] We are now at the stage where macromolecular crystallography is switching back to a more "bottom-up" role, by providing high resolution structures that serve as the building blocks to interpret lower resolution images provided by electron and other microscopies to stitch together molecular-level depictions of cells.

X-ray analyses of complex biological systems present two challenges: (i) the biochemical problem of tackling large complexes, membrane proteins, etc., where sample purification and preparation of suitable crystals create major barriers; and (ii) the technical problems associated with the crystal structure determination. Preparative biochemistry issues are particularly acute for multimolecular assemblages with numerous, perhaps transiently associated, subunits, and the consequent challenges of preparing material that retains a full complement of subunits. Although daunting, these obstacles can be overcome, as with the structural analysis of RNA polymerase.[16] Membrane proteins pose a distinct set of challenges; in particular, the same properties that enable membrane proteins to function in the heterogeneous milieu of cell membranes also have profound consequences for their structural analysis. Reflecting the small membrane surface area to cellular volume ratio, membrane proteins are typically present in low abundance, and solubilization from the membrane requires the use of detergents or other amphiphiles that may not faithfully mimic the lipid bilayer. Although there are notable exceptions,[17] eukaryotic membrane proteins have been particularly problematic in terms of the ability to recombinantly produce material appropriate for structural studies.

For systems such as macromolecular complexes or membrane proteins, the amount of material available for an X-ray analysis may create significant problems. In this regard, microfluidic methods that manipulate volumes down to nanoliters have significant potential to revolutionize the handling, purification and crystallization of biomolecules.[18] An estimate of the minimal amount of material required for an X-ray structure may be obtained from the size of a crystal ($\sim 10^3 \, \mu^3$) that can be used for data collection at a synchrotron microfocus beamline. A crystal of this size contains ~ 1 ng of material, which is equivalent to 0.1 nl of a 10 mg/ml solution. Screening a thousand crystallization conditions would consequently require 100 nl of protein solution, which contains 1 μg protein. While this undoubtedly represents the "best-case" scenario, it is still orders of magnitude less material than the 10–100 mg that have been ball-park targets over the past few decades.

The technical challenges in an X-ray structure determination will generally reflect the diffraction quality of the crystals, which encompasses such characteristics as resolution, mosaicity, radiation sensitivity, etc. In fortunate cases, these characteristics can be improved by appropriate sample manipulations. The ability to successfully collect diffraction data from a crystal will ultimately depend, among other factors, on the ability to resolve neighboring reflections, which in turn is dependent upon the unit cell dimensions. All things being equal, larger assemblages will crystallize in larger unit cells — the exact relationship will depend on the details of the crystallographic symmetry, the extent (if any) of non-crystallographic symmetry and the packing density. The largest crystal structures that have been solved to date can have primitive unit cell axes over 1000 Å (Ref. 19) and asymmetric unit volumes greater than $10^8 \, \text{Å}^3$ (Fig. 2). An isotropic object with this volume would have overall dimensions of ~ 500 Å. The data collection challenges are determined in part by the unit cell dimensions and, for objects of this size, even moderate resolution structures will have several

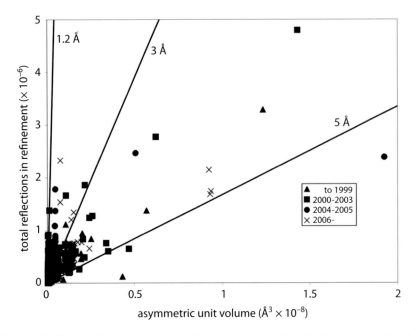

Fig. 2. Distribution of the asymmetric unit volume and number of reflections used in the refinement for very large X-ray structures deposited in the Protein Data Bank. The structures are classified into four sequential time periods with roughly equal numbers of structures (~ 8,000–10,000) solved by all methods. These groups cover the periods from the beginning to 1999, within 2000–2003, 2004–2005 and 2006–2007, and are designated by triangles, squares, circles and crosses, respectively. Assuming a solvent content of ~ 50% typical for water soluble proteins,[43] the molecular weight of the contents of an asymmetric unit with a volume equal to $10^8 \, \text{Å}^3$ would be ~ 40×10^6. Although the total number of structures solved each year is increasing (Fig. 1), this is not the case for very large structures, likely because these are tackled on a case-by-case basis and are not amenable to more high-throughput approaches. ~ 70% of the 37,655 structures used in this analysis had asymmetric unit volumes below $0.002 \times 10^8 \, \text{Å}^3$ and were excluded from this analysis, as were structures solved at resolutions above $5 \, \text{Å}$. For reference, the theoretical number of reflections to limiting resolutions of $1.2 \, \text{Å}$ (atomic resolution), $3 \, \text{Å}$ and $5 \, \text{Å}$ are indicated as a function of the asymmetric unit volume. As the actual X-ray reflection sets used for refinement are typically incomplete, points corresponding to structures solved at a given resolution generally fall below the theoretical curves.

million observed reflections. While the largest crystal structures that have been solved are of viruses with substantial non-crystallographic symmetry (that significantly facilitates the structure determination), asymmetric objects with molecular weights of several million (e.g., intact ribosomes[20,21]) have been solved in atomic detail.

As this survey indicates, the upper limits of presently solved crystal structures ($\sim 1000\,\text{Å} = 0.1\,\mu$) are approaching sizes comparable to those of cells ($\sim 1\,\mu$ for bacteria). Provided suitable crystals can be obtained, undoubtedly crystals with even larger cell dimensions could be studied, particularly given the capabilities of third generation synchrotron sources and advances in microfocus optics. The limit would correspond to recording the continuum X-ray patterns from single objects; while this may appear fanciful, calculations suggest this is indeed possible using 4th generation pulsed X-ray sources, such as the Linac Coherent Light Source (LCLS) or the X-ray free electron laser (XFEL) under development at SLAC and DESY, respectively, where the scattering pattern may be recorded before the sample is destroyed through radiation damage.[22] The overarching goal is to push the limits of X-ray methods to bridge high-resolution structures with developments in cryo-electron microscopy to realize the development of atomic scale models of living systems.

Beyond the challenges provided by large molecular size, there are significant components of biological systems that lack the long-range order requisite for crystallographic analyses. In these cases, X-ray analysis can proceed in one of two directions: preparation of suitably modified forms of a sample that are amenable to crystallization, or implementation of more generalized scattering methods. Two systems, of many, that illustrate these considerations are amyloids and cell membranes. Amyloids are fibril-type aggregates that form from a remarkably diverse set of proteins that have apparently little in common, at least in terms of structure, except for the ability to form these fibrils.[23] A variety of studies, including the high resolution structure of an amyloid forming peptide solved from microcrystals,[24] are consistent with a β-sheet secondary structure arrangement of the polypeptide chain in these structures. Despite the ubiquity of these elements and the local order in the β-sheet, more detailed structural arrangements have been problematic due to the lack of long-range tertiary order that precludes formation of high quality X-ray crystals of amyloids from full-length proteins. The structural organization of biological membranes reflect similar considerations. The general arrangement of the lipid components into a bilayer is well understood and may be considered as equivalent to the membrane's "secondary structure." The typical representation of a membrane bilayer as composed of a set of "lollypop" structures, though, is clearly misleading in terms of both structure and dynamics. Understanding how all the components pack together,[25] including the distribution and interactions of the protein components and the myriad different types of lipids, to form the cell membrane remains a great challenge.

3. Structure Determination at Ever Higher Resolution

Accompanying the explosion in the Protein Data Bank has been the associated increase in the number of structures refined at atomic resolution (Fig. 1), defined as resolutions beyond 1.2 Å.[26] The highest resolution structure deposited to date in the PDB is crambin, solved at 0.54 Å resolution,[27] which is comparable to the resolution used to refine small molecule structures. Structural analysis at atomic resolution is essential to define the structure and stereochemistry of novel cofactors, modifications and ligands, such as the structure of the nitrogenase FeMo-cofactor at 1.16 Å resolution.[28] These analyses not only define the topology of the structure in terms of the heavy (i.e., non-hydrogen) atoms, but can also identify hydrogen atoms in macromolecular structures (such as direct visualization of the Watson–Crick hydrogen bonds between base pairs in B-form DNA[29]), establish the oxidation states of metals in metalloclusters,[30] and illuminate the details of the bonding electron density in peptides.[31] An as yet unrealized, but tantalizing, possibility is the high resolution characterization of active sites, particularly metallocenters, in terms of experimental charge densities[32] that would give clues to the electronic structure of catalytic centers within proteins.

While the rules of covalent bonding are generally well understood, particularly for organic and biological molecules, the native tertiary structures of macromolecules reflect the crucial role of non-covalent interactions. In contrast to covalent interactions that are strong and highly directional, non-covalent interactions are weak and non-directional. What they lack in strength, however, they make up in numbers, with the result that many-body effects are significant for the treatment of both energetic and entropic contributions to the thermodynamics of these systems.[33] As a consequence, potential functions used to evaluate the contributions of non-covalent interactions to macromolecular energies are imprecise, and incorrect structures can have similar energies to the correct ones. A particularly intriguing prospect for ultra-high resolution crystallography is to help quantitatively characterize non-covalent interactions through precise location of all the components of the system — both macromolecule and solvent — to improve the modeling of these effects through comparison of predicted and observed positions. An important advantage of this approach is that both the experimental work and the theoretical analysis can be done on a structurally well characterized system — a protein crystal. As discussed below, applications to electrostatic and solvation interactions appear particularly appealing.

The electrostatic potential $\Phi(r')$ at a point r' in a crystal may be calculated by Coulomb's law from the surrounding charge distribution. Following the derivation by Bertaut,[34] extended to a screened Coulomb potential to take into account the presence of surrounding electrolytes, the potential may be evaluated as a lattice sum calculated by the Fourier transform:

$$\Phi(r') = \frac{z_i e}{r'} e^{-\kappa r'} = \frac{1}{\pi V} \sum_{\mathbf{h}} \frac{F_{\text{tot}}(\mathbf{h})}{h^2 + \alpha^2} e^{-2\pi i \mathbf{h} \cdot \mathbf{r}'} \tag{1}$$

where the total structure factor, F_{tot}, contains both the negative contribution from the electrons (given by the macromolecular structure factor for which the amplitude is measured in the diffraction experiment), as well as the positive contribution from the nuclei, and $2\pi\alpha = \kappa$ where κ is the reciprocal of the Debye length. With the h^2 term in the denominator, high resolution data are not required to calculate the potential, although this is needed for the electric field and other derived quantities.[32] Although this type of electrostatic analysis has apparently not been exploited for macromolecules, the beauty of this approach is that the electrostatic potential may be calculated from experimental data and compared to the Poisson–Boltzmann computational models.[35] An intriguing direction would be to experimentally analyze the electric field across membranes in the presence of an applied electric field. These gradients are estimated to be ~ 10 MeV/m (0.1 V across 100 Å); while experimental characterization of the trans-membrane potential is a challenging problem, it is crucial to do so in view of the bioenergetic significance of this property.

High resolution structural analyses may also be of value in detailing the solvation structure surrounding crystalline macromolecules. The solvent properties of water are essential for life, both to provide a medium that allows proper and timely associations between biomolecules, and through solvation interactions that are central to the ability of macromolecules to adopt and maintain stable, well defined tertiary structures. The chemical simplicity of water belies the complexity of the physical chemical description of these processes. One approach to evaluating models for protein–water interactions would be to computationally simulate the solvent structure in macromolecular crystals where the crystal structure is accurately known,[36] including (particularly) the solvent distribution; the availability of multiple high resolution crystal structures with clearly defined solvent structure (including low occupancy sites) would provide excellent test systems for this purpose.

In addition to frontier-type research, high resolution structural analyses will also be crucial as a component of more general research programs addressing macromolecular function, mechanism and design. This role for X-ray crystallography corresponds to Bernal's "applied X-ray analysis" and is comparable to that of a small molecule X-ray facility solving structures generated by synthetic chemists. (Indeed, one could argue that many of the developments in macromolecular crystallography follow comparable developments in small molecule crystallography by a few decades.) High resolution structures are crucial in this arena, as they provide the chemical-level definition of structure and non-covalent interactions (hydrogen bonds, contacts, torsional angle conformations) that are fundamental to pursuing molecule design, recognition and catalytic processes. At Caltech, an integrated approach to crystallography spanning the small to large molecule limits is being developed as a "Molecular Observatory" to facilitate the efforts of the broader chemical and biological communities to take advantage of contemporary technologies in X-ray crystallography.

4. Structural Dynamics of Macromolecules

Biological macromolecules are not static entities, and atomic-level characterization of the dynamics of internal motions, solvation and catalysis, as well as larger scale processes such as protein folding and membrane fusion, gets to the heart of our ability to understand their function in biological systems. X-ray crystallography has made relatively few contributions to the study of these types of dynamic processes; instead the strength of X-ray crystallography arises from the ability to precisely define the atomic structures for systems of essentially any size (provided suitable crystals can be obtained and the phase problem solved). As traditionally implemented, an X-ray structure represents a time and space average over the $\sim 10^{12}$ unit cells (within several orders of magnitude) present in a crystal lattice. The dynamic properties of the system cannot be generally accessed and only inferences concerning the probability distribution for atomic positions around the equilibrium value can be deduced from the shape of the electron density profiles.

An exception to this generalization is provided by time resolved crystallography studies, where the structural evolution of a system is followed by recording the diffraction pattern as a function of time following initiation of a process in the crystal. The most successful studies involve heroic analyses of photo-initated reactions that have yielded time resolutions down to ~ 100 ps.[37–39] There are relatively few reactions that can be photo-initiated in crystals, however. A more general approach to starting a reaction is by reactant addition, but molecular diffusion through the crystal restricts the time resolution to minutes. These constraints on rapid reaction initiation, together with the inability of crystal lattices to accommodate larger conformational changes, have undoubtedly restricted the more widespread application of time-resolved crystallography.

These technical issues notwithstanding, the characterization of structural dynamics by X-ray methodologies represents a long-term challenge with significant opportunities. While other approaches, most notably NMR[40] and ultrafast electron diffraction,[41] have great advantages for studying the dynamics of biological systems, X-ray methods combine the ability to obtain global structural information (as can electron diffraction) with the ability to study samples in solution (as can NMR). A serious restriction, touched on above, is the incompatibility of crystal lattices to accommodate large structural changes. In this respect, the availability of fourth generation sources, such as the LCLS and XFEL currently under construction, that will provide high intensity, pulsed X-rays (time resolution in the femtosecond region) are eagerly awaited for the unique experimental capabilities they will provide to study the dynamics of non-crystalline samples in biology and other disciplines.

Bimolecular association reactions represent a promising opportunity for the characterization of structural dynamics at the molecular level, particularly as these processes set a physical limit to the time scale of events in biological systems. Due to the large numbers of species, the small volumes involved and the finite solute capacity of water, the concentration of any one molecular species inside a cell (with the exception of water and a few ionic components) must necessarily be low. The molecular species in this cellular matrix are continually interacting to carry out the chemical processes of life: enzymes must bind the correct substrate for conversion to product while simultaneously discriminating between incorrect, but chemically similar, alternate molecules at comparably low concentrations; regulatory molecules must interact with the correct receptors in signaling pathways; proteins and nucleic acids must be synthesized and replicated with high fidelity. All these processes need to occur within an appropriate timescale to maintain a functioning cellular organization, and this time scale is ultimately limited by the association kinetics of the interacting molecules.

The concept of a diffusion controlled reaction set is central to the analysis of these interactions, as it sets an upper limit for the rate at which two molecules can associate. For the bimolecular association reaction:

$$A + B \underset{k_{-1}}{\overset{k_1}{\rightleftharpoons}} AB \; . \tag{2}$$

The diffusion controlled rate, k_D, derived from the diffusion equation governing the collision of non-interacting spherical molecules, is given by the expression[42]:

$$k_D = \frac{2RT}{3000\eta} \frac{(r_A + r_B)^2}{r_A r_B} \tag{3}$$

where R is the gas constant, T is the absolute temperature, r_A and r_B are the molecular radii of molecules A and B; and η is the viscosity of water. The diffusion controlled rate constant is independent of size for the interaction of identical molecules and equals $7 \times 10^9 \, M^{-1} s^{-1}$. For the interaction of different size molecules, higher rates can be achieved. Rates that exceed this value are proposed to involve additional attractive interactions or reduced dimensionalities, while lower rates invoke steric factors to correct for the fraction of the surface represented by the binding site. Experimentally, protein–protein association rates are found to span the range from 10^4 to above $10^9 \, M^{-1} s^{-1}$, while enzyme-substrate association rates range from 10^6

to $10^8\,M^{-1}s^{-1}$ (Ref. 42). While the upper rates are close to that expected for diffusion controlled processes, the lower values are substantially below this limit. The half-life, $t_{1/2} = 1/(k(A))$, for the second order association of two identical molecules present at a concentration of $1\,\mu M$ will vary from 10^2 to $10^{-3}\,s$ as k ranges from 10^4 to $10^9\,M^{-1}s^{-1}$, respectively; the corresponding values for a species at an initial concentration of $1\,nM$ will range from 10^5 to $1\,s$, respectively. This variation in half-life could have profound consequences for the response time of a biological system.

Why is there such a range in the rates of bimolecular association reactions? The rate of diffusional collision is governed by well understood physical principles that must be valid, at least until the interacting molecules reach proximity. The influence of long-range electrostatic interactions or the restricted size of the binding sites will influence the rates, but these are unlikely to introduce variations of more than 1–2 orders of magnitude from the diffusion controlled value.[42] Consequently, a span of 5 orders of magnitude suggests that the simple kinetic model presented above may be inadequate to describe the association of two molecules in all cases. Such differences in association rates must reflect the contribution of other events, such as desolvation or protein conformational adjustments, to the binding mechanism. The objective of high resolution, time-resolved structural studies will be to characterize the dynamics of these processes in atomic detail, to understand the fundamental determinants of the association rate between biomolecules.

5. Concluding Remarks

Bernal's general view that X-ray analysis will tackle more and more complex structures while becoming a powerful tool in other fields of research is as relevant for speculating about the next half century of biological X-ray analysis as it was in 1945. The easiest prediction is that more and more complex structures will succumb to an assault by X-ray crystallography, with nuclear pores, kinetochores and chromosomal fragments likely targets over the next decade or two, along with various transiently associated complexes (appropriately stabilized), eukaryotic membrane proteins, and ultimately, transiently associated complexes of eukaryotic membrane proteins. These achievements will be propelled by new generations of X-ray sources, as well as developments in biological and biophysical methodologies and computation. From the perspective of "pure" X-ray analysis, perhaps the biggest advances will be the structural characterization of non-crystalline samples with faster and faster time resolution. Molecularly explicit models for the structures and dynamics of cells and higher order assemblages, with predictive capabilities over time scales ranging from femtoseconds to days or longer (corresponding to the biological counterpart of weather forecasting), is a worthy objective for the next half century of structural biology. While these problems "will not yield to X-ray analysis alone" (just as Bernal predicted would be the case for protein structure[7]), X-ray analysis will undoubtedly continue to play an important role in deciphering biological systems in terms of structure and dynamics with atomic detail. The implications of these developments for understanding living processes at this level will undoubtedly provide us with ample future opportunities to wander the streets in excited contemplation.

Acknowledgments

Discussions with B.T. Hsu on the electrostatic analysis, and with W.M. Clemons, R. Phillips and J.B. Howard are greatly appreciated, as are the contributions of group members past and present.

References

1. Bernal JD, Crowfoot D. (1934) X-ray photographs of crystalline pepsin. *Nature* **133:** 794–795.
2. Hodgkin DC, Riley DP. (1968) Some ancient history of protein X-ray analysis. In: Rich A & Davidson N (eds), *Structural Chemistry and Molecular Biology.* W.H. Freeman, San Francisco.

3. Ferry G. (1998) *Dorothy Hodgkin: A Life*. Granta Books, London.

4. Bernal JD. (1939) Structure of proteins. *Nature* **143:** 663–667.

5. Kendrew JC, Dickerson RE, Strandberg BE, Hart RB, Davies DR, Phillips DC, Shore VC. (1960) Structure of myoglobin: 3-dimensional Fourier synthesis at 2 Å resolution. *Nature* **185:** 422–427.

6. Berman HM, Westbrook J, Feng Z, Gilliland G, Bhat TN, Weissig H, Shindyalov IN, Bourne PE. (2000) The Protein Data Bank. *Nuc Acids Res* **28:** 235–42.

7. Bernal JD. (1945) The future of X-ray analysis. *Nature* **155:** 713–715.

8. Green DW, Ingram VM, Perutz MF. (1954) The structure of haemoglobin. 4. Sign determination by the isomorphous replacement method. *Proc Roy Soc* **A225:** 287–307.

9. Ban N, Nissen P, Hansen J, Moore PB, Steitz TA. (2000) The complete atomic structure of the large ribosomal subunit at 2.4 Å resolution. *Science* **289:** 905–920.

10. Wimberly BT, Brodersen DE, Clemons WM, Morgan-Warren RJ, Carter AP, Vonrhein C, Hartsch T, Ramakrishnan V. (2000) Structure of the 30S ribosomal subunit. *Nature* **407:** 327–339.

11. Schluenzen F, Tochilj A, Zarivach R, Harms J, Gluehmann M, Janell D, Bashan A, Bartels H, Agmon I, Franceschk F, Yonath A. (2000) Structure of functionally activated small ribosomal subunit at 3.3 Å resolution. *Cell* **102:** 615–623.

12. Pauling L, Niemann C. (1939) The structure of proteins. *J Amer Chem Soc* **61:** 1860–1867.

13. Harrison SC. (2004) Whither structural biology? *Nature Struct Mol Biol* **11:** 12–15.

14. Pauling L, Corey RB. (1951) Configurations of polypeptide chains with favored orientations around single bonds: Two new pleated sheets. *Proc Natl Acad Sci USA* **37:** 729–740.

15. Pauling L, Corey RB, Branson HB. (1951) The structure of proteins: Two hydrogen-bonded helical configurations of the polypeptide chain. *Proc Natl Acad Sci USA* **37:** 205–211.

16. Cramer P, Bushnell DA, Kornberg RD. (2001) Structural basis of transcription: RNA polymerase II at 2.8 Å resolution. *Science* **292:** 1863–1876.

17. Long SB, Campbell EB, MacKinnon R. (2005) Crystal structure of a mammalian voltage-dependent Shaker family K$^+$ channel. *Science* **309**: 897–903.

18. Hansen C, Quake SR. (2003) Microfluidics in structural biology: Smaller, faster... better. *Curr Opin Struct Biol* **13**: 538–544.

19. Gan L, Speir JA, Conway JF, Lander G, Cheng N, Firek BA, Hendrix RW, Duda RL, Liljas L, Johnson JE. (2006) Capsid conformational sampling in HK97 maturation visualized by X-ray crystallography and cryo-EM. *Structure* **14**: 1655–1665.

20. Schuwirth BS, Borovinskaya MA, Hau CW, Zhang W, Vila-Sanjurjo A, Holton JM, Cate JH. (2005) Structures of the bacterial ribosome at 3.5 Å resolution. *Science* **310**: 827-834.

21. Selmer M, Dunham CM, Murphy FV, Weizlbaumer A, Petry S, Kelley AC, Weir JR, Ramakrishnan V. (2006) Structure of the 70S ribosome complexed with mRNA and tRNA. *Science* **313**: 1935–1942.

22. Neutze R, Wouts R, van der Spoel D, Weckert E, Hajdu J. (2000) Potential for biomolecular imaging with femtosecond X-ray pulses. *Nature* **406**: 752–757.

23. Chiti F, Dobson CM. (2006) Protein misfolding, functional amyloid, and human disease. *Ann Rev Biochem* **75**: 333–366.

24. Nelson R, Sawaya MR, Balbirnie M, Madsen AO, Riekel C, Grothe R, Eisenberg D. (2005) Structure of the cross-beta spine of amyloid-like fibrils. *Nature* **435**: 773–778.

25. Engelman DM. (2005) Membranes are more mosaic than fluid. *Nature* **438**: 578–580.

26. Dauter Z. (2003) Protein structures at atomic resolution. *Meth Enz* 288–337.

27. Jelsch C, Teeter MM, Lamzin V, Pichon-Pesme V, Blessing RH, Lecomte C. (2000) Accurate protein crystallography at ultra-high resolution: Valence electron distribution in crambin. *Proc Natl Acad Sci USA* **97**: 3171–3175.

28. Einsle O, Tezcan FA, Andrade SLA, Schmid B, Yoshida M, Howard JB, Rees DC. (2002) The nitrogenase MoFe-protein at 1.16 Å resolution: A central ligand in the FeMo-cofactor. *Science* **297**: 1696–1700.

29. Kielkopf C, Ding S, Kuhn P, Rees DC. (2000) Conformational flexibility of B-DNA at 0.74 Å resolution: d(CCAGTACTGG)$_2$. *J Mol Biol* **296:** 787–801.

30. Einsle O, Andrade SLA, Dobbek H, Meyer J, Rees DC. (2007) Assignment of individual metal redox states in a metalloprotein by crystallographic refinement at multiple X-ray wavelengths. *J Amer Chem Soc* **129:** 2210–2211.

31. Lecomte C, Guillot B, Muzet N, Pichon-Pesme V, Jelsch C. (2004) Ultra-high resolution X-ray structures of proteins. *Cell Mol Life Sci* **61:** 774–782.

32. Coppens P. (1997) *X-ray Charge Densities and Chemical Bonding.* Oxford University Press, New York.

33. Moore PB. (2007) Let's call the whole thing off: Some thoughts on the protein structure initiative. *Structure* **15:** 1350–1352.

34. Bertaut F. (1952) L'énergie électrostatique de réseaux ioniques. *J Phys Radium* **13:** 499–505.

35. Warwicker J, Watson HC. (1982) Calculation of the electric potential in the active-site cleft due to α-helix dipoles. *J Mol Biol* **157:** 671–679.

36. Hagler AT, Moult J. (1978) Computer simulation of the solvent structure around biological molecules. *Nature* **272:** 222–226.

37. Aranda R, Levin EJ, Schotte F, Anfinrud PA, Phillips GN. (2006) Time-dependent atomic coordinates for the dissociation of carbon monoxide from myoglobin. *Acta Cryst* **D62:** 776–783.

38. Coppens P, Vorontsov II, Graber T, Gembicky M, Kovalevsky AY. (2004) The structure of short-lived excited states of molecular complexes by time-resolved X-ray diffraction. *Acta Cryst* **A61:** 162–172.

39. Moffat K. (2001) Time-resolved biochemical crystallography: A mechanistic perspective. *Chem Rev* **101:** 1569–1581.

40. Palmer AG. (2004) NMR characterization of the dynamics of biomacromolecules. *Chem Rev* **104:** 3623–3640.

41. Zewail AH. (2006) 4D ultrafast electron diffraction, crystallography and microscopy. *Annu Rev Phys Chem* **57:** 65–103.

42. Fersht A. (1999) *Structure and Mechanism in Protein Science.* W.H. Freeman, New York.

43. Matthews BW. (1968) Solvent content of protein crystals. *J Mol Biol* **33:** 491–497.

Reinterpreting the Genetic Code
Implications for Macromolecular Design, Evolution and Analysis

*David A. Tirrell**

The genetic code, elucidated in the 1960s through the work of Nirenberg, Ochoa, Khorana and their coworkers, provides a set of molecular instructions for translating nucleic acids into proteins. Codons are assigned to amino acids through high-fidelity charging of transfer RNAs, and through accurate base-pairing between charged tRNAs and messenger RNA. Over the last decade, cells have been outfitted with modified translational machinery that enables the participation of an expanded set of amino acids in protein synthesis. These developments have stimulated a unified view of the chemistry of natural and synthetic macromolecules, and provided a basis for powerful new approaches to materials design, protein evolution, biological imaging, and proteome-wide analysis of cellular processes.

1. Introduction

Macromolecular chemistry has traditionally been divided into two fields, with biochemists and biophysicists working primarily on proteins and nucleic acids while polymer chemists and materials scientists concerned themselves with synthetic polymers.[1] Natural and synthetic macromolecules are profoundly different from one another; proteins and nucleic acids are uniform in length and sequence and well folded, whereas synthetic polymers are heterogeneous and for the most part adopt random-coil conformations.

*Division of Chemistry and Chemical Engineering, California Institute of Technology, Pasadena, California, USA, e-mail: tirrell@caltech.edu.

(a)   (b)

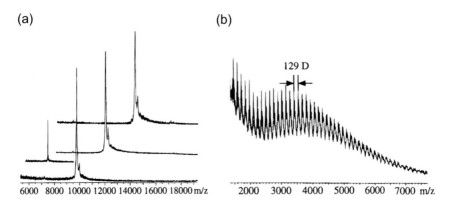

Fig. 1. MALDI-TOF mass spectra of monodisperse PLGAs (a) and polydisperse PLGA (DP ≈ 92, PDI = 1.2) (b). Satellite signals that appear as high molecular weight shoulders in part a are due to a matrix-protein conjugate. The mass difference (129) between the major peaks in part b corresponds to the mass of the PLGA repeating unit. Reprinted with permission from Yu SM, Soto CM, Tirrell DA. Nanometer-Scale Smectic Ordering of Genetically Engineered Rod-Like Polymers: Synthesis and Characterization of Monodisperse Derivatives of Poly(γ-benzyl α,L-glutamate). *J Am Chem Soc* **122:** 6552–6559. Copyright 2000, American Chemical Society.

Perhaps the clearest demonstration of the striking differences between natural and synthetic polymers is provided by a comparison of the chain-length distributions produced by biological and chemical polymerization processes. Figure 1 shows mass spectra of poly(glutamic acid)s (PLGAs) prepared either by gene expression (Fig. 1a) or by ring-opening polymerization (Fig. 1b).[2] The gene products are uniform, with chain lengths specified by the lengths of the corresponding coding sequences, while the chemical polymerization product is a complex mixture of chain-length variants.

The complexity of synthetic polymer mixtures increases to unfathomable levels when considerations of comonomer sequence and stereochemical structure are taken into account. For example, growth according to Bernoullian statistics of a "simple" binary copolymer of chain length 1000 yields more than 10^{300} possible sequence variants! In contrast, tight control of sequence and stereochemistry in the synthesis of proteins and nucleic acids confers on these long-chain molecules the informational, catalytic and recognition properties that are central to life.

The molecular complexity of synthetic polymers has not precluded their scientific study or their widespread use. The physical properties of polymer mixtures are intriguing and important, and their commercial impact spans nearly every aspect of modern materials technology. At the same time, synthetic methods that afford improved control of macromolecular architecture have had profound scientific and technological consequences; for example, the control of topology and stereochemistry enabled by Ziegler-Natta catalysis led directly to the introduction of the advanced polyolefins that now constitute the most important class of polymeric materials.

The advent of recombinant DNA technology made it possible to approach macromolecular chemistry in fundamentally new ways, and to draw together the study of natural and synthetic polymers. No longer is it necessary to choose between design (previously the purview of the synthetic chemist) and control (the hallmark of the proteins and nucleic acids). Instead, one can design new macromolecules (or classes of macromolecules) and make the desired products with essentially complete control of the molecular architecture. We adopted this approach in the late 1980s and focused our early efforts on control of crystallization behavior,[3] liquid crystal structure,[4] and viscoelastic properties in macromolecular systems.[5]

We soon recognized that the impact of this approach to macromolecular chemistry would be enhanced by access to an expanded set of monomeric building blocks. The genetic code specifies the nucleic acid sequences that direct the incorporation of twenty different amino acids into growing polypeptide chains, and the translational apparatus maintains a high level of fidelity in the process. To what extent might it be possible to "re-interpret" the genetic code to enable the programmed incorporation of new amino acid constituents into proteins?

Reports of incorporation of amino acid analogues into cellular proteins pre-date the elucidation of the structure of DNA and the nature of the genetic code. The first such report appears to be that of Levine and Tarver, who in 1951 described the incorporation of ethionine (a homologue of the natural amino acid methionine) into proteins in the rat.[6] A few years later, Cowie and Cohen described the translational activity of selenomethionine in bacterial cells,[7] a phenomenon that, through the work of Hendrickson and coworkers on multiwavelength anomalous diffraction, has had revolutionary consequences for protein crystallography.[8] Since 1990, expression and analysis of selenomethionyl proteins has become the method of choice for solution of protein crystal structures, and has enabled the rapid advances that underlie current initiatives in structural genomics.

Our initial motivation in exploring new, "non-canonical" amino acids was quite different, in that we wished to lay the groundwork for creating macromolecules with unusual physical and chemical properties. We therefore focused much of our effort on fluorinated amino acids and their potential role in increasing the hydrophobic character of peptides and proteins, and on the introduction of reactive side chains carrying alkene, alkyne, azide, aryl halide, and ketone functional groups, which would render proteins susceptible to an expanded set of chemical transformations.

Because all of the triplet codons in the genetic code are assigned specific roles in protein synthesis, the introduction of a non-canonical amino acid into a protein requires assignment of one or more codons to the amino acid of interest. The simplest approach to this problem is to use the codons assigned to one of the canonical amino acids, and to replace the canonical amino acid – in whole or in part – by a structural analogue. This approach results, in most cases, in introduction of the analogue at multiple sites – a desirable outcome if one is interested in creating new macromolecules with unusual properties. An alternative approach, used primarily to label proteins at a single site, is to recruit a stop codon or a four-base codon to direct insertion of the non-canonical amino acid. The two approaches are complementary, and are often described as "residue-specific" and "site-specific," respectively. This chapter will focus on residue-specific methods, which have been of primary interest in our laboratory over the past decade.

Beautiful work on site-specific methods has been reported by Furter,[9] Schultz,[10] Sisido[11] and others, and has been reviewed recently.[12]

2. Codon Assignment – Interpreting the Genetic Code

Within months of the 1953 publication of the double-helical structure of DNA,[13] there emerged the working hypothesis that the path from genes to proteins involves RNA templates for protein synthesis.[14] The transient "messenger RNA" (mRNA) was identified in 1960, and by 1966, the genetic code had been elucidated through the work of Nirenberg, Ochoa, Khorana and their coworkers.[15] Amino acids are activated by aminoacyl-tRNA synthetases and charged to cognate transfer RNAs (tRNAs); the resulting aminoacyl-tRNAs recognize mRNA through base-pairing interactions between their anticodon loops and complementary triplet codons in the message. In a classic confirmation of Crick's "adaptor hypothesis," Benzer and coworkers showed in 1962 that cysteine tRNA mischarged with alanine (Ala-tRNA^{Cys}) delivers alanine to growing polypeptide chains *in vitro* in response to cysteine codons.[16] The mischarged tRNA was prepared by reduction of Cys-tRNA^{Cys} by Raney nickel. In a companion experiment, Chapeville showed that oxidation of Cys-tRNA^{Cys} resulted in incorporation of cysteic acid in response to cysteine codons,[17] a result with clear implications for the programmed insertion of non-canonical amino acids into proteins.

3. Assigning Codons to Non-Canonical Amino Acids – Design of the Cellular Machinery

The observation that mischarged tRNAs can deliver their amino acid cargoes to growing polypeptide chains illustrates the critical role of the aminoacyl-tRNA synthetases in ensuring the fidelity of protein synthesis in cells. At the same time, it suggests that changes in the activity or specificity of the synthetases, coupled with the design of complementary, non-canonical amino acid substrates, might enable cells to make proteins from altered or expanded sets of building blocks.

3.1. Over-Expression of Wild-Type Synthetases

The simplest approach to altering synthetase activity involves increasing the intracellular concentration of the wild-type enzyme through over-expression. We adopted this approach first in our work on analogues of methionine (Met, **1**), after we observed that *trans*-crotylglycine (Tcg, **2**), unlike homoallylglycine (Hag, **3**) and homopropargylglycine (Hpg, **4**), did

not support protein synthesis in bacterial cells depleted of Met.[18] We found through *in vitro* kinetic assays that the *E. coli* methionyl-tRNA synthetase (MetRS) activates Tcg, but that the rate of activation is several orders of magnitude lower than that for Met. Apparently, activation of Tcg in the cell is too slow to support observable protein synthesis in the absence of Met. Simple introduction of a plasmid-borne copy of a gene encoding wild-type MetRS enabled the expression of proteins in which nearly all of the Met sites were occupied by Tcg. The approach has proven to be quite general, and has enabled the incorporation of hexafluoroleucine,[19] (*2S,3R*)-*4,4,4-trifluorovaline* (Tfv),[20] and a wide variety of proline (Pro) analogues[21] into recombinant proteins in bacterial cells. Remarkably, Tfv can be assigned either to valine codons or to isoleucine codons, depending on which synthetase is over-expressed (and which amino acid is depleted) in the cell.[20]

A single RNA message can be translated in different ways by controlling the cellular pools of synthetases and amino acids.

3.2. Design of Mutant Synthetases

For analogues that are activated very slowly – or not at all – by any of the wild-type synthetases, it is often possible to prepare mutant synthetases with the required activity. Hennecke and coworkers took an important first step in this direction when they showed that a single mutation (A294G) in the active site of the *E. coli* phenylalanyl-tRNA synthetase (PheRS) enabled incorporation of *p*-chlorophenylalanine (pCF) into cellular proteins.[22] The A294G mutant and related variants were subsequently shown to activate Phe analogues carrying azide, alkyne, and other functional groups,[23,24,25] and Conticello and coworkers have used a similar approach to modify the activity of the *E. coli* prolyl-tRNA synthetase (ProRS).[21]

3.3. Combinatorial Approaches to Mutant Synthetases

The design of mutant enzymes can be a difficult and uncertain task, and prudent investigators often opt for combinatorial strategies that allow rapid evaluation, either by screening or by selection, of large populations of enzyme variants. Several different combinatorial approaches have been used to identify aminoacyl-tRNA synthetases that activate and charge non-canonical amino acids.

5

In some cases, it is possible to look directly for the appearance of the non-canonical amino acid in cellular proteins. In collaboration with Carolyn Bertozzi and her coworkers, we used this approach in a search for MetRS variants that activate azidonorleucine (Anl, **5**).[26] We transformed an *E. coli* culture with a MetRS library randomized at four positions adjacent to the amino acid binding site, and with a cell-surface

6

reporter protein that could be expressed under control of an inducible promoter. Cell-surface azides were detected through the azide-alkyne cycloaddition, which was used to deliver a biotin probe (**6**) to cells bearing MetRS variants active toward Anl. Finally, treatment of the culture with a dye-labeled avidin reagent allowed us to isolate the cells of interest by fluorescence-activated cell sorting. Three MetRS mutants active toward Anl were identified in this way. Because all three shared a common mutation that converted leucine at position 13 into glycine, we prepared and examined the single-site L13G mutant both *in vitro* and *in vivo*. The single-site mutant proved to be the most efficient of the four, and enabled the synthesis of good yields of protein in which Met was replaced essentially quantitatively by Anl.

Schultz and coworkers have developed several powerful combinatorial methods for identifying synthetases of altered specificity.[10] These methods rely on the introduction of "orthogonal pairs" of synthetases and suppressor tRNAs (i.e., pairs that do not cross-react with the translational machinery of the host), and are most directly relevant to site-specific incorporation of non-canonical amino acids into recombinant proteins. Because these methods involve "negative" as well as "positive" selection, they can be used to eliminate variants that activate any of the canonical amino acids at substantial rates.

3.4. Design of Amino Acid Substrates

For many purposes, the best approach of all is to design amino acid analogues that exhibit good translational activity without requiring modification of the protein synthesis machinery of the host cell. Such analogues must not be excluded from the cell, and must be activated by the endogenous aminoacyl-tRNA synthetases fast enough to support a useful rate of protein synthesis (where the definition of "useful" varies, depending on the objective). It is not, in our experience, necessary that the activation rate of the analogue be comparable to that of the canonical amino acid. For example, we have achieved near-quantitative replacement of Met by Hpg[27] (or by the azide analogue azidohomoalanine, Aha, **7**)[28] without loss of protein yield, despite the fact that the value of k_{cat}/K_m for MetRS is reduced more than 100-fold for both analogues as compared to the authentic amino acid substrate.

4. Non-Canonical Amino Acids in Macromolecular Design, Evolution and Analysis

The combination of approaches described above has expanded greatly the number and diversity of amino acids known to exhibit translational activity in microbial and mammalian cells. Figure 2 shows a representative set of non-canonical amino acids examined in our laboratory; a more complete summary is presented in the excellent book by Budisa.[12] The color code in Fig. 2 indicates the measures required for translational activity; amino acids shown in blue are active in wild-type cells, those drawn in black require extra copies of the wild-type synthetase, and those shown in other colors require the introduction of mutant synthetases.

Development of these and other non-canonical amino acids required many years of work in many laboratories. The result of that effort was not only the amino acids themselves, but also an understanding of the methods needed to expand the amino acid pool and a conviction that further expansion could be accomplished as required. Much of the most recent work in the field has been directed toward the use of non-canonical amino acids to develop

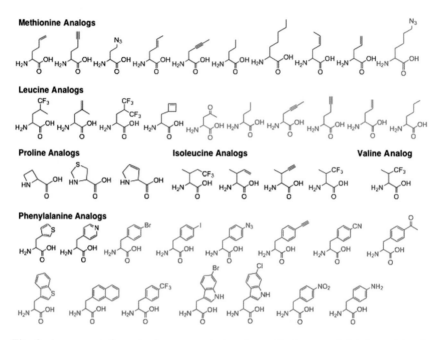

Fig. 2. A representative set of non-canonical amino acids incorporated into proteins in microbial and mammalian cells.

new approaches to macromolecular science and engineering, with special emphasis on protein design, evolution and analysis.

4.1. Protein Design

As described in the introduction to this chapter, the development of recombinant DNA methods has created unprecedented opportunities for the design of new macromolecules of defined sequence. Especially exciting is the prospect of macromolecular materials in which both physical and biological properties are subject to tight control through gene design. We have referred to such materials as "artificial extracellular matrix (ECM) proteins" because of their analogy to the extracellular matrices that play such central roles in determining the structure and function of tissues and organs.

Our first artificial ECM proteins were assembled from elements of elastin and fibronectin, and drew heavily on prior work by Urry and coworkers,[29] who showed that simple repeating polypeptides related to elastin could be formed into robust protein elastomers, and by Ruoslahti,[30] Hubbell[31] and others, who showed that short peptides derived from fibronectin retain their affinity for cell-surface adhesion receptors even when presented in non-native contexts.

Artificial genes encoding multiple elastin and fibronectin domains are readily assembled and expressed in bacterial cells, but like natural tropoelastin, elastin-derived aECM proteins must be crosslinked in order to serve effectively as matrices for organizing cells and tissues.

Crosslinking through lysine side chains is effective for many purposes, and has yielded matrices that exhibit sequence-dependent adhesion of mammalian cells.[32] But spatial and temporal control of crosslinking, needed for patterning or for the preparation of gradient materials, is more readily accomplished by photochemical means. In order to enable photochemical crosslinking of artificial ECM proteins under mild conditions, we introduced the photosensitive amino acid *p*-azidophenylalanine (pAzF, **8**), which undergoes rapid and efficient loss of molecular nitrogen upon mild irradiation.[33] The resulting nitrene intermediate forms intermolecular crosslinks, probably through several parallel reaction pathways.

Transformation of bacterial cells with genes encoding the ECM protein of interest and the A294G PheRS variant described above, allows the preparation of photocrosslinkable protein matrices in good yield. Because pAzF competes with Phe in decoding Phe codons, one can prepare not just one protein from a single coding sequence, but rather a family of related proteins of controlled composition. Control of the amino acid pool (and in particular, of the ratio of concentrations of Phe and pAzF) enables control of the extent of incorporation of pAzF and of the density of crosslinks

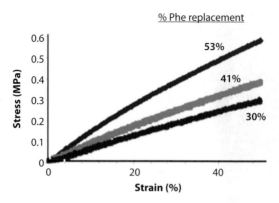

Fig. 3. Tensile stress-strain curves for hydrated artificial ECM proteins crosslinked by photolysis of pAzF side chains. Reprinted with permission from Carrico IS, Maskarinec SA, Heilshorn SC, Mock M, Liu JC, Nowatzki PJ, Franck C, Ravichandran G, Tirrell DA. Lithographic Patterning of Photoreactive Cell-Adhesive Proteins. *J Am Chem Soc* **129:** 4874–4875. Copyright 2007, American Chemical Society.

formed upon irradiation. And because crosslink density controls the elastic modulus of the molecular network, the stiffness of the protein matrix can be varied within wide limits. Figure 3 shows stress-strain curves acquired for a representative set of photocrosslinked ECM proteins, in which the modulus varies from 0.5 to 1.4 MPa, depending on the pAzF content.[33] Such materials constitute new tools for the study of cell-matrix interactions and of the interplay of physical and biochemical signals in determining cellular behavior. In the future, artificial ECM proteins and related materials may enable new approaches to surgery and regenerative medicine.

Photochemical crosslinking also enables lithographic patterning and the preparation of protein networks characterized by spatial gradients in the elastic modulus.[33] Figure 4 shows mammalian cells cultured on a patterned ECM protein substrate; cell adhesion is observed only in those regions of the substrate that were subject to irradiation. Removal of the soluble protein after photolithography confines the cells to a predetermined pattern. Gradients in the elastic modulus have been prepared by spatial variation in the irradiation dose.

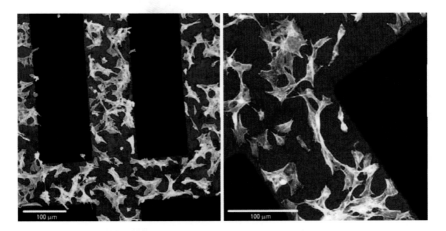

Fig. 4. Confocal microscopy of Rat-1 fibroblasts attached to photopatterned artificial ECM protein. Immunostaining with anti-T7 (blue) demonstrates colocalization of protein and cells (stained with rhodamine phalloidin (yellow)). Scale bars represent 100 μm. Reprinted with permission from Carrico IS, Maskarinec SA, Heilshorn SC, Mock M, Liu JC, Nowatzki PJ, Franck C, Ravichandran G, Tirrell DA. Lithographic Patterning of Photoreactive Cell-Adhesive Proteins. *J Am Chem Soc* **129**: 4874–4875. Copyright 2007, American Chemical Society.

4.2. Protein Evolution

As described earlier in this chapter, our initial motivation in developing new amino acids stemmed from our interest in designing proteins with novel physical, chemical and biological properties. We started from the perspective of the synthetic polymer chemist, by considering the properties of interest and the molecular architectures that give rise to such properties, designing analogous structures that could be prepared via expression of artificial genes, and adding non-canonical amino acids as needed. But the availability of an expanded amino acid pool also raises interesting questions about the extent to which *natural* proteins might acquire new molecular properties if they could be assembled from altered sets of amino acid constituents.

The straightforward approach of effecting complete replacement of one of the canonical amino acids by a non-canonical analogue seemed likely to fail for most proteins. Protein folding is often sensitive to minor changes in amino acid sequence; amino acid replacement at multiple sites would be expected to compromise protein structures and protein folding pathways that have evolved in the context of the canonical amino acids. But what if one could "re-evolve" proteins in the laboratory in new amino acid contexts? Is it possible to retain proper folding and function while "evolving out" canonical amino acids and "evolving in" new ones? And if it is possible, what kinds of changes in structure, dynamics and function might be realized in this way?

As a first test system for exploring these questions, we examined the replacement of leucine by 5,5,5-trifluoroleucine (Tfl, **9**) in several peptides and proteins. Fluorocarbons are, in general, more hydrophobic than their hydrocarbon counterparts, although the origins of this effect remain puzzling. Experiments in our laboratory,[34] and in those of Kumar[35] and Marsh,[36] have shown that introduction of fluorinated amino acids into coiled-coil and four-helix bundle peptides can cause marked increases in the stability of such peptides with respect to thermal and chemical denaturation, as expected from the central role of hydrophobic forces in stabilizing helix bundles.

We expected a different outcome when we substituted Tfl for leucine in chloramphenicol acetyltransferase (CAT).[37] Unlike the simple helical peptides discussed above, CAT adopts an intricately folded trimeric structure in which each of the three polypeptide chains consists of 219 amino acids, including 13 leucine residues (Fig. 5).[38] Global replacement of Leu by Tfl (which in practice was limited to ca. 75% of the Leu sites) caused a marked decrease in the thermal stability of CAT; the half-life for thermal inactivation ($t_{1/2}$) of the enzyme at 60°C was reduced 20-fold upon fluorination.

Laboratory evolution was explored as a means of adapting CAT to the introduction of fluorinated side chains. Random mutagenesis was used to prepare a library of approximately 2000 sequence variants, and the library was screened for CAT mutants that (in fluorinated form) retained activity

front   back

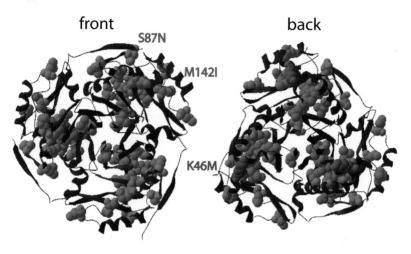

Fig. 5. Structural model of the CAT trimer showing the 3 stabilizing mutations in L2-A1. Two of the mutations (S87N and M142I) from generation 1 are depicted in orange; the third mutation (K46M) from generation 2 is represented in pink. Residues highlighted in blue are the leucine/TFL residues of CAT; the chloramphenicol substrate is highlighted in red. Reproduced with permission from Montclare JK, Tirrell DA. Evolving Proteins of Novel Composition. *Angew Chem Int Ed* **45:** 4518-4521. Copyright 2006, Wiley-VCH Verlag GmbH & Co. KGaA.

after incubation at 60°C. Two rounds of mutagenesis and screening yielded a fluorinated variant containing three amino acid substitutions (S87N, M142I and K46M), for which $t_{1/2}$ was improved 27-fold as compared to the fluorinated form of the parent enzyme. The activity and stability of the evolved, fluorinated mutant were virtually indistinguishable from those of wild-type CAT in its conventional (leucine) form.

Several aspects of this experiment deserve comment. First, it proved to be remarkably straightforward to adapt CAT to the introduction of multiple copies of the non-canonical amino acid; just two rounds of mutagenesis and three amino acid replacements were required. Second, none of the mutations lies close to the active site of the enzyme; all are at least 15 Å from the chloramphenicol binding site, and none makes contact with any of the Leu/Tfl residues of the protein. Finally, and most importantly, no leucine codons were lost from the coding sequence; the evolved mutant is heavily fluorinated, and the mutations acquired through evolution compensate for the introduction of many copies of the fluorinated amino acid.

A similar experiment examined the laboratory evolution of the green fluorescent protein (GFP) subsequent to replacement of Leu by Tfl.[39] Fluorination of GFP caused misfolding and loss of fluorescence, but the emission from cells expressing the fluorinated variant was restored (enhanced approximately 650-fold) through eleven rounds of evolution. Twenty amino acid substitutions were acquired, and there was a net loss of four Leu residues (of 19; six were lost and two others were added to the sequence). The folding behavior of the evolved, fluorinated variant was essentially identical to that of the leucine form of the parent sequence.

In each of these cases, further study will be required to determine the mechanisms by which sequence changes compensate for the introduction of fluorinated side chains, but the implications of these results are encouraging with respect to the prospects for evolving proteins of novel composition, structure and function. Studies of this kind may also advance the more ambitious goals of evolving organisms with novel genetic codes, and of elucidating the mechanisms by which code evolution occurs.[40,41]

4.3. Proteomic Analysis

If we are to understand complex biological phenomena such as differentiation, development, and learning, we must understand the underlying changes in the synthesis, degradation and assembly of the proteins of the cell. Scientists who wish to probe such changes face a daunting problem: new proteins are always made in complex mixtures of pre-existing proteins that are present in much larger numbers. How can one identify and visualize just the newly synthesized proteins, or just a subset of proteins of specific interest, in such mixtures? Gaining spatial, temporal and chemical selectivity in protein imaging and analysis will allow us to address critically important biological questions in fundamentally new ways.

As an example, consider the phenomenon of synaptic plasticity, the remodeling of the synapse in response to stimulation. Remodeling of synaptic connections is believed to play a central role in learning and memory, and is known to require the synthesis of new proteins. Is it possible to identify new proteins as they are made in the neuron and to visualize them as they are transported to their cellular destinations?

Recent developments suggest that it should be possible. In a collaborative effort with Erin Schuman and her coworkers, we have shown that newly synthesized proteins can be selectively labeled in cells by co-translational incorporation of amino acid analogues that carry carefully designed, reactive side chains (Fig. 6).[42,43] For example, methionine analogs bearing azide or alkyne side chains can be used to "tag" proteins as they are made, and the tagged proteins can be labeled subsequently with affinity reagents or fluorescent dyes. Affinity-labeled proteins can be purified by chromatographic methods and sequenced by tandem mass spectrometry, while dye-labeled proteins can be visualized directly by fluorescence microscopy. Thus one can establish both the location and the identity of the newly synthesized proteins in the cell. Susceptibility to amino acid tagging

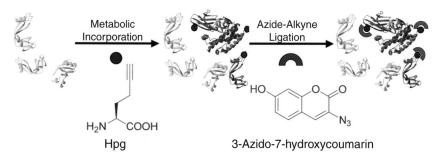

Fig. 6. Selective dye-labeling of newly synthesized proteins in cells. Metabolic incorporation of the methionine surrogate Hpg is followed by ligation to an azidocoumarin dye. Reproduced with permission from Beatty KE, Liu JC, Xie F, Dieterich DC, Schuman EM, Wang Q, Tirrell DA. Fluorescence Visualization of Newly Synthesized Proteins in Mammalian Cells. *Angew Chem Int Ed* **45:** 7364–7367. Copyright 2006, Wiley-VCH Verlag GmbH & Co. KGaA.

is determined not by the identity of the protein, but rather by the spatial and temporal character of its synthesis.

Strategies similar to that outlined in Fig. 6 have been implemented not only in cell culture, but also in complex tissue systems such as hippocampal slices. Hippocampal slices are widely used in neurobiological studies because they maintain much of the essential architecture of the tissue. Figure 7 compares hippocampal slices labeled under conditions that allow (Fig. 7a, left) or inhibit (Fig. 7b, right) the synthesis of new proteins. Robust labeling of new proteins is apparent in Fig. 7a, indicating that active protein synthesis is underway. The absence of signal in Fig. 7b illustrates the selectivity of the method with respect to labeling of newly synthesized proteins.

It should also be possible to achieve further selectivity in amino acid tagging. Because, as described earlier in this chapter, certain amino acid surrogates require mutant aminoacyl-tRNA synthetases for activation, amino acid tagging can be made dependent upon synthetase expression. The required synthetases can be genetically encoded in the cells of interest,

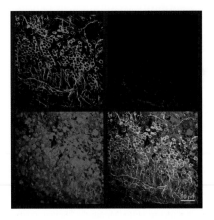 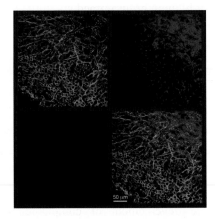

Fig. 7. Imaging of hippocampal slices in the absence (a, left) or presence (b, right) of a protein synthesis inhibitor. Lower left panel in each figure shows the image formed by dye-labeling of newly synthesized proteins. Top panels show labeling by neuronal markers Map 2 and bassoon; lower right panels are overlays of the other three panels. Figure courtesy of Daniela Dieterich and Erin Schuman.

and expressed only upon activation of specific promoters. Approaches of this kind will allow unique and powerful insights into the nature of the proteomic responses to cellular signals of various kinds. For example, in systems such as the hippocampal slices shown in Fig. 7, it should be possible to place the synthetase under control of a promoter that is activated only in neurons, and not in adjacent glial cells (or vice versa). Labeling can then be restricted to the cell type of interest, and the intrinsic noise in such experiments can be reduced substantially. In complex microbial populations (e.g., in biofilms) the synthetase can be encoded in only one genome, and labeling can be restricted to just one of the tens or hundreds of species that typically comprise such populations. Host-pathogen interactions, responses to extreme environments (e.g., heat or cold shock), cellular transdifferentiation, and many other critically important biological processes, also become susceptible to fundamentally new methods of experimental investigation. The most ambitious prospect is that of imaging and analysis via amino acid tagging in whole animals, an approach that should become feasible through the design of amino acid surrogates that can be fed to animals and activated only upon programmed synthetase expression.

5. Conclusions

It has been known for more than fifty years that non-canonical amino acids can be incorporated into cellular proteins. The most important application of this phenomenon stems from the work of Hendrickson and coworkers, who in 1990 introduced the use of selenomethionine as a powerful tool for determination of protein structure. In the last decade, the design of amino acid analogues carrying fluorinated and reactive side chains, coupled with the development of complementary aminoacyl-tRNA synthetases, has expanded the field greatly, and has created important new opportunities for macromolecular chemistry, biomaterials science, protein evolution, biological imaging and proteome-wide analysis of cellular processes.

Acknowledgments

Research at Caltech on the chemistry of non-canonical amino acids has been supported by grants from the National Institutes of Health (GM62523), the Institute for Collaborative Biotechnologies, the Beckman Institute and the Joseph J. Jacobs Institute for Molecular Engineering for Medicine. The author is grateful to the many coworkers and collaborators who contributed to the development of this chemistry.

References

1. Morawetz H. (1985) *Polymers: The Origins and Growth of a Science.* Dover Publications, New York.
2. Yu SM, Soto CM, Tirrell DA. (2000) Nanometer-Scale Smectic Ordering of Genetically Engineered Rod-Like Polymers: Synthesis and Characterization of Monodisperse Derivatives of Poly(γ-benzyl α,L-glutamate). *J Am Chem Soc* **122**: 6552–6559.
3. Krejchi MT, Atkins EDT, Waddon AJ, Fournier MJ, Mason TL, Tirrell DA. (1994) Chemical Sequence Control of ß-Sheet Assembly in Macromolecular Crystals of Periodic Polypeptides. *Science* **265**: 1427–1432.
4. Yu SM, Conticello V, Zhang GH, Kayser C, Fournier MJ, Mason TL, Tirrell DA. (1997) Smectic Ordering in Solutions and Films of a Rod-like Polymer Owing to Monodispersity of Chain Length. *Nature* **389**: 167–170.
5. Petka WA, Harden JL, McGrath KP, Wirtz D, Tirrell DA. (1998) Reversible Hydrogels from Self-Assembling Artificial Proteins. *Science*, **281**: 389–392.
6. Levine M, Tarver H. (1951) Studies on ethionine. 3. Incorporation of ethionine into rat proteins. *J Biol Chem* **192**: 835–850.
7. Cowie DB, Cohen GN. (1957) Biosynthesis by Escherichia-Coli of active altered proteins containing containing selenium instead of sulfur. *Biochim Biophys Acta* **26**: 252–261.
8. Hendrickson WA, Horton JR, LeMaster DM (1990) Selenomethionyl proteins produced for analysis by multiwavelength anomalous

diffraction (MAD): a vehicle for direct determination of three-dimensional structure. *EMBO J* **9:** 1665–1672.

9. Furter R. (1998) Expansion of the genetic code: Site-directed p-fluoro-phenylalanine incorporation in *Escherichia coli*. *Protein Science* **7:** 419–426.

10. Wang L, Xie J, Schultz PG. (2006) Expanding the genetic code. *Annu Rev Biophys Biomol Struct* **35:** 225–249.

11. Hohsaka T, Sisido M. (2002) Incorporation of non-natural amino acids into proteins. *Curr Opin Chem Biol* **6:** 809–815.

12. Budisa N. (2006) *Engineering the Genetic Code: Expanding the Amino Acid Repertoire for the Design of Novel Proteins*. Wiley-VCH, Weinheim.

13. Watson JD, Crick FHC. (1953) Molecular structure of nucleic acids - a structure for deoxyribose nucleic acid. *Nature* **171:** 737–738.

14. Watson JD, Hopkins NH, Roberts JW, Steitz JA, Weiner AM. *Molecular Biology of the Gene*, p. 81. (1987) Benjamin/Cummings, Menlo Park.

15. Nirenberg M. (2004) Historical review: Deciphering the genetic code – a personal account. *Trends Biochem Sci* **29:** 46–54.

16. Chapeville F, Lipmann F, von Ehrenstein G, Weisblum B, Ray WJ Jr, Benzer S. (1962) On the role of soluble ribonucleic acid in coding for amino acids. *Proc Natl Acad Sci USA* 1086–1092.

17. Chapeville F. (1962) Oxidation of cysteine to cysteic acid on SRNA with retention of specificity in amino acid incorporation. *Fed Proc* **21:** 414d.

18. Kiick KL, van Hest JCM, Tirrell DA. (2000) Expanding the Scope of Protein Biosynthesis by Altering the Methionyl-tRNA Synthetase Activity of a Bacterial Expression Host. *Angew Chem Int Ed* **39:** 2148–2152.

19. Tang Y, Tirrell DA. (2001) Biosynthesis of a Highly Stable Coiled-coil Protein Containing Hexafluoroleucine in an Engineered Bacterial Host. *J Am Chem Soc* **123:** 11089–11090.

20. Wang P, Fichera A, Kumar K, Tirrell DA. (2004) Alternative Translations of a Single RNA Message: An Identity Switch of 2S,3R-4,4,4-Trifluorovaline Between Valine and Isoleucine Codons. *Angew Chem Int Ed* **43:** 3664–3666.

21. Kim W, George A, Evans M, Conticello VP. (2004) Cotranslational incorporation of a structurally diverse series of proline analogues in an *Escherichia coli* expression system. *ChemBioChem* **5**: 928–936.

22. Kast P, Hennecke H. (1991) Amino acid substrate specificity of *Escherichia coli* phenylalanyl-tRNA synthetase altered by distinct mutations. *J Mol Biol* **222**: 99–124.

23. Kirshenbaum K, Carrico IS, Tirrell DA. (2002) Biosynthesis of Proteins Incorporating a Versatile Set of Phenylalanine Analogs. *ChemBioChem* **3**: 235–237.

24. Datta D, Wang P, Carrico IS, Mayo SL, Tirrell DA. (2002) A Designed Phenylalanyl-tRNA Synthetase Variant Allows Efficient in vivo Incorporation of Aryl Ketone Functionality into Proteins. *J Am Chem Soc* 124, 5652–5653.

25. Bentin T, Hamzavi R, Salomonsson J, Roy H, Ibba M, Nielsen PE. (2004) Photoreactive bicyclic amino acids as substrates for mutant *Escherichia coli* phenylalanyl-tRNA synthetases. *J Biol Chem* **279**: 19839–19845.

26. Link AJ, Vink MKS, Agard NJ, Prescher JA, Bertozzi CR, Tirrell DA. (2006) Discovery of Aminoacyl-tRNA Synthetase Activity Through Cell-Surface Display of Noncanonical Amino Acids. *Proc Natl Acad Sci USA* **103**: 10180–10185.

27. van Hest JCM, Kiick KL, Tirrell DA. (2000). Efficient Incorporation of Unsaturated Methionine Analogues into Proteins *in vivo*. *J Am Chem Soc* **122**: 1282–1288.

28. K.L. Kiick, E. Saxon, C.R. Bertozzi and D.A. Tirrell. (2002) Incorporation of Azides into Recombinant Proteins for Chemoselective Modification by the Staudinger Ligation. *Proc Natl Acad Sci USA* **99**: 19–24.

29. Urry DW. (1993) Molecular machines: How motion and other functions of living organisms can result from reversible chemical changes. *Agnew Chem Int Ed Engl* **32**: 819–841.

30. Ruoslahti E, Pierschbacher MD. (1987) New perspectives in cell adhesion: RGD and integrins. *Science* **238**: 491–497.

31. Hubbell JA, Massia SP, Desai NP, Drumheller PD. (1991) Endothelial cell-selective materials for tissue engineering in the vascular graft via a new receptor. *Bio/Tech* **9**: 568–572.

32. Liu JC, Heilshorn SC, Tirrell DA. (2004) Comparative Cell Response to Artificial Extracellular Matrix Proteins Containing the RGD and CS5 Cell Binding Domains. *Biomacromolecules* **5**: 497–504.

33. Carrico IS, Maskarinec SA, Heilshorn SC, Mock M, Liu JC, Nowatzki PJ, Franck C, Ravichandran G, Tirrell DA. (2007) Lithographic Patterning of Photoreactive Cell-Adhesive Proteins. *J Am Chem Soc* **129**: 4874–4875.

34. Son S, Tanrikulu IC, Tirrell DA. (2006) Stabilization of bzip Peptides through Incorporation of Fluorinated Aliphatic Residues. *ChemBioChem* **7**: 1251–1257.

35. Bilgicer B, Kumar K. (2004) De novo design of defined helical bundles in membrane environments. *Proc Natl Acad Sci USA* **101**: 15324–15329.

36. Lee H-Y, Lee K-H, Al-Hashimi HM, Marsh ENG. (2006) Modulating protein structure with fluorous amino acids: Increased stability and native-like structure conferred on a 4-helix bundle protein by hexafluoroleucine. *J Am Chem Soc* **128**: 337–343.

37. Montclare JK, Tirrell DA. (2006) Evolving Proteins of Novel Composition. *Angew Chem Int Ed* **45**: 4518–4521.

38. Leslie AGW, Liddell JM, Shaw WV. (1986) Crystallization of a type III chloramphenicol acetyl transferase. *J Mol Biol* **188**: 283–285.

39. Yoo TH, Link AJ, Tirrell DA. (2007) Evolution of a Fluorinated Green Fluorescent Protein. *Proc Natl Acad Sci USA* **104**: 13887–13890.

40. Hughes RA, Ellington AD. (2005) Mistakes in translation don't translate into termination. *Proc Natl Acad Sci USA* **102**: 1273–1274.

41. Bacher JM, Waas WF, Metzgar D, de Crécy-Lagard V, Schimmel P. (2007) Genetic code ambiguity confers a selective advantage on *Acinetobacter baylyi*. *J Bacteriol* **189**: 6494–6496.

42. Dieterich DC, Link AJ, Graumann J, Tirrell DA, Schuman EM. (2006) Selective identification of newly synthesized proteins in mammalian cells using bioorthogonal noncanonical amino acid tagging (BONCAT). *Proc Natl Acad Sci USA* **103**: 9482–9487.

43. Beatty KE, Liu JC, Xie F, Dieterich DC, Schuman EM, Wang Q, Tirrell DA. (2006) Fluorescence Visualization of Newly Synthesized Proteins in Mammalian Cells. *Angew Chem Int Ed* **45**: 7364–7367.

chapter 9

Designing Ligands to Bind Tightly to Proteins

George M. Whitesides*, *Phillip W. Snyder,*
Demetri T. Moustakas, and Katherine A. Mirica

1. Introduction

In a scientific reductionist's view, a cell is a compartment containing a large ensemble of reacting molecules. Molecules enter the cell from outside; metabolic waste products leave it; the reactions occurring inside the membrane generate the new molecules required both for the cell to replicate itself, and for it to respond to changes in its environment. The intense activity of the cell at the molecular scale is almost entirely controlled by catalysts (most of which are proteins, but among which are a few very important catalytic nucleic acids, or aggregates of proteins and nucleic acids), or by other proteins that perform specific functions (for example, facilitating passage of ions or molecules across the cell membrane, or sensing the environment outside the cell).

A central task faced by the cell is that of ensuring the correct level of specificity in its reactions. That specificity, in turn, is accomplished through what is called "molecular recognition": loosely, the structure-specific association of one molecule with another.[1–10] In an important sense, molecular recognition is the most fundamental process in biochemistry, as

*Department of Chemistry and Chemical Biology, Harvard University, Cambridge MA 02138, USA, e-mail: gwhitesides@gmwgroup.harvard.edu.

its fidelity enables the exquisite control the cell exercises over its composition and chemical activity.

Molecular recognition is typically a non-covalent event: the two or more interacting molecules form a structured aggregate without forming a covalent bond. Enzymatic catalysis is also a kind of molecular recognition: that is, the preferential recognition (stabilization) of the transition state relative to the ground state of the reactants. Molecular recognition in biology almost always involves releasing some or all of the water (or other) molecules normally in contact with the interacting surfaces, and bringing these surfaces together; it may also involve changes in the conformation of the molecules, and in other details (for example, in the extent of ionization of functional groups, and in the coordination of ions or molecules).[11–13] The interactions involved are typically individually comparable to thermal energy (kT) — from tenths of a kcal/mol to a few kcal/mol.[14,15]

Given the ubiquity and enormous importance of molecular recognition in biology, one might assume that we chemists would have "figured it out": that is, rationalized the interactions involved, and developed methods of predicting interactions involving new structures. In fact, we have not. The prototypical problems of molecular recognition — understanding how and why a ligand (which might be another protein) binds to a protein, or understanding how a polypeptide folds into a protein — were well-posed but unsolved problems 50 years ago, and largely remain so today.[16–22]

The importance of molecular recognition is abundantly clear: it is the basis for a reductionist understanding of life; it is essential for the rational design of ligands to bind to proteins of known structure, and therefore an important contributor to the early stages of drug design. Molecular recognition in water is also a subject that is technically very difficult. Chemists — in thinking about molecular recognition — have tended to work on the much easier problems posed by systems soluble in aprotic organic solvents (especially chloroform and methylene chloride, chosen largely because they are excellent solvents for NMR spectroscopy and relatively simple liquids).[1–3,8,23,24] These problems have produced an abundance of interesting science, but very little that is relevant to molecular recognition in water. In fact, they may have contributed to a misunderstanding of the problem: because it is possible to develop workable rules of thumb for molecular recognition in certain systems (particularly those involving hydrogen bonding) in these solvents that do not explicitly include solvent or solvation, they have given the sense that molecular recognition is a problem involving only two players — "host" and "guest" — rather than three — "host," "guest," and "solvent." In the study of molecular recognition in chloroform (and the like) the solvent is largely ignored, as the penalty of desolvating surface area of the host and guest is negligible. Molecular recognition in water is *much* more complicated, and it remains the case that: (i) we cannot rationalize why a ligand does or does not bind to a particular protein, other than after-the-fact or very qualitatively; (ii) we cannot predict the structures of new ligands that will bind tightly to a protein, other than by close analogy with known ligands; (iii) we cannot even predict the effects of seemingly minor perturbations to the structure of known, characterized ligands on their binding.

2. What are the Difficulties?

One should perhaps simplify the question, and ask "What *is* the difficulty?" to which the short answer would be "water."[25-27] The story is more complicated than that, of course, and all of the interacting parties — protein, ligand, and medium — contribute to the difficulty of understanding protein-ligand interactions in aqueous media. Still, liquid water is an astonishing material, and after decades of intense study, it remains mysterious, unexplained in many of its properties, and unpredictable in many of its interactions. Water has a number of characteristics that collectively make it uniquely complicated: (i) It is small, has a large dipole moment, and is capable of donating and accepting hydrogen bonds. Together, these characteristics give water a high cohesive energy density and a large heat capacity relative to other molecules of similar molecular weight; (ii) Water has a high cohesive energy density: it associates with itself through networks of hydrogen bonds in ways that simply are not understood. Phrases such as "structured water," "ice-like water," "near-surface water," and "flickering cluster" speak to the elusive nature of the structure of this network.[27] Even the median number of hydrogen bonds in which a water molecule is involved is still undefined: after years of assuming the number was four, it now seems to be closer to three[14,28,29]; (iii) Its small molar volume and extensive structure underscore the importance of configurational entropy for the structure of water — this structure depends strongly on temperature; (iv) Just as the structure of the hydrogen bond network depends strongly on temperature, so too do its properties: the temperature dependence of the dielectric constant is especially important in estimating electrostatic interactions; (v) It has a very high surface tension, a parameter that probably reflects — in some way that is not well understood — the change between the mean strength of the hydrogen bonds of water molecules at the interface between water and air, and that of water molecules in bulk water, and the order of "surface" and "bulk" water. The surface tension is important in estimating energetics in "cavity" models for solubility, and in rationalizing the hydrophobic effect; (vi) Water is an excellent solvent for ions and polar molecules, species so common within the cellular milieu. It associates strongly with ions by ion-dipole interactions and with polar-protic functionality by hydrogen bonding. In both cases, the associations are enthalpically favorable but entropically unfavorable; (vii) The properties of water still include surprises that are difficult to understand: for example, studies of zeta potential suggest that hydroxide ion apparently preferentially partitions to the water-non-polar interface.[30,31] We do not know why.

The fundamental interactions among species dissolved in water, and their interactions with water, are centrally important in trying to understand phenomena such as the hydrophobic effect, and electrostatic interactions in water.[14,27,32,33] These phenomena — which seem to reflect the ensemble properties of water, large molecular surfaces, and networks of charges — provide conceptual and technical difficulties that are distinct from those

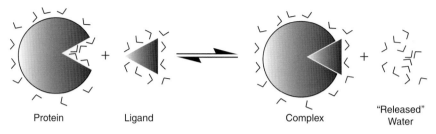

Fig. 1. The recognition of a ligand by a protein in water. Formation of a protein-ligand complex requires the dehydration of the molecular surface area of both the ligand and the active site of the protein.

associated with simple pair-wise interactions of molecules. In particular, the current model of the hydrophobic effect as a phenomenon dominated by the entropy of formation of water — whatever its structure — at the interface between non-polar materials and water is probably oversimplified in its neglect of different types of interactions (for example, dispersion interactions) between non-polar but polarizable molecules or functional groups. Similarly, efforts to provide a unified rationalization of biochemical interactions in terms of electrostatic interactions are too simple to be general and predictive.[11,34–36,37] Although these efforts must, in some sense, be correct (chemistry is largely electrostatic interactions operating within the constraints of quantum mechanics), these treatments tend to be very approximate. Biochemical systems tend to be too large to treat in detail using *ab initio* methods, and estimation of entropies requires running molecular dynamics simulations for periods of time that remain impractical.[38–47]

2.1. "Swatting Mosquitoes"

The picture that has emerged in thinking about biochemical phenomena is, from the point of view of simplicity, wholly unsatisfactory. It is, in a qualitative sense, a problem of systems containing large numbers of molecules with many small energetic interactions — each poorly defined in thermodynamic terms. The enthalpy of binding is a sum of both unfavorable terms (desolvation and configurational restriction) and favorable terms (numerous electrostatic interactions between host and guest) of similar magnitude. Entropic terms — which remain even more difficult to estimate — make equally important contributions: differences between the bound and unbound states in translation, rotation, conformation, vibration and solvent ordering are (at least) poorly defined and in all cases impossible to measure directly.[12,13,17,18,48–50] It is a problem that has the feeling of swatting mosquitoes: kill one, and so many remain that it seems to make no difference.

Science moves most rationally toward the solution of a complex problem when that problem can be broken down into parts with a clearly differentiated hierarchy in importance. When there are many contributors of equal importance, it is difficult to know what problem to focus on, and difficult to know when one has been solved until all have been solved simultaneously. The hypothesis is evident: the problem of molecular recognition in water will not resolve itself into one interaction sufficiently dominant to form the basis for simple rules of thumb. Although far from resolving the issue, this realization suggests that computation and simulation will be *required* for solving this problem. Computers do better than people at keeping track of many small numbers in a problem with many parts.

2.2. "If You can't Measure It, You can't Manage It"

The difficulty in understanding binding of proteins and ligands in water is such that progress requires the application of all plausible tools. Fundamentally, the field of molecular recognition in water requires both reliable, accurate, and interpretable experimental data, and relevant, predictive theory. The scientific method — the best template we have for the systematic resolution of a complex problem — proceeds from observation to theory to experimental test of theory. In molecular recognition, we are still in the situation of needing data on which to build and test an adequate theory.

The objective of our program in understanding the molecular recognition of ligands by proteins in aqueous buffer has been to provide good experimental information, against which to calibrate theory and simulation.

A number of experimental and computational techniques have been developed in the last several decades that, collectively, have made major steps toward a process for the rational design of ligands. (i) *Thermodynamics.* On the one hand, many techniques make it possible to measure binding constants — often with high accuracy — and thus to estimate values of $\Delta G°$. On the other, separation of $\Delta G°$ into values of $\Delta H°$ and $T\Delta S°$ has been more problematic. Measurement of $\Delta G°$ as a function of temperature — van't Hoff analysis — is now accepted to be fundamentally too error-prone for use in biochemistry. Too many factors — the structure of proteins, the conformations of ligands, the properties of water — change with temperature to make it possible to isolate parameters that are specific to the molecular recognition event.[51–54] Unfortunately, many thermodynamic data in the literature have been obtained by van't Hoff analysis, and are consequently too unreliable to be useful.[55] Fortunately, a technique that has recently become widely used as a result of the introduction of good commercial apparatus is isothermal titration calorimetry (ITC), which solves many of the problems of estimating enthalpies by measuring heat directly at constant temperature. (We discuss this technique in more detail in the next section). (ii) *Structure.* X-ray crystallography (and to an extent, NMR spectroscopy) have provided enormously useful information concerning the structure of representative soluble proteins, both with and without bound ligands. The development of synchrotron light sources has dramatically increased the quality and decreased the cost of obtaining these structures. High-energy light sources and new NMR techniques have extended the information available using them to include structural information concerning some of the water molecules at the surface of the protein (and particularly in the region of the active site).[12,48,56–58] Site-specific mutagenesis makes it routine to change the amino acid sequence in proteins to test hypotheses that depend on this sequence.[36,59–62] These advances, although invaluable, still leave critical gaps. Most of the water molecules in a protein are not defined by crystallography, and no information is available concerning the water molecules solvating the complementary binding face of the ligand. Large changes in structure of a protein that might occur on binding a ligand can be characterized by X-ray crystallography, but small ones — especially small ones distributed over many amino acids — cannot be, and even if they could, it is not clear what to do with the information. "Structure" has been extremely useful in defining what *not* to do to get binding (that is, the rational design of ligands that do *not* bind to a protein is now straightforward), but much less useful in designing ligands that *do* bind. Emerging techniques in NMR spectroscopy may someday supplement our understanding of dynamics in molecular recognition, but these techniques are still too highly specialized to be applied broadly.[48,58,63] (iii) *"Molecular Association"*. Association of molecules occurs in a very wide range of processes in chemistry. Phase transitions are forms of association. The properties of materials often reflect association of molecules. Subjects such as molecular recognition have been extensively studied in non-aqueous solutions. All of this information is relevant to the protein-ligand problem, but has not been integrated into a unified theory of molecular association. (iv) *Water*. Water has been extensively studied experimentally and theoretically, with results that have provided some (but certainly not complete) clarity.[14,27,32] The structure of water at or close to interfaces is a particularly contentious subject, and complex aqueous media — water containing dissolved organic molecules and ions — is nearly *terra incognita*.[27]

2.3. Theory, Computation, and Simulation

Theory and simulation is not *our* primary business, but understanding the theory with which the experimental data must accord is essential for the design of a useful experimental program. The availability of powerful computers and networks of computers has increased the ability of simulations to handle large numbers of atoms, but the potential functions used in these calculations are still approximate and of uncertain accuracy. Density functional theory (DFT) — although still a semi-empirical method — provides estimates of energies with high accuracies for certain classes of problems.[64] The many advances in theory and simulation still, however, fall short of what is needed in an important way. Molecular recognition in water is a many-atom problem with contributions of similar magnitude from enthalpy and entropy. The most widely used computational method for estimating entropies in complex systems of many molecules is the quasi-harmonic approximation method.[41] This method uses data from a molecular dynamics trajectory of a biomolecular system to parametrize a quasi-harmonic Hamiltonian for the system, from which the entropy of the system can be estimated. Simulations of realistic systems of protein, ligands, and medium still require so much time that it is impractical to run them over sufficient intervals to generate useful estimates of entropies. (And again, even if they could be run for very long times, or if their speed could be increased by many orders of magnitude, it is not clear how accurate the resulting entropies would be, as everything ultimately depends on potential functions of uncertain accuracy.) Although it is possible in some cases to build empirical effective potential functions that estimate $\Delta G°$, the ultimate accuracy of all theoretical methods has to be judged against their ability to predict the results of experiments before they are carried out. By that criterion, for whatever reasons, theory has not yet reached a level where it provides results that are broadly useful to designers of ligands.

3. What is New? Back to the Future with Isothermal Titration Calorimetry (ITC)

There are several glasses through which to regard molecular recognition: we think of it primarily in terms of thermodynamics — the difference in free energies between protein and ligand separately in solution, and protein and ligand associated in solution. Ultimately, one would wish to predict this difference for new structures and specific media in order to design ligands that bind tightly; at present, the task is to rationalize the thermodynamic data that are available (or can be obtained) using proteins and ligands of known structure.

Thermodynamics is an approach to the study of molecular recognition that has been used perhaps more than any other.[6,53,65–69] The unreliability of the separation of $\Delta G°$ into $\Delta H°$ and $T\Delta S°$ has, however, been such that it has been difficult to develop a set of interpretable thermodynamic parameters on which to build a theory. Isothermal calorimetry has changed this unfortunate situation for the better (although it remains to be seen if a useful theory will emerge as a consequence).[70,71] ITC is, of course, a technique that has been a part of calorimetry for many decades, but commercial instruments that have the accuracy to measure heats of association of ligands and proteins have been available for only about two decades. These instruments still require a substantial amount of protein (~1 mg for an experiment, and ~50 mg for a complete set of experiments), but preparing pure protein on this scale is no longer impossibly arduous, given the techniques of genetic engineering and protein over-expression.

Commercial instruments now being introduced will provide the ability to generate data entirely automatically, and are thus appropriate for providing the substantial quantities of replicate experiments that may be necessary to obtain estimates of statistical significance for experiments that require reliable identification of trends in small numbers. These data could provide the basis for investigations of phenomena that have previously eluded explanation because the magnitude of their effect on the thermodynamics of binding is small. The thermodynamic solvent isotope effect, for example, which is the difference between thermodynamics of binding in light and heavy waters, has been difficult to address experimentally because of the small magnitude of the effect (typically just larger than the error of measuring enthalpies of binding).[72,73] Similarly, the incremental heat capacity of binding is *known* to be a hallmark of the hydrophobic effect, but the relationship between this parameter and the structure of water near hydrophobic surface area is far from well established. Resolving these issues will also come from expanded structure-thermodynamic relationships: automated calorimetry should facilitate the collection of data for large numbers of ligands of different structures, perhaps in different media and at different temperatures. Such aggressive programs directed at understanding the thermodynamics of molecular recognition have, until now, been impractical.

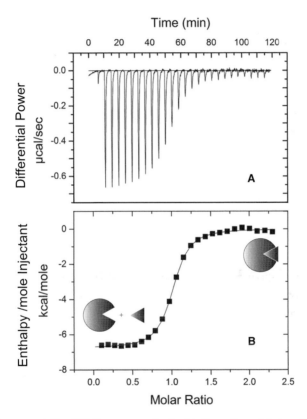

Fig. 2. (A) A representative ITC thermogram for the binding of a protein to a ligand. Small quantities of ligand are periodically injected into a protein solution, and the heat evolved by each injection (ΔH) is recorded. As the concentration of ligand in the sample cell approaches saturation, the magnitude of the heat evolved by each injection decreases. **(B)** A binding curve is obtained by plotting the ΔH measured for each injection against the ratio of ligand to protein at the time of the injection. This binding curve yields the K_d, ΔH and n, the binding stoichiometry.

ITC is normally carried out by adding sub-stoichiometric aliquots of ligand to a solution of protein, and measuring the heat evolved.[74] Analysis of this heat with a known model of binding provides three useful parameters: (i) a direct measure (the total integrated heat evolved on binding) of the enthalpy of binding at constant temperature; (ii) the binding constant; and (iii) the stoichiometry of binding. Because the experiment is carried out at constant temperature, most of the potential artifacts of van't Hoff analysis disappear.

In ITC, the evolved heat at constant pressure gives the enthalpy of binding ($\Delta H°$) directly. The dissociation constant gives $\Delta G°$ [Eq. (1)]; this parameter can usually be obtained by some entirely independent measurement (enzymatic activity, or spectrophotometry). The difference between them thus provides $T\Delta S°$ [Eq. (2)]. ITC offers, for the first time, the ability to generate high-quality thermodynamic data characterizing binding of ligands to proteins under conditions relevant to biological chemistry.

$$\Delta G° = -RT \ln(K_d) \tag{1}$$

$$\Delta G° = \Delta H° - T\Delta S° . \tag{2}$$

4. A Model System: Carbonic Anhydrase and Arylsulfonamides

Most of our experimental work has involved the enzyme carbonic anhydrase (CA, E.C. 4.2.1.1), and studied the association of derivatives of benzenesulfonamide (as the anion: $ArSO_2NH^-$) to the zinc ion in the active site of this enzyme. The choice of this system is based primarily on simplicity. The system of CA and arylsulfonamides offers: (i) an enzyme that is available commercially and inexpensively, and that has high structural stability (and thus few complications from changes in its structure — plasticity — on binding the arylsulfonamide anion); (ii) a class of active-site directed inhibitors — arylsulfonamides — that is easy to prepare synthetically, and whose binding at the active site of the CA is exceptionally constant and well-characterized in geometry; (iii) an active site that includes a substantial hydrophobic surface, and thus provides a test-bed for studying hydrophobic interactions; (iv) accessible molecular surface proximate to the active site that is non-polar, and thus useful in other examinations of hydrophobic effects. We have summarized the information concerning CA as a model for physical-organic and biophysical studies of protein-ligand binding in an extensive review article.[75]

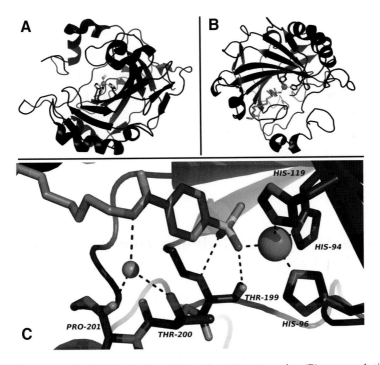

Fig. 3. Structures of carbonic anhydrase II bound to **(A)** water and to **(B)** a para-substituted benzenesulfonamide ligand with an oligoethylene glycol "tail."[76] **(C)** A more detailed view of the polar interactions between the protein and the ligand. The coordination bond between the nitrogen of the sulfonamide and the Zn(II) cofactor is shown as a dashed line. Hydrogen bonds between the nitrogen of the sulfonamide and the side-chain hydroxyl of THR-199 of the protein are shown as dashed lines as well. A network of hydrogen bonds between the amide of the *para*-substituted "tail" and two protein residues (PRO-201 and THR-200) are shown as dashed lines.

We have used CA as a model in a number of studies. The most relevant to considerations of thermodynamics simplifies the system even further by using it primarily as the basis for perturbational studies. In these studies, we leave the structure of the enzyme and the benzenesulfonamide group unchanged, vary only the structure of the substituents in the para-position of the benzene ring, and examine the influence of those variations on the thermodynamics of binding; these variations reflect interactions of the para-substituents with the surface of the protein proximate to, but not directly in, the active site. We believe this system is the simplest and best defined that can be devised for realistic studies of protein-ligand binding. We have carried out a number of studies using it; probably the most honest description of these studies is that they have consistently produced results that were — to us — *unexpected*: we draw the conclusion that we cannot presently accurately predict the results of even these kinds of perturbational experiments.

We have carried out a number of studies of the binding of arylsulfonamides to CA.[76–85] These include examinations of the influence of Coulombic interactions on binding (using protein charge ladders), of the relation of the structure of the aryl group and binding constant, and of hydrophobic substituents on the aryl ring and binding constants; all have given interesting results that have clarified the interactions that contributed to binding. Detailed interpretation of these results has also remained difficult.

In principle, the most interpretable types of studies should be the perturbation studies that have examined the influence of para-substituents on the aryl ring on the binding. Since these studies concern only binding outside of the immediate active site, and since they do not appear to influence the binding site region itself, we had expected them to minimize changes to the structure of the protein, and thus to provide the simplest case we could develop for studying molecular recognition in a system of protein and ligand. The results of these studies have been uniformly surprising: that is, they have not been in accord with current theory (or at least with *our* understanding of current theory). The interpretation of these experiments is often complicated, and we give only one example to show the kinds of information that ITC yields, and the difficulty in reconciling this information with simple rationalizations of molecular recognition in this system. The papers we reference (and in the case of electrostatic effects, a review) give further views of these subjects.[11,75]

4.1. Binding of p-R-C$_6$H$_4$SO$_2$NH$_2$ to Carbonic Anhydrase

We prepared a series of benzenesulfonamides with conformationally flexible ("floppy") oligomeric chains in the para-position (oligomers of ethylene glycol, glycine, and sarcosine).[78,79,85] These systems are important in one current approach to the rational design of ligands: this approach identifies a secondary binding site close to the active site (by NMR spectroscopy, by crystallography, and/or by computation), and attempts to construct a tightly binding inhibitor by linking groups that bind simultaneously at the active and the secondary sites. The question of interest is: "What should be the structure of the linker to maximize the value of $\Delta G°$ for the bivalent ligand?" Answering the question of the nature of bivalent interactions, and of the strategy that maximizes binding through bivalency, is a separate and engaging problem, but one for which it is clearly important to understand the interaction of the primary ligand and the linker (*without* the group targeted to the secondary binding site) with the protein.[49,50] We, therefore, prepared benzenesulfonamides with $R = (EG)_n$, $(Gly)_n$, and $(Src)_n$, measured values of heat evolved upon binding to CA (and thus $\Delta H°$) and $\Delta G°$, and obtained $T\Delta S°$ from these two. The results of these studies were entirely unexpected.[85]

What we had anticipated was to observe either a trend to tighter binding with longer oligomeric chains (reflecting favorable enthalpy of interaction of the floppy chain with the surface of the protein) or a trend to weaker binding (reflecting loss in conformational entropy due to this interaction, or simply to a restriction in the volume of space accessible to the chain). In fact, we observed — for all three series — *no* change: the binding constant was independent of the structure of the floppy chains. This independence was *not* due to the absence of interaction of the chains with the surface of the CA: both $\Delta H°$ and $T\Delta S°$ varied with the chain length.

This result was thus unexpected in three ways: (i) The binding constant was essentially independent of the length and structure of the floppy chain; (ii) There *were* changes in $\Delta H°$ and $T\Delta S°$, but they precisely cancelled; (iii) The changes in $\Delta H°$ and $T\Delta S°$ were opposite to those we had expected: that is, as the chains in each series became long, the contribution to the enthalpy of binding we attributed to the chain became *less favorable*, and the contribution to the entropy of binding *less unfavorable*. How do we explain these very puzzling results? We do not, in fact, presently know. We have a rationalization, but it is very much *post facto*.[85] We are confident that these results are not experimental error, however, and we note that a number of other studies that we have carried out with CA have generated related but equally puzzling results. We conclude that we simply cannot use our current, qualitative, understanding of molecular recognition involving proteins in water in a way that successfully rationalizes magnitudes and contributions of enthalpy and entropy to free energy.

4.2. Entropy-Enthalpy Compensation

We also do not understand the apparent entropy-enthalpy compensation (EEC), the tendency for structural changes in ligands and proteins to have little or no effect on the free energy of binding because of compensating contributions from enthalpic and entropic terms. The occurrence of EEC is one of the puzzles waiting to be resolved in the study of molecular recognition in water. EEC has been claimed in many experimental investigations of the thermodynamics of biological systems. In studies in which van't Hoff analysis generated the values of $\Delta H°$ and $T\Delta S°$, the potential for compensating errors, and for artifacts in the procedures, was so great that EEC was largely (and we believe correctly) dismissed as a credible phenomenon.[52,54,55] When ITC is used to generate the data that underpin EEC, the potential for artifact is smaller (although still not zero). So: is EEC a real phenomenon, or is it an artifact? There are plausible, if qualitative, arguments that it should exist, and we are inclined to believe that it is real.[6,86] If it is, it hints at other relationships in the thermodynamics of biochemical systems whose basis we do not presently see clearly.

5. Assessment

Our studies of benzenesulfonamides binding to carbonic anhydrase provide the simplest biophysical system that we can construct with which to study the physical-organic chemistry of binding of a low molecular weight ligand to a globular protein. The combination of ITC (for thermodynamics) and X-ray crystallography (for structure) provides a solid basis for characterizing the entropies and enthalpies of binding, and for trying to relate trends in these values to trends in structure, and for comparisons with the results of simulations. The situation from these studies suggests that there is still a substantial amount of work — both experimental and conceptual — to be done before this simplest system can be claimed to be understood. We can rationalize some of its features.[75] For example: (i) The largest single contribution to binding is probably the interaction of the Zn^{+2} ion in the active site, the $^{-}HNSO_3$-R group to which it binds, and the network of hydrogen bonds — some directly between protein and sulfonamide anion, and some involving water — in the active site. (ii) The major influence of the structure of the arylsulfonamide on the binding is through its influence on the *pKa* of the H_2NSO_2Ar group. (iii) Strongly hydrophobic substituents in the para position of the benzene ring increase binding through the hydrophobic effect. Others we cannot rationalize: (i) We do not understand the contributions of substituents to the enthalpy and entropy of binding; (ii) We do not understand the apparent importance of EEC in these (and other biochemical) systems; (iii) Our estimations of entropy and enthalpy in interactions that would plausibly be ascribed to the hydrophobic effect do not suggest a system dominated by entropy (as one would expect from the most commonly accepted explanations of the hydrophobic effect; (iv) We do not understand the role of water — its release on binding of inhibitor, its role in forming hydrogen bonds in unbound and bound states, its role in mediating electrostatic effects — at all; and (v) We do not understand how to take into account the conformational mobility of the ligand (free in solution, and bound to the protein), or the plasticity of the protein (although we speculate that protein plasticity is small — or at least smaller than for most other proteins — for CA).

So, although we (and others) are making good progress in rationalizing the binding of arylsulfonamides to CA, the progress is more at the level of understanding what it is that we do not understand, than it is in understanding. We are — in the phraseology of the military — converting "unknown unknowns" into "known unknowns," but we still have a long road to travel before converting "known unknowns" into "knowns."

6. What Do We Need?

We close this chapter with a wish list. What is it that could make an important contribution to the rational design of ligands? This list is not unique, and other researchers in this field might come up with other subjects, but there would be broad agreement on most of these items. (i) *Reliable experiments in interpretable systems.* Although we have spoken enthusiastically of the virtues of ITC, it is still too difficult (slow and labor-intensive) to provide the large numbers of data that would be helpful in unraveling the relationships between structure and binding in this difficult field. It would also be extremely useful to have new techniques to measure thermodynamic parameters. There are, for example, no good ways of measuring thermodynamic parameters for materials with very low solubilities (e.g., hydrophobic materials — often the ones of the greatest interest). (ii) *Simulation.* It seems inevitable that computation and simulation will play a very important part in the future of this field: computers are the only practical way to add many small contributions — often with opposite signs — with reasonable accuracy. Computation is still too slow, and especially too slow to provide estimates of entropies. (iii) *Theory.* There is very little about molecular recognition in water that we can truly claim to understand theoretically. The plasticity of proteins. The role and structure of water. The nature of the hydrophobic effect. The character of electrostatic effects. The tradeoff between exact fits ("lock-in-key") with favorable enthalpy and unfavorable entropy, and loose fits ("horse-in-stall") with less favorable enthalpy and less unfavorable configurational entropy. All, and more, are understood in only general terms. (iv) *Experiments in realistic media.* Much of biophysical chemistry is carried out in simple, dilute buffers. The inside of the cell — where water rarely achieves a thickness of more than a few monolayers — is a much more complex medium: approximately 300 g/L of organic materials, approximately 1 M ionic strength, high concentrations of zwitterions.[27] It would be most valuable to have some comparison between thermodynamic parameters in simple and complex media. (v) *Understanding of extra-thermodynamic relationships.* Is EEC real? How do we tell? If so, what is its origin, and why is it so common in biochemical systems. (vi) *Understanding of protein plasticity.* It is evident that the conceptual model of a protein as a rigid object, with a definite shape, is — in general — wrong. Many proteins are probably more like structured liquid crystals than the rigidly defined structures so elegantly suggested by crystallography. How does this plasticity influence binding? Time-resolved spectroscopic methods (electron spin resonance and infrared spectroscopies, for example) might give insight into the problems associated with protein dynamics on very short time scales, but these techniques are far from being common approaches in the biophysics community.[57,58,87–89] (vii) *New types of experiments.* New tools open new doors. Molecular recognition involving proteins is still starved for relevant data. We need everything: tools to define the structure of water, and to watch proteins at atomic resolution in real time, and to measure entropy changes, and to explore complex media. The list of measurements we would *like* to make is almost endless. (viii) *Better understanding of fundamentals.*

Finally, at the core, we need to understand the fundamentals of non-covalent interactions in water better. What is a hydrogen bond? The hydrophobic effect? Structured water at interfaces? Ionic atmospheres around proteins? Protein conformational mobility/plasticity? It may or may not be necessary to have a profound understanding of these and other fundamental interactions to build working engineering solutions to the problem of designing ligands to fit tightly to proteins, but it cannot hurt to have them.

Acknowledgments

This work is supported by the National Institutes of Health (GM030367, GM051559). D.T.M. was supported by fellowship AI068605. K.A.M. acknowledges support from Eli Lilly.

References

1. Lehn JM. (1990) Perspectives in supramolecular chemistry — from molecular recognition towards molecular information-processing and self-organization. *Angew Chem* **29**: 1304–1319.
2. Cram DJ. (1992) Molecular container compounds. *Nature* **356**: 29–36.
3. Hof F, Craig SL, Nuckolls C, Rebek J. (2002) Molecular encapsulation. *Angew Chem Int Ed* **41**: 1488–1508.
4. Houk KN, Leach AG, Kim SP, Zhang XY. (2003) Binding affinities of host-guest, protein-ligand, and protein-transition-state complexes. *Angew Chem Int Ed* **42**: 4872–4897.
5. Meyer EA, Castellano RK, Diederich F. (2003) Interactions with aromatic rings in chemical and biological recognition. *Angew Chem Int Ed* **42**: 1210–1250.
6. Williams DH, Stephens E, O'Brien DP, Zhou M. (2004) Understanding noncovalent interactions: Ligand binding energy and catalytic efficiency from ligand-induced reductions in motion within receptors and enzymes. *Angew Chem Int Ed* **43**: 6596–6616.
7. Hof F, Diederich F. (2004) Medicinal chemistry in academia: Molecular recognition with biological receptors. *Chem Commun* 477–480.

8. Reed CA. (2005) Molecular architectures — guest editorial. *Acc Chem Res* **38:** 215–216.

9. Butlin NG, Meares CF. (2006) Antibodies with infinite affinity: Origins and applications. *Acc Chem Res* **39:** 780–787.

10. Hirsch AKH, Fischer FR, Diederich F. (2007) Phosphate recognition in structural biology. *Angew Chem Int Ed* **46:** 338–352.

11. Gitlin I, Carbeck JD, Whitesides GM. (2006) Why are proteins charged? Networks of charge-charge interactions in proteins measured by charge ladders and capillary electrophoresis. *Angew Chem Int Ed* **45:** 3022–3060.

12. Homans SW. (2007) Water, water everywhere — except where it matters? *Drug Discov Today* **12:** 534–539.

13. Homans SW. (2007) Dynamics and thermodynamics of ligand-protein interactions, Bioactive conformation i, Vol. 272, Topics in Current Chemistry, pp. 51–82.

14. Dill KA, Truskett TM, Vlachy V, Hribar-Lee B. (2005) Modeling water, the hydrophobic effect, and ion solvation. *Annu Rev Biophys Biomol Struct* **34:** 173–199.

15. Paulini R, Muller K, Diederich F. (2005) Orthogonal multipolar interactions in structural chemistry and biology. *Angew Chem Int Ed Engl* **44:** 1788–1805.

16. Kauzmann W. (1959) Some factors in the interpretation of protein denaturation. *Adv Protein Chem* **14:** 1–63.

17. Jencks WP, Page MI. (1974) Orbital steering, entropy, and rate accelerations. *Biochem Biophys Res Commun* **57:** 887–892.

18. Page MI, Jencks WP. (1971) Entropic contributions to rate accelerations in enzymic and intramolecular reactions and chelate effect. *Proc Natl Acad Sci USA* **68:** 1678–1683.

19. Tanford C. (1962) Contribution of hydrophobic interactions to stability of globular conformation of proteins. *J Am Chem Soc* **84:** 4240–&.

20. Mirsky AE, Pauling L. (1936) On the structure of native, denatured, and coagulated proteins. *Proc Natl Acad Sci USA* **22:** 439–447.

21. Makhatadze GI, Privalov PL. (1993) Contribution of hydration to protein-folding thermodynamics. 1. The enthalpy of hydration. *J Mol Biol* **232:** 639–659.

22. Privalov PL, Makhatadze GI. (1993) Contribution of hydration to protein-folding thermodynamics. 2. The entropy and gibbs energy of hydration. *J Mol Biol* **232**: 660–679.

23. Rebek J. (2005) Simultaneous encapsulation: Molecules held at close range. *Angew Chem Int Ed* **44**: 2068–2078.

24. Whitesides GM, Simanek EE, Mathias JP, Seto CT, Chin DN, Mammen M, Gordon DM. (1995) Noncovalent synthesis — using physical-organic chemistry to make aggregates. *Acc Chem Res* **28**: 37–44.

25. Pal SK, Zewail AH. (2004) Dynamics of water in biological recognition. *Chem Rev* **104**: 2099–2123.

26. Jayaram B, Jain T. (2004) The role of water in protein-DNA recognition. *Annu Rev Biophys Biomol Struct* **33**: 343–361.

27. Ball P. (2008) Water as an active constituent in cell biology. *Chem Rev* **108**: 74–108.

28. Head-Gordon T. (1995) Is water structure around hydrophobic groups clathrate-like? *Proc Natl Acad Sci USA* **92**: 8308–8312.

29. Head-Gordon T, Hura G. (2002) Water structure from scattering experiments and simulation. *Chem Rev* **102**: 2651–2669.

30. Beattie JK, Djerdjev AM. (2004) The pristine oil/water interface: Surfactant-free hydroxide-charged emulsions. *Angew Chem Int Ed* **43**: 3568–3571.

31. Beattie JK, Djerdjev AM, Franks GV, Warr GG. (2005) Dipolar anions are not preferentially attracted to the oil/water interface. *J Phys Chem B* **109**: 15675–15676.

32. Chandler D. (2005) Interfaces and the driving force of hydrophobic assembly. *Nature* **437**: 640–647.

33. Arnett EM, McKelvey DR. (1969) Solvent isotope effect on thermo-dynamics of nonreacting solutes. In: Coetzee JF & Ritchie CD (eds), *Solute-Solvent Interactions*, pp. 343–398. Marcel Dekker, New York.

34. Caravella JA, Carbeck JD, Duffy DC, Whitesides GM, Tidor B. (1999) Long-range electrostatic contributions to protein-ligand binding estimated using protein charge ladders, affinity capillary electrophoresis, and continuum electrostatic theory. *J Am Chem Soc* **121**: 4340–4347.

35. Kangas E, Tidor B. (1999) Charge optimization leads to favorable electrostatic binding free energy. *Phys Rev E* **59**: 5958–5961.

36. Kangas E, Tidor B. (2000) Electrostatic specificity in molecular ligand design. *J Chem Phys* **112:** 9120–9131.
37. Sheinerman FB, Norel R, Honig B. (2000) Electrostatic aspects of protein-protein interactions. *Curr Opin Struct Biol* **10:** 153–159.
38. Kollman PA, Massova I, Reyes C, Kuhn B, Huo S, Chong L, Lee M, Lee T, Duan Y, Wang W, Donini O, Cieplak P, Srinivasan J, Case DA, Cheatham TE, 3rd. (2000) Calculating structures and free energies of complex molecules: Combining molecular mechanics and continuum models. *Acc Chem Res* **33:** 889–897.
39. Massova I, Kollman P. (2000) Combined molecular mechanical and continuum solvent approach (mm-pbsa/gbsa) to predict ligand binding. *Persp Drug Discov* **18:** 113–135.
40. Swanson JM, Henchman RH, McCammon JA. (2004) Revisiting free energy calculations: A theoretical connection to mm/pbsa and direct calculation of the association free energy. *Biophys J* **86:** 67–74.
41. Schafer H, Mark AE, Gunsteren WFv. (2000) Absolute entropies from molecular dynamics simulation trajectories. *J Chem Phys* **113:** 7809–7817.
42. Karplus M, Kushick JN. (1981) Method for estimating the configurational entropy of macromolecules. *Macromolecules* **14:** 325–332.
43. Gilson MK, Given JA, Bush BL, McCammon JA. (1997) The statistical-thermodynamic basis for computation of binding affinities: A critical review. *Biophys J* **72:** 1047–1069.
44. Gilson MK, Zhou HX. (2007) Calculation of protein-ligand binding affinities. *Annu Rev Biophys Biomol Struct* **36:** 21–42.
45. Ruvinsky AM, Kozintsev AV. (2005) New and fast statistical-thermodynamic method for computation of protein-ligand binding entropy substantially improves docking accuracy. *J Comput Chem* **26:** 1089–1095.
46. Wang J, Deng Y, Roux B. (2006) Absolute binding free energy calculations using molecular dynamics simulations with restraining potentials. *Biophys J* **91:** 2798–2814.
47. Woo HJ, Roux B. (2005) Calculation of absolute protein-ligand binding free energy from computer simulations. *Proc Natl Acad Sci USA* **102:** 6825–6830.

48. Homans SW. (2005) Probing the binding entropy of ligand-protein interactions by nmr. *ChemBioChem* **6**: 1585–1591.

49. Lundquist JJ, Toone EJ. (2002) The cluster glycoside effect. *Chem Rev* **102**: 555–578.

50. Mammen M, Choi SK, Whitesides GM. (1998) Polyvalent interactions in biological systems: Implications for design and use of multivalent ligands and inhibitors. *Angew Chem Int Ed* **37**: 2755–2794.

51. Wold S, Exner O. (1973) Statistics of enthalpy-entropy relationship. 4. Temperature-dependent activation parameters. *Chemica Scripta* **3**: 5–11.

52. Krug RR, Hunter WG, Grieger RA. (1976) Statistical interpretation of enthalpy-entropy compensation. *Nature* **261**: 566–567.

53. Liu YF, Sturtevant JM. (1997) Significant discrepancies between van't Hoff and calorimetric enthalpies. 3. *Biophys Chem* **64**: 121–126.

54. Cornish-Bowden A. (2002) Enthalpy-entropy compensation: A phantom phenomenon. *J Biosci* **27**: 121–126.

55. Beasley JR, Doyle DF, Chen LX, Cohen DS, Fine BR, Pielak GJ. (2002) Searching for quantitative entropy-enthalpy compensation among protein variants. *Proteins: Struct Funct Genet* **49**: 398–402.

56. Mukherjee M, Dutta K, White MA, Cowburn D, Fox RO. (2006) Nmr solution structure and backbone dynamics of domain iii of the e protein of tick-borne langat flavivirus suggests a potential site for molecular recognition. *Protein Sci* **15**: 1342–1355.

57. Frederick KK, Marlow MS, Valentine KG, Wand AJ. (2007) Conformational entropy in molecular recognition by proteins. *Nature* **448**: 325–U3.

58. Igumenova TI, Frederick KK, Wand AJ. (2006) Characterization of the fast dynamics of protein amino acid side chains using nmr relaxation in solution. *Chem Rev* **106**: 1672–1699.

59. Horovitz A, Fersht AR. (1990) Strategy for analyzing the cooperativity of intramolecular interactions in peptides and proteins. *J Mol Biol* **214**: 613–617.

60. Hunt JA, Ahmed M, Fierke CA. (1999) Metal binding specificity in carbonic anhydrase is influenced by conserved hydrophobic core residues. *Biochemistry* **38**: 9054–9062.

61. McCall KA, Fierke CA. (2004) Probing determinants of the metal ion selectivity in carbonic anhydrase using mutagenesis. *Biochemistry* **43:** 3979–3986.

62. Nair SK, Calderone TL, Christianson DW, Fierke CA. (1991) Altering the mouth of a hydrophobic pocket — structure and kinetics of human carbonic anhydrase-ii mutants at residue val-121. *J Biol Chem* **266:** 17320–17325.

63. Agarwal PK, Billeter SR, Rajagopalan PTR, Benkovic SJ, Hammes-Schiffer S. (2002) Network of coupled promoting motions in enzyme catalysis. *Proc Natl Acad Sci USA* **99:** 2794–2799.

64. Parr RG, Yang W. (1995) Density-functional theory of the electronic structure of molecules. *Annu Rev Phys Chem* **46:** 701–728.

65. Bingham RJ, Findlay JBC, Hsieh SY, Kalverda AP, Kjeliberg A, Perazzolo C, Phillips SEV, Seshadri K, Trinh CH, Turnbull WB, Bodenhausen G, Homans SW. (2004) Thermodynamics of binding of 2-methoxy-3-isopropylpyrazine and 2-methoxy-3-isobutylpyrazine to the major urinary protein. *J Am Chem Soc* **126:** 1675–1681.

66. Freire E. (1994) Statistical thermodynamic analysis of differential scanning calorimetry data — structural deconvolution of heat-capacity function of proteins, Numerical computer methods, pt b, Vol. 240, *Methods in Enzymology*, pp. 502–530.

67. Chervenak MC, Toone EJ. (1994) A direct measure of the contribution of solvent reorganization to the enthalpy of ligand-binding. *J Am Chem Soc* **116:** 10533–10539.

68. Christensen T, Gooden DM, Kung JE, Toone EJ. (2003) Additivity and the physical basis of multivalency effects: A thermodynamic investigation of the calcium edta interaction. *J Am Chem Soc* **125:** 7357–7366.

69. Malham R, Johnstone S, Bingham RJ, Barratt E, Phillips SEV, Laughton CA, Homans SW. (2005) Strong solute-solute dispersive interactions in a protein-ligand complex. *J Am Chem Soc* **127:** 17061–17067.

70. Leavitt S, Freire E. (2001) Direct measurement of protein binding energetics by isothermal titration calorimetry. *Curr Opin Struct Biol* **11:** 560–566.

71. Ladbury JE, Chowdhry BZ. (1996) Sensing the heat: The application of isothermal titration calorimetry to thermodynamic studies of biomolecular interactions. *Chem Biol* **3**: 791–801.

72. Connelly PR, Thomson JA, Fitzgibbon MJ, Bruzzese FJ. (1993) Probing hydration contributions to the thermodynamics of ligand-binding by proteins — enthalpy and heat-capacity changes of tacrolimus and rapamycin binding to fk506 binding-protein in d2o and h2o. *Biochemistry* **32**: 5583–5590.

73. Oas TG, Toone EJ. (1997) Thermodynamic solvent isotope effects and molecular hydrophobicity. *Adv Biophys Chem* **6**: 1–52.

74. Wiseman T, Williston S, Brandts JF, Lin LN. (1989) Rapid measurement of binding constants and heats of binding using a new titration calorimeter. *Anal Biochem* **179**: 131–137.

75. Krishnamurthy VM, Kaufman GK, Urbach AR, Gitlin I, Gudiksen KL, Weibel DB, Whitesides GM. (In press) Carbonic anhydrase as a model for biophysical and physical-organic studies of proteins and protein-ligand binding. *Chem Rev.*

76. Boriack PA, Christianson DW, Kingery-Wood J, Whitesides GM. (1995) Secondary interactions significantly removed from the sulfonamide binding pocket of carbonic anhydrase ii influence inhibitor binding constants. *J Med Chem* **38**: 2286–2291.

77. Gao JM, Qiao S, Whitesides GM. (1995) Increasing binding constants of ligands to carbonic-anhydrase by using greasy tails. *J Med Chem* **38**: 2292–2301.

78. Jain A, Huang SG, Whitesides GM. (1994) Lack of effect of the length of oligoglycine-derived and oligo(ethylene-glycol)-derived para-substituents on the affinity of benzenesulfonamides for carbonic-anhydrase-ii in solution. *J Am Chem Soc* **116**: 5057–5062.

79. Jain A, Whitesides GM, Alexander RS, Christianson DW. (1994) Identification of 2 hydrophobic patches in the active-site cavity of human carbonic-anhydrase-ii by solution-phase and solid-state studies and their use in the development of tight-binding inhibitor. *J Med Chem* **37**: 2100–2105.

80. Carbeck JD, Colton IJ, Gao J, Whitesides GM. (1998) Protein charge ladders, capillary electrophoresis, and the role of electrostatics in biomolecular recognition. *Acc Chem Res* **31**: 343–350.

81. Grzybowski BA, Ishchenko AV, Kim C-Y, Topalov G, Chapman R, Christianson DW, Whitesides GM, Shakhnovich EI. (2002) Combinatorial computational method gives new picomolar ligands for a known enzyme. *Proc Natl Acad Sci USA* **99:** 1270–1273.

82. Krishnamurthy VM, Bohall BR, Kim C-Y, Moustakas DT, Christianson DW, Whitesides GM. (2007) Thermodynamic parameters for the association of fluorinated benzenesulfonamides with bovine carbonic anhydrase ii. *Chem-Asia J* **2:** 94–105.

83. Krishnamurthy VM, Bohall BR, Kim CY, Moustakas DT, Christianson DW, Whitesides GM. (2007) Thermodynamic parameters for the association of fluorinated benzenesulfonamides with bovine carbonic anhydrase ii. *Chem-Asia J* **2:** 94–105.

84. Krishnamurthy VM, Semetey V, Bracher PJ, Shen N, Whitesides GM. (2007) Dependence of effective molarity on linker length for an intramolecular protein-ligand system. *J Am Chem Soc* **129:** 1312–1320.

85. Krishnamurthy VM, Bohall BR, Semetey V, Whitesides GM. (2006) The paradoxical thermodynamic basis for the interaction of ethylene glycol, glycine, and sarcosine chains with bovine carbonic anhydrase ii: An unexpected manifestation of enthalpy/entropy compensation. *J Am Chem Soc* **128:** 5802–5812.

86. Dunitz JD. (1995) Win some, lose some — enthalpy-entropy compensation in weak intermolecular interactions. *Chem Biol* **2:** 709–712.

87. Ishikawa H, Finkelstein IJ, Kim S, Kwak K, Chung JK, Wakasugi K, Massari AM, Fayer MD. (2007) Neuroglobin dynamics observed with ultrafast 2d-ir vibrational echo spectroscopy. *Proc Natl Acad Sci USA* **104:** 16116–16121.

88. Ishikawa H, Kim S, Kwak K, Wakasugi K, Fayer MD. (2007) Disulfide bond influence on protein structural dynamics probed with 2d-ir vibrational echo spectroscopy. *Proc Natl Acad Sci USA* **104:** 19309–19314.

89. Park S, Kwak K, Fayer MD. (2007) Ultrafast 2d-ir vibrational echo spectroscopy: A probe of molecular dynamics. *Laser Phys Lett* **4:** 704–718.

Chapter 10

Biology by the Numbers

Rob Phillips[*]

Science was once called "natural philosophy" and had as its purview the scientific study of all of nature. Increasing specialization led to a splintering of natural philosophy into a number of separate disciplines and one of the outcomes of this trend was that physics emerged largely as the study of inanimate matter. This peculiar state of affairs imposed an unnatural barrier which largely prevented physicists from seeing the study of living matter as part of their core charter. Similarly, the style of analysis favored in the life sciences was often descriptive, isolating much of the biological mainstream from quantitative descriptions as the rule rather than the exception. An exciting outcome of the biological revolution of the last 50 years is that the study of living matter is emerging as a true interdisciplinary science that will enrich traditional physics and biology alike. I examine some of the philosophical underpinnings of physical biology and then illuminate these ideas through several case studies that highlight the interplay between quantitative data and the models set forth to greet them. One of the interesting outcomes of an analysis from the physical biology perspective is that topics that seem very distant biologically are next door neighbors in physical biology.

1.1. Introduction

Science has always been driven by the invention of new ways of observing and measuring the world around us. When Galileo turned his "tube" to the vault of the heavens, he discovered the phases of Venus and the moons of Jupiter.[1] By connecting telescopes to spectrometers or detectors that can "see" at wavelengths other than the visible, our view of the universe has been transformed again and again. Over the last 50 years while we have

[*]Department of Applied Physics, California Institute of Technology, Pasadena, CA 91125,
e-mail: phillips@pboc.caltech.edu.

garnered an unprecedented view of inanimate matter at the scales of both the very large and the very small, another revolution has taken place in the study of living matter. Indeed, in just over 50 years, biologists have gone from uncertainty as to whether the molecules of the gene are protein or nucleic acid[2] to an era in which high school students can reprogram bacteria to do their own bidding. As a result of the technological advances that have come on the heels of fundamental biological understanding, it is becoming increasingly possible to study living matter in a quantitative way, resulting in the science of what one might call "physical biology."

One of the key tenets of physical biology is that quantitative data demands quantitative models. More specifically, as evidenced by casual perusal of almost any scientific journal reporting on the life sciences, data from biological experimentation is routinely quantitative, presented in the form of graphs illustrating the kinds of functional dependence that are the lifeblood of the interplay between theory and experiment in the physical sciences. One of the central arguments of physical biology as I see it is that theory must be practiced at a level commensurate with the experimental state of the art. That is, if those performing experiments are going to go to all of the trouble of generating reproducible, quantitative data, then the field must demand that theory be practiced in a way that can respond to this data with more than words and cartoons.

To get an impression of the kind of data that can inspire a new kind of interplay between theory and experiment in biology, Fig. 1 shows recent experiments involving the physics of genome management. In these two cases (both of which will be fleshed out in more detail later in the essay), there is an interplay between the informational and physical properties of DNA as the central molecule of heredity.[3] In the first case, the physical properties of DNA as a charged, semiflexible polymer place physical constraints on the way that viruses can pack their genomes within the tight confines of a viral capsid.[4] As a result of these physical effects, the virus has to resort to the services of an ATP-driven molecular motor in order to fully pack its genome. In the second case, DNA loops are formed when a transcription factor binds simultaneously to two sites near a promoter.[5] The outcome of the measurement in this case is that the extent of gene expression depends sensitively upon the length of the DNA loop between the two sites.[6] The second of these experiments convincingly demonstrates that reproducible, quantitative data can be gotten from the *in vivo* setting as well as from their *in vitro* counterparts.

Both of these cases implicitly involve one of the key injunctions of physical biology, namely, the existence of some "tuning variable" that can be varied systematically both experimentally and in the models that are put forth to greet the data. As will be shown later in the chapter, in the viral case, the genome length itself can be used as a tuning variable, whereas in the gene expression example, it is similarly the DNA length, but this time the distance between two binding sites. Generally, it is not the absolute magnitudes of the numbers that show up in these experiments (and the corresponding models) that are important. Rather, it is the ability to

(A)

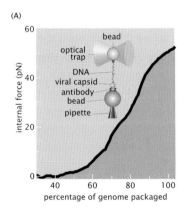

(B)

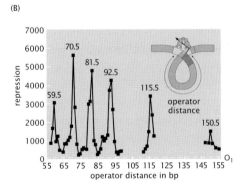

Fig. 1. Quantitative data demands quantitative models. Data from two experiments involving the *physical* properties of DNA. **(A)** Force build up during packing of a viral genome into the viral capsid. The inset shows a schematic of the optical tweezers experiment used to measure the force buildup during DNA packaging.[4] **(B)** Repression in the *lac* operon as a function of the distance between two Lac repressor binding sites. The inset shows a schematic of the binding of Lac repressor to DNA.[6]

understand how these numbers vary as some key tuning variable is varied. For example, as shown in Fig. 1(B), by tuning the length of the DNA loop in one basepair increments, the relative strength of repression is varied considerably. Though the *absolute* value of repression may be out of reach of simple model building, the scaling with length provides a chance to test our understanding of the regulatory process.

Traditionally, much of biological understanding has been captured through subtle cartoons that represent a careful decision about which features of a problem are important and which are not. On the other hand, this kind of cartoon-level understanding is inadequate as a response to well characterized quantitative data like that described above. In the remainder of the chapter, I will argue how we can go beyond pictorial and verbal descriptions and use quantitative analysis as a test of understanding that is even more stringent. This kind of thinking has already been a limited but powerful part of the scientific study of life for more than a century, with one of the most powerful examples being the analysis of propagation of action potentials with giants like Helmholtz measuring their velocity and Hodgkin and Huxley dissecting their mechanism.[7–10] One of the outcomes of this kind of analysis is that it can sharpen the kinds of questions that can be asked about a given problem and suggest new lines of experimental investigation that cannot even be conceived in the absence of a quantitative framework.

1.2. Order of Magnitude Biology

One of the main obsessions that is passed along to students when they are first learning science is to check their units. In freshman chemistry, equations with long strings of unit conversions routinely navigate students between the number of grams of a particular reactant mixed in some solution and the energy released in the resulting reaction measured in kilojoules, for example. The reason teachers harp on students to "check their units" is that these unit checks constitute the first line of defense in the sanity check to make sure that the results make intuitive sense.

A less formulaic, but deeper class of sanity checks is offered by the arithmetic of order of magnitude estimates. The idea of such estimates is to find out if the magnitudes of the quantities in question are reasonable. I remember a student once computing the deflection of a bridge due to the weight of a train crossing the span and finding a result that was 10 to some large power (maybe 10^8) with units measured in meters! Clearly this result failed the order of magnitude sanity check, but the homework was still turned in. Aside from ferreting out outright errors, the more powerful use of simple estimates is their ability to tell us if we have the right factors in play.

There is a long tradition of the value of order of magnitude estimates in the service of both physics and biology. Indeed, one of the stories that epitomized the physicists of the mid-20th century concerns Enrico Fermi who asked his students in a graduate class to estimate the number of piano tuners in Chicago. However, this approach based upon getting numbers out at the end was a serious part of an approach to physics that has been argued as one of the core reasons for the strength of American physics after the Second World War. The argument goes "In the United States, theory and experiments were always housed under the same roof in one department. And perhaps more so than everywhere else, in the United States physics was about numbers, and theories were deemed to be algorithms for getting the numbers out."[11] More recently, Edward Purcell and Victor Weisskopf kept this tradition alive in the pages of the *American Journal of Physics* where they performed all sorts of interesting estimates in a set of papers that appeared over several years.[12–15]

This tradition continues today with one of the most useful classes in the physics curriculum at Caltech, entitled "Order of Magnitude Physics." What is especially fun about this course is that the instructors (primarily Peter Goldreich and Sterl Phinney) have an open question policy in which they are willing to make an order of magnitude estimate of anything — the heights of mountains, the deflection of the Keck telescope mirrors under their own weight or the energy consumption of the human brain. My argument here is that this is definitely more than fun and games (though it definitely is fun) and serves as a self-conscious way to begin to understand what a given problem is about.[16] One of my favorite examples of the power and subtlety of order of magnitude estimates is the simple calculation that permits an order of magnitude determination of the lattice parameter, cohesive energy and bulk modulus of a metallic solid by treating it as an "electron gas."[17,18]

Perhaps the most famous example of this kind of *eureka* moment resulting from simple estimates was experienced by Newton when he realized that the ratio of the acceleration of a falling body near the Earth and the acceleration of the moon as it "falls" towards the Earth are inversely proportional to the square of their distances from the center of the Earth. It is probably not an exaggeration to say that this simple estimate was foundational in the discovery of the law of universal gravitation.[19] Newton himself noted "I deduced that the forces which keep the Planets in their Orbs must be reciprocally as the squares of their distances from centers about which they revolve: & thereby compared the force required to keep the Moon in her Orb with the force of gravity at the surface of the Earth, & found them answer pretty nearly."[19] To see the math unfold for ourselves, we need simply use the familiar formula $s = \frac{1}{2}gt^2$ and compare the distance fallen by an object at the surface of the Earth to the distance that the moon falls in the same time at its distance of nearly 60 Earth radii away from the center of the Earth. What we find is that in one second, an object right near the surface of the Earth will fall roughly 4.9 m, while the moon will "fall" 1.3×10^{-3} m. The ratio of these two distances is roughly 1/3700, the "pretty nearly" to 1/3600 that Newton speaks of.[20] To see a more detailed and beautiful exposition of Newton's "Moon Test" see pp. 357–361 of Chandrasekhar.[21]

It is exciting to hear Newton carry out this estimate in his own words.[22] Frankly, this estimate still just fills me with the thrill of seeing a deep and wonderful result revealed on the basis of a simple calculation and a few elementary facts. Proposition IV, Theorem IV of Newton's *Principia* says: "if we imagine the moon, deprived of all motion, to be let go, so as to descend towards the earth with the impulse of all that force by which it is retained in its orb, it will in the space of one minute of time, describe in its fall $15\frac{1}{12}$ *Paris* feet... Wherefore, since that force, in approaching to the earth, increases in the proportion of the inverse square of the distance, and, upon that account, at the surface of the earth, is $60 \cdot 60$ times greater than at the moon, a body in our regions, falling with that force, ought in the space of one minute of time, to describe $60 \times 60 \times 15\frac{1}{12}$ *Paris* feet; and, in the space of one second of time, to describe $15\frac{1}{12}$ of those feet. And with this very force we actually find that bodies here upon earth do really descend; for a pendulum oscillating seconds in the latitude of Paris will be 3 *Paris* feet, and 8 lines $\frac{1}{2}$ in length, as Mr. Huygens has observed. And the space which a heavy body describes by falling in one second of time is to half the length of this pendulum as the square of the ratio of the circumference of a circle to its diameter (as Mr. Huygens has also shown), and is therefore 15 *Paris* feet, 1 inch, 1 line $\frac{7}{9}$. And therefore the force by which the moon is retained in its orbit becomes, at the very surface of the earth, equal to the force of gravity which we observe in heavy bodies there. And therefore the force by which the moon is retained in its orbit is that very same force which we commonly call gravity."[22] That is an estimate!

A fundamental tenet of the physical biology approach, in my view, is the need to further strengthen this tradition. Barbara McClintock's interesting biography is entitled *A Feeling for the Organism* and builds the case that the best way to study biological systems is to really have a feeling for the organism.[23] From the perspective of physical biology, this feeling for the organism amounts to having a sense of the sizes of molecules, organelles, cells and organisms, an intuition for the rates of biological processes and a developed quantitative sense of the forces and energies relevant to biology.

One of the most interesting and simple ways to get an intuitive feel for the cell is to perform the cellular census.[24-26] This serves as a prime case study in order of magnitude biology. There are many different ways to come at such an estimate, all of which essentially converge on the same qualitative picture: the cell is highly crowded.[27-29] One simple way to see this is to start with the fact that the cytoplasm is characterized by a protein density of roughly 300 mg/mL.[30] Alternatively, we can build our order of magnitude estimates around the fact that the cell is roughly 30% by mass macromolecule with roughly half of that coming from proteins. If we assume a "typical" protein with a mass of 300 kDa, it is left as an exercise to the reader to demonstrate that this implies roughly three million proteins in an *E. coli* cell.[31] This estimate can be used in turn to calculate that the mean spacing between proteins is less than 10 nm. The key point here is that simply by knowing a few elementary facts it is possible to perform an order of magnitude analysis of a vast array of features of cells and organisms. The outcome of such thinking is shown in Fig. 2.

We can complement the simple estimates on the crowded nature of the cytoplasm by reflecting similarly on the cell surface. Here, too, there are many different ways to come to an estimate of the mean spacing between proteins on the cell surface. One of the simplest ideas is to use the fact that roughly 1/3 of genomes code for membrane proteins. In light of this rule of thumb, for a bacterium such as *E. coli*, this suggests that there are

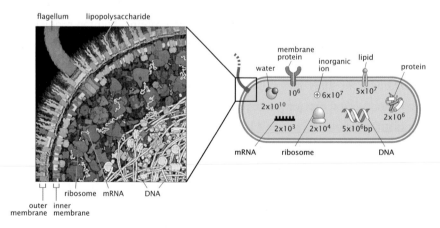

Fig. 2. Order of magnitude census of a bacterial cell. Goodsell's illustration of a bacterium shows the crowded nature of both the cytoplasm and the cell membrane.[27,32] The numbers on the right are *estimates* of the number of various classes of molecule in a bacterium.

roughly 10^6 membrane proteins distributed roughly equally on the inner and outer membranes. A second way to come to the same basic conclusion is to appeal to recent measurements of the ratio of membrane protein mass to phospholipid mass. For example, in the *E. coli* membrane, for every 1 g of phospholipid, there is roughly 2 g of protein.[33] This result can be used in turn to deduce the mean spacing between proteins. The upshot is that these kinds of order of magnitude estimates suggest a mean spacing between membrane proteins of less than 5 nm. One of the insights that emerges from such an estimate is that it is likely that there are strong physical interactions between membrane proteins which can alter their function.[34,35]

As noted above, these ideas are not new to biology. One of my favorite examples is offered in the beautiful Spiers Memorial Lecture of W.T. Astbury, entitled "X-Ray Adventures Among the Proteins."[36] Astbury starts his paper in the kind of lively and personal style that has been systematically and sadly removed from almost all modern scientific prose with the words: "All scientific research is an intellectual adventure of course, and scientific thrills are among the best and highest kind of thrills. But nowhere do we feel this sense of adventure so much as when investigating living things and the complex bodies that take part in the life process. Far and away the most important and complex of these bodies are the proteins, and the problem of their structure and properties is, I think, the greatest scientific adventure of our times."[36]

Though we take the sequences, structures and functions of proteins and their macromolecular partners for granted, Astbury's 1938 lecture demonstrates that the way was not always clear. The part of the paper that especially intrigued me and is pertinent to the discussion of order of magnitude biology centers on his analysis of whether fibrous proteins (such as keratin) "are constructed to a common plan" with globular proteins. Astbury undertakes a simple estimate of the masses of the amino acid residues, noting "It is possible to calculate it from X-ray data and the density, however, without making use of the amino-acid proportions given by the chemistry."[36] The reader is encouraged to examine Astbury's paper to see how simple estimates were used as both a sanity check and a deductive tool as the attempt was being made to figure out what proteins are really like (again noting that just because something is now taught as a triviality in high school does not imply that it was always so obvious). The story of the long road to our understanding of proteins is described in Tanford and Reynolds' excellent book.[37]

An even more dramatic example of the way in which simple estimates can be biologically illuminating is to examine the biophysical underpinnings of fidelity. Biological polymerization during DNA replication, transcription and translation is very high with error rates lower than 10^{-4}.[38] From an order of magnitude perspective, one question that immediately comes to mind is whether or not such high fidelity can be the result of thermodynamic specificity alone. For example, as was explored by both Hopfield[38] and Ninio,[39] if one computes the relative probability of binding the correct vs the incorrect tRNA during translation as a result of codon-anticodon pairing, this implies much higher error rates than are actually observed. The fact that the estimate does *not* jibe with measured error rates served as the basis of their analysis of kinetic mechanisms of discrimination.

The argument of this section is that, as Astbury (and many others) were inspired to do naturally, order of magnitude estimates can go a long way towards instructing us if we are on the right track in our thinking.

1.3. Case Studies in Physical Biology

After the first step of carrying out the relevant order of magnitude estimates, the next stage in one's analysis is often a more systematic examination of the problem from the point of view of both theory and experiment. In this section, I will describe three case studies that have been important to me personally that illustrate how such thinking might go. A different author could come up with a different set of examples (and the existence of many different examples is the whole point!) that would illustrate precisely the same point, namely, that the time is ripe to demand a rich and quantitative interplay between theory and experiment. Further, the essence of this new era of model building should explicitly aim for the kind of simplification and abstraction that intentionally attempts to ignore every detail of a given problem.

How Viruses Make New Viruses. Many of the most famous model systems from the history of molecular biology seem to have lost their luster in the biology community at large. However, in many cases, it is just these systems which have yielded so much biologically already that are optimally poised to serve as case studies in physical biology. Bacteriophage are among the most celebrated of model systems and were once referred to by Delbruck as the "atom of biology."[40] As already revealed in Fig. 1.1, bacterial viruses have served as a provocative model system for carrying out the kind of rich interplay between theory and experiment that is characteristic of physical biology. The measurements shown earlier in the paper considered the intriguing process of DNA packing. An interesting counterpart to the packing process is the ejection process whereby viral DNA is released from the bacteriophage once it has attached to the bacterium that is under siege.

We reasoned that by using genome length and salt conditions as tuning parameters to dial in the pressure associated with the packed DNA, the rate of DNA ejection from these viruses would be altered. One of the most powerful tools in the physical biology arsenal are those coming from single-molecule biophysics. These experiments complement traditional bulk experiments by revealing the unique features of the trajectories of individual molecules. A beautiful single-molecule assay for viral ejection had already been developed[41] and we decided to tailor this assay to pose a well-formulated quantitative question: How does the rate of ejection depend upon genome length and on the salt conditions? As shown in Fig. 3, it is possible to watch individual phage particles eject their DNA one molecule at a time by fluorescently labeling the DNA as it emerges from the viral capsid. As noted above, the idea of the experiment is to examine how the rate of ejection depends upon tuning parameters such as the genome length. In the experiment shown here, two different genome lengths (the wild-type length

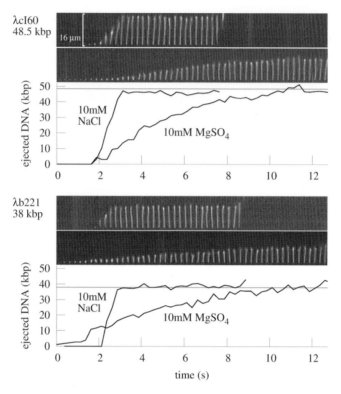

Fig. 3. Viral DNA ejection. Single-molecule measurements of the ejection process for two different choices of the viral genome length. The images show sequences of images of the same ejection event from one virion.[42]

of 48.5 kbp and a strain with a much shorter genome length of 37.5 kbp) are used to control the driving force for ejection.

Interestingly, in our own theoretical musings on this problem, we had a simple model of the ejection process based upon the idea that the friction experienced by the DNA is independent of how much DNA remains in the capsid.[43] Roughly at the same time our theoretical paper on the topic came out, my student Paul Grayson came to me with his experimental data showing that our model was overly simplistic. In particular, the friction experienced by the DNA as it is ejected from the capsid depends upon how much DNA remains in the capsid. However, from my point of view, the theory had done its job by suggesting these experiments in the first place. Now we are thinking hard about the nature of the friction associated with the packaged DNA and how it determines the rate of ejection. The very question itself is only meaningful when posed quantitatively.

How Cells Decide. One of the great episodes in the history of modern biology was the discovery of the idea of gene regulation. The simple idea is that there are certain genes whose job is to control the expression of other genes.[2,44] It is an interesting twist of history that the particular examples that led to the elucidation of the operon concept both involve DNA looping, a situation in which transcription factors bind at two sites on the DNA simultaneously and loop the intervening DNA. As shown in Fig. 1(B), beautiful, quantitative measurements reveal the subtle way in which transcriptional regulation depends upon physical factors such as the deformability of looped DNA.[6]

The study of transcriptional regulation has become impressively quantitative. It is now possible to say how much expression occurs at a given time in the cell cycle and where within the cells. As a result of these quantitative advances, it is incumbent upon those who are responding to this quantitative data to do so at a level that is similarly quantitative. One powerful tool for thinking about this class of problems is the "thermodynamic models" of gene regulation which make the simplifying but falsifiable assumption that the binding of RNA polymerase to the promoter of interest can be thought of as an equilibrium process.[45,46] Within this class of models, one computes the probability that the promoter is occupied and the predicted effect on gene expression is captured through a quantity known as the regulation factor.[47] This function gives the fold-change in gene expression as a function of key tuning variables such as repressor and activator concentrations, binding strengths of the transcription factor binding sites and quantities such as the free energy of looping. A wide range of regulation functions are shown in Fig. 4 which shows how different regulatory architectures give rise to different functional forms for the fold-change.

Regulation factors for several different regulatory motifs.

Case	Regulation factor (F_{reg})	
1. Simple repressor	$(1+r)^{-1}$	$\left(1+\dfrac{[R]}{K_R}\right)^{-1}$
2. Simple activator	$\dfrac{1+ae^{-\frac{\varepsilon_{ap}}{k_BT}}}{1+a}$	$\dfrac{1+\dfrac{[A]}{K_A}f}{1+\dfrac{[A]}{K_A}}$
3. Activator recruited by a helper (H)	$\dfrac{1+a\dfrac{1+he^{-\frac{\varepsilon_{ha}}{k_BT}}}{1+h}e^{-\frac{\varepsilon_{ap}}{k_BT}}}{1+a\dfrac{1+he^{-\frac{\varepsilon_{ha}}{k_BT}}}{1+h}}$	$\dfrac{1+\dfrac{[H]}{K_H}+\dfrac{[A]}{K_A}f+\dfrac{[A][H]}{K_AK_H}f\omega}{1+\dfrac{[H]}{K_H}+\dfrac{[A]}{K_A}+\dfrac{[A][H]}{K_AK_H}\omega}$
4. Repressor recruited by a helper (H)	$\left(1+\dfrac{1+he^{-\frac{\varepsilon_{hr}}{k_BT}}}{1+h}r\right)^{-1}$	$\dfrac{1+\dfrac{[H]}{K_H}}{1+\dfrac{[H]}{K_H}+\dfrac{[R]}{K_R}+\dfrac{[R][H]}{K_RK_H}\omega}$
5. Dual repressors	$(1+r_1)^{-1}(1+r_2)^{-1}$	$\left(1+\dfrac{[R_1]}{K_{R_1}}\right)^{-1}\left(1+\dfrac{[R_2]}{K_{R_2}}\right)^{-1}$
6. Dual repressors interacting	$\left(1+r_1+r_2+r_1r_2e^{-\frac{\varepsilon_{r_1r_2}}{k_BT}}\right)^{-1}$	$\left(1+\dfrac{[R_1]}{K_{R_1}}+\dfrac{[R_2]}{K_{R_2}}+\dfrac{[R_1][R_2]}{K_{R_1}K_{R_2}}\omega\right)^{-1}$
7. Dual activators interacting	$\dfrac{1+a_1e^{-\frac{\varepsilon_{a_1}p}{k_BT}}+a_2e^{-\frac{\varepsilon_{a_2}p}{k_BT}}+a_1a_2e^{-\frac{\varepsilon_{a_1}p+\varepsilon_{a_2}p+\varepsilon_{a_1a_2}}{k_BT}}}{1+a_1+a_2+a_1a_2e^{-\frac{\varepsilon_{a_1}p+\varepsilon_{a_2}p}{k_BT}}}$	$\dfrac{1+\dfrac{[A_1]}{K_{A_1}}f_1+\dfrac{[A_2]}{K_{A_2}}f_2+\dfrac{[A_1][A_2]}{K_{A_1}K_{A_2}}f_1f_2\omega}{1+\dfrac{[A_1]}{K_{A_1}}+\dfrac{[A_2]}{K_{A_2}}+\dfrac{[A_1][A_2]}{K_{A_1}K_{A_2}}\omega}$
8. Dual activators cooperating via looping	$\dfrac{1+a_1e^{-\frac{\varepsilon_{a_1}p}{k_BT}}+a_2e^{-\frac{\varepsilon_{a_2}p}{k_BT}}+a_1a_2e^{-\frac{\varepsilon_{a_1}p+\varepsilon_{a_2}p+F_{loop}}{k_BT}}}{(1+a_1)(1+a_2)}$	$\dfrac{1+\dfrac{[A_1]}{K_{A_1}}f_1+\dfrac{[A_2]}{K_{A_2}}f_2+\dfrac{[A_1]}{K_{A_1}}+\dfrac{[A_2]}{K_{A_2}}f_1f_2\omega}{\left(1+\dfrac{[A_2]}{K_{A_2}}\right)\left(1+\dfrac{[A_1]}{K_{A_1}}\right)}$
9. Repressor with two DNA binding units and DNA looping	$\left(1+r_m+\dfrac{r_m}{1+r_a}e^{-\frac{\Delta\varepsilon_{r_a}d+F_{loop}}{k_BT}}\right)^{-1}$	$\dfrac{1+\dfrac{[R]}{K_a}}{\left(1+\dfrac{[R]}{K_m}\right)\left(1+\dfrac{[R]}{K_a}\right)+\dfrac{[R][L]}{K_mK_a}}$
10. N non-overlapping activators and/or repressors acting independently on RNAP	$F_{reg1}\cdot F_{reg2}\cdots F_{regN}$	$F_{reg1}\cdot F_{reg2}\cdots F_{regN}$

Fig. 4. Quantitative treatment of gene regulation. Each entry in the table corresponds to a different regulatory architecture. The mathematical expressions give the predicted "regulation function" for each motif and predict the fold-change in gene expression as a function of key parameters such as the concentration of transcription factors, the free energy of looping and the strengths of DNA-protein interactions.[47]

One of the interesting case studies that can be considered is that of DNA looping. As with the case of viral packing, attacking this problem from the physical biology perspective suggests that by tuning the DNA bendability (by controlling the sequence of the looped region), it is possible to probe transcriptional regulation in a way that explicitly appeals to model predictions like those shown in Fig. 4. Experiments can be performed both *in vitro* (such as single-molecule looping experiments[48]) and *in vivo* single-cell measurements which directly test the predicted regulation factors and how they depend upon DNA flexibility, for example.

How Cells Detect Force. As another interesting example, this time from the cell biology and biophysics of membranes, we turn now to the subject of mechanosensation. Cells have a vast array of molecular components that couple to external forces.[49] One of the most palpable examples from everyday life is our sense of hearing which is based upon mechanosensors known as stereocilia which jiggle about in the hair cells of our inner ear as a result of sound waves impinging upon them. However, mechanosensation is ubiquitous and one of the most powerful model systems for exploring this phenomenon is offered by bacteria.

It has become possible to carry out a case study in physical biology on these problems because of the confluence of structural and functional information about the channels that mediate mechanosensation in the case of bacterial mechanosensitive channels. In this case, X-ray crystallography, crosslinking studies and other tools have resulted in an atomic-level picture of the structure in the closed state and hypotheses for what the channel looks like when open.[50–53] These results are complemented by functional measurements in which the current that passes through the channel is measured as a function of the load to which the surrounding membrane is subjected. Interestingly, by examining the gating properties of channels that are reconstituted in membranes with different lipid molecules, it has been found that the gating tension appears to depend sensitively on the properties of the surrounding membrane. Evidence for this effect is shown in Fig. 5. Though the details of how one might think about this kind of data using

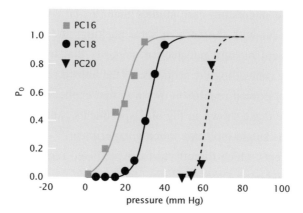

Fig. 5. Gating of mechanosensitive channels. The graph shows the probability the channel will be open (P_0) as a function of the suction pressure applied in the pipette. The critical pressure at which the mechanosensitive channel of large conductance opens depends upon the length of the lipid tail lengths in the surrounding membrane. The three curves correspond to lipid tail lengths of 16, 18 and 20.[58]

physical models are spelled out elsewhere,[54–57] for my purposes here the point of this data is to demonstrate yet again the way in which *scaling* of some observable property with some tunable parameter works out. In this case, the lipid tail length serves to alter the gating tension.

All three of the case studies highlighted here share the common feature that a particular parameter has been tuned that can be incorporated into our theory analysis and that elicit different biological function. By necessity, when biological questions are formulated in these precise quantitative terms, the questions are sharpened and the notion of what it means to understand a given phenomenon is tightened. Similarly, in each one of these case studies, it is possible to take the cartoon-level description of each problem and to recast it in mathematical terms.

Mathematicizing the Cartoons. Like most ideas, the notion of translating visual representations into a corresponding mathematical formalism is not new. One of the greatest episodes in the history of physics was the realization of the unity of electricity, magnetism and optics. Much of the experimental and conceptual foundation for this revolution was ushered in by the experiments of Michael Faraday, captured in his *Experimental Researches in Electricity*.[59] As part of that revolution, Faraday introduced the concept of a "field" in a way that stuck and became one of the key features of the modern equipment of physics.

The experimental insights of Faraday concerning the relation between electricity and magnetism were described verbally and characterized conceptually by cartoons. It is into Faraday's world of lines of force permeating space that James Clerk Maxwell entered, equipped as he was with the mathematical tools of the theoretical physicist. By his own admission, Maxwell attached enormous importance to Faraday's experimental successes, commenting in the preface to his own classic *A Treatise on Electricity and Magnetism*, "before I began the study of electricity I resolved to read no mathematics on the subject till I had first read through Faraday's *Experimental Researches in Electricity*."[60] As a result of his reading of Faraday, Maxwell perceived that the key conceptual elements of the model of the electromagnetic field were already in place and that what was required was to translate the cartoons and verbal descriptions of Faraday into the more familiar mathematical language of post Newtonian physics. Indeed, Maxwell himself characterized his successes thus: "I had translated what I considered to be Faraday's ideas into a mathematical form."[60]

Of course, analogies are almost never perfect, but I like to think that much of the beautiful and hard work that has gone into the biology of the last several hundred years is bringing us to the same sort of critical moment that was faced by Maxwell, and which was similarly illustrated by Kepler once Tycho Brahe had made his careful, quantitative measurements on the motion of Mars. In both of these cases, qualitative observations had given way to quantitative measurements. It was only in light of these measurements that it became possible to finally crack problems such as the motion of celestial bodies and the dynamics of electromagnetic fields. By way of contrast, these measurements raised the bar on what would constitute convincing theoretical understanding of these problems. The amazing explosion of understanding of all kinds about living organisms, much of which is captured diagrammatically in ways that show how different components are linked both informationally and physically serve as an invitation to the kind of translation into mathematical form described by Maxwell.

1.4. On the Virtues of Being Wrong

One of the challenges sure to be faced by those attempting to practice "physical biology" is a skepticism about the usefulness and correctness of theory and models. This skepticism comes in many different forms, ranging from those who argue that theory offers nothing that the data cannot already say for themselves, through those who are bothered by the omission of some feature of the problem, to those who dismiss models because they are "wrong" because they do not "agree" with some feature of the data. To take stock of the significance of these objections, we need to stop and examine what the goals of building simple models really are.

1.4.1. Some "Wrong" Ideas We Teach in School

I have recently heard it said that the goal of theory in biology is to be wrong. Though that might sound like an odd claim, I know just what was meant. One of the pillars of the physics curriculum is the study of thermodynamics and its microscopic partner science, statistical mechanics. Within the thermodynamic and statistical mechanical canon, one of the most important topics is the study of specific heats, i.e., the amount of heat that is required to raise the temperature of a substance by one degree. Interestingly, classical physics has a very concrete and universal prediction for the specific heat of crystalline solids that goes under the name of the law of Dulong and Petit which says that the specific heat is $3R$, where R is the universal gas constant.[61]

Interestingly, though this result is reasonably universal as a first cut at the high temperature specific heat of solids, the low temperature behavior is another matter altogether. Indeed, as the temperature drops, so too does the specific heat in direct contradiction with the law of Dulong and Petit. This anomaly was one of the great challenges to the ideas of classical physics and resulted in one of the many key contributions of Einstein to the development of quantum theory.[62] But here too, Einstein's model made drastic simplifying assumptions that resulted in inconsistencies with the scaling of the low-temperature specific heat with temperature. Should we dismiss Dulong-Petit and Einstein simply because they do not agree with the data over the entire range of measured values? As evidenced by a half-century or more of teaching in the physical sciences, the answer is clearly no. Similar statements can be made about the Lorentz model of optical absorption,[63] the Bohr model of the atom, the electron gas picture of metallic solids or the Ising model of magnetism.[64] However, in each of these cases the virtues of these models have far outweighed their vices and there is a marked lack of nuance in simply dismissing them as "wrong."

Not surprisingly, the utility of hypotheses that have later been shown to be incomplete (or even truly wrong, such as the phlogiston hypothesis which attributed a material reality to heat) has played an important role in biology as well. My argument is that a more generous interpretation of the words "right" and "wrong" would go a long way towards making the interplay between quantitative theory and experiment work better in the biological setting.

1.4.2. Of Soups and Sparks

A fascinating episode in the recent history of biology that makes the same point as the example about specific heats of solids described above is the nature of synaptic transmission. How do different neurons communicate with one another? Strongly held and distinct views on the nature of synaptic connections go back to Ramon y Cajal and Golgi and were played out in the mid-20th century as the controversy of soups and sparks.[65] In particular, the question being debated was whether or not communication between neurons was electrical (along the lines of the action potential) or rather involved chemical messengers. Both hypotheses had their proponents and were used as the basis for drawing conclusions about how various experiments would turn out. It seems like serious overreaching to me to dismiss this historical episode as one in which one group had it right and the other did not, for such throwaway remarks ignore the value of suggestive hypotheses as part of the iterative search for a better picture of how nature works.

This view of the role of theory in biology was stated succinctly by Francis Crick who noted: "The job of theorists, especially in biology, is to suggest new experiments."[66] A model is necessarily a distillation and is used to illustrate something such as a way of thinking, or how a system depends upon some parameter or another. One of the highest compliments that can be paid a model is to say that it suggests further experimentation. Ultimately, even if the hypothesis turns out to be flawed or to lack sufficient generality, it should still be viewed as one of the engines that drives scientific progress.

The history of science seems to me to send the message that often when our preconceived notions are inconsistent with experiments, these are halcyon moments when we are maximally poised to learn something new. One of the teachers from whom I have learned the most is E.T. Jaynes, one of the pioneers who showed the connection between ideas from information theory and statistical mechanics. Jaynes was also a pleasure to be around with not only because of his smarts, but because he seemed to me to live a maxim we have always passed along to our kids when out skiing — if you do not fall, you are not trying hard enough or skiing on terrain that is steep enough. It is exciting to be associated with ideas that are no longer on the bunny slopes of academia, where the risk of falling is always present, but the probability of learning something is almost sure. Jaynes summarized his views on calculations being wrong thus: "if our calculation should indeed prove to be 'unsound,' the result would be far more valuable to physics than a 'successful' calculation!"[67] As noted above, the modern quantum theory owes its existence to at least two such "unsound" results, namely, the failure of the classical theory of the specific heats of solids and the so-called ultraviolet catastrophe which was a similar anomaly of classical physics in explaining blackbody radiation.

Interestingly, recent measurements on the packing of DNA in bacteriophage ϕ-29 have cast doubt on the ability of models like those described in this chapter to properly account for the observed forces.[68] In the spirit of this section, we look upon these recent challenges to the theory as precisely the kind of interplay between theory and experiment that helps move understanding forward. The immediate reaction is not one of trying to find some way to tweak the parameters in the models so that the data is "fit" once again. Rather, it is to look into the underlying assumptions of the model and to see which assumptions should be relaxed or reconsidered.

1.5. A Look Ahead

A skeptical response to the thrust of this chapter can be framed from at least several different perspectives. First, if understanding living matter (i.e., cells and organisms and the molecules that make them tick) is the objective, will invoking the kind of rich interplay between quantitative theory and experiment advocated here teach us anything fundamentally new about the living world? Second, from the standpoint of the physical sciences, is the study of living matter going to teach us anything new about the physical laws that animate the world? Both of these are useful questions and we examine them in turn.

Will demanding a quantitative interplay between theory and experiment reveal anything new, deep or interesting about the living world? Ultimately, the only way to really find out is to try it and see. However, I would bet that the answer is yes. To see why, we need look no further than the examples highlighted in this chapter. For example, it is not an exaggeration to say that the quantitative discrepancies between the classical description of specific heats and measured low-temperature values was one of the main seeds of the modern quantum theory. Similarly, attention to details in the features of atomic spectra led to the elucidation of key ideas in modern physics such as spin. In this sense, my thinking is guided by historical analogies like these that have repeatedly demonstrated that new insights are revealed by demanding a quantitative accounting of some phenomenon of interest. It is precisely the *failure* of the simple models that provided the experimental and theoretical impetus for the next generation of discoveries.

Equally interesting is the question of whether new insights into physics itself will be provided by the study of living matter. The themes of energy, information and geometry are some of the most constant threads running through the physical sciences, and yet, they serve as the cornerstone of many biological phenomena as well. There is something still mysterious and wonderful about the complexity of living organisms and it still feels like a long way from our understanding of the "complexity" revealed by ordinary matter to the richness of the living world. Further, I cannot shake a lingering disappointment in the impotence of conventional physics in the face of systems that are far from equilibrium and my own sense is that the best place for physics to look for clues about how to think about nonequilibrium is in the setting of living matter. The idea that other sciences have something to gain from biology was stated far more eloquently as a query in the title of a paper by Joel Cohen: "Is mathematics biology's next microscope, only better? Is biology mathematics' next physics only better?"[69] What Cohen is noting is the historic synergy between mathematics and physics where ideas from one would enrich the other and then back again. In particular, for several centuries, applications in physics were one of the main driving forces in the development of mathematics and advances in mathematics permitted the expansion of physics to new domains. Cohen is betting on a similar synergy with biology at the center this time.

Ultimately, one of the main arguments of this chapter is that the hard work and deep insights of biologists on a variety of model systems makes this an opportune time to reexamine these systems with an eye to the type of interplay between theory and experiment more familiar from physics. By demanding a more stringent (i.e., quantitative) definition of what it means to understand a system, as with many previous episodes (such as the specific heat example given earlier in the chapter), we are sure to learn new things about the workings of nature. Further, by carrying out this program, physical biology will strengthen the roots of both physics and biology themselves.

Educating the Next Generation of Life Scientists and Engineers. Another way in which physical biology must surely touch the future of both physics and biology concerns the way we educate our students. The traditional physics curriculum is based upon a canon of topics that includes mechanics, electricity and magnetism, statistical mechanics and quantum mechanics. Will the applications of the ideas of physics to living matter become a part of this canon, or will physics remain stuck with the bizarre adiabatic wall that separates the study of the inanimate from living matter? Similarly, will tomorrow's biology students come to view their training in physics and mathematics as a reasonable price of entry into the study of biology? In a report released by the National Academy of Sciences, entitled "Bio2010," it was argued that the future of education in the life sciences must include a nod to the more quantitative aspects of biology.[70]

Applied Biology. One of the most amazing features of science is in the form of unanticipated technological spin offs. For every fundamental science, ultimately, there seems to always be an engineering partner. Fundamental understanding of the quantum mechanics of confined electrons gave rise to parts of a vast array of engineering disciplines (materials science, electrical engineering, computer science, etc.). Ironically, one of the poster problems learned by every beginning quantum mechanic, namely, the particle in a box, has become a technological reality. Similarly, fundamental studies of the mechanics of gases, liquids and solids literally serve as the scientific underpinnings used in engineering the new Airbus A380 or the Boeing "Dreamliner." Mechanical engineers use these foundational ideas to examine the structural mechanics of wings. Anyone who has ever sat in the window seat of a large airplane such as a 777 or a 747 will know that the wing tips can suffer deflections in excess of a meter and it is our understanding of the elasticity of structures that ensures that these wing tip deflections are nothing more than uncomfortable. Similarly, engineering fluid mechanics provides insights into the generation of wing tip vortices and drag that can be used to increase fuel efficiency. Materials scientists use fundamental understanding of the thermodynamics of alloys to construct enormous single-crystal turbine blades to prevent the kind of fracture at material interfaces that were the cause of air disasters. In each of these cases, enlightened empiricism has been superseded by rational engineering based upon rigorous scientific underpinnings. We should expect a similar transformation of the engineering outgrowths of biology.

The mindset represented by "physical biology" is one in which the construction of the fundamental biological infrastructure to fuel an engineering discipline is front and center. The February 1989 issue of *Physics Today* was a special feature on the many contributions of Richard Feynman to modern physics. Several features of this issue caught my eye. One was that Feynman considered his teaching to be his paramount achievement (but that is another story). A second intriguing feature of this issue was a photograph of the blackboard of Feynman at the time of his death (see Fig. 6) which had various interesting notes that he had written to himself. Of the many interesting things written on his blackboard, the one that caught my attention perhaps more than all the others was "What I

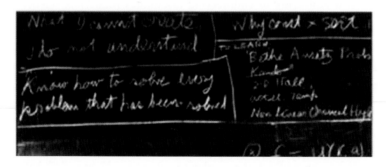

Fig. 6. Blackboard of Richard Feynman. Besides the quote mentioned in the text, the board also says "Know how to solve every problem that has been solved" and provides a list of things "to learn."

cannot create I do not understand." Physical biology begins to move in the direction of that definition of understanding by demanding that quantitative data is met with predictive, quantitative models. More to the point, the emergence of synthetic biology in which new networks, cells and organisms are constructed from scratch demonstrates that the study of living matter is well on its way to measuring up to Feynman's definition of understanding.

Acknowledgments

Many of the sentiments and ideas expressed in this essay have arisen as a result of my conversations with Jane' Kondev, Julie Theriot, Nigel Orme and Hernan Garcia in our work on our book *Physical Biology of the Cell.* I am grateful to each of them for the pleasure of working together and for having taught me so much. I should note, however, that many of the views expressed here did not pass muster with my friends and they should not be held responsible for any of my opinions that are found objectionable. Steve Quake and Jon Widom have both patiently instructed me on the interface between physics and biology. My referencing throughout the essay has not been exhaustive and is intended to provide an entry into the literature for the interested reader. I am grateful to the many others who have assisted me in my adventures in quantitative journalism as I have gone from one lab to another to learn about many of the amazing discoveries being made in the study of living matter. It is also a great pleasure to acknowledge the support of the NIH Director's Pioneer Award.

References

1. Galilei G, Drake S. (1990) *Discoveries and opinions of Galileo: including The starry messenger (1610), Letter to the Grand Duchess Christina (1615), and excerpts from Letters on sunspots (1613), The assayer (1623),* Anchor Books, New York.
2. Judson HF. (1996) *The Eighth Day of Creation.* Cold Spring Harbor Laboratory Press, Cold Spring Harbor, New York.

3. Garcia HG, Grayson P, Han L, Inamdar M, Kondev J, Nelson PC, Phillips R, Widom J, Wiggins PA. (2007) Biological consequences of tightly bent DNA: The other life of a macromolecular celebrity. *Biopolymers* **85(2):** 115–130.

4. Smith DE, Tans SJ, Smith SB, Grimes S, Anderson DL, Bustamante C. (2001) The bacteriophage straight phi29 portal motor can package DNA against a large internal force. *Nature* **413(6857):** 748–752.

5. Müller-Hill B. (1996) *The lac Operon: A Short History of a Genetic Paradigm*. Walter de Gruyter, Berlin; New York.

6. Müller J, Oehler S, Muller-Hill B. (1996) Repression of *lac* promoter as a function of distance, phase and quality of an auxiliary *lac* operator. *J Mol Biol* **257(1):** 21–29.

7. Hodgkin AL, Huxley AF. (1945) Resting and action potentials in single nerve fibres. *J Physiol* **104(2):** 176–195.

8. Hodgkin AL. (1951) The ionic basis of electrical activity in nerve and muscle. *Biol Rev Cambridge Philosoph Soc* **26(4):** 339–409.

9. Hodgkin AL, Huxley AF. (1952) A quantitative description of membrane current and its application to conduction and excitation in nerve. *J Physiol* **117(4):** 500–544.

10. Hodgkin AL, Huxley AF. (1952) Propagation of electrical signals along giant nerve fibers. *Proc R Soc Lond B Biol Sci* **140(899):** 177–183.

11. Schweber SS. (1994) *QED and the Men who Made it: Dyson, Feynman, Schwinger, and Tomonaga*. Princeton Series in Physics, Princeton University Press, Princeton, NJ.

12. Purcell EM. (1977) Life at low Reynolds number. *Am J Phys* **45:** 3.

13. Purcell EM. (1983) The back of the envelope. *Am J Phys* **51(1):** 11.

14. Bernstein HJ, Weisskopf VF. (1987) About liquids. *Am J Phys* **55(11):** 974–983.

15. Nauenberg M, Weisskopf VF. (1978) Why does the sun shine? *Am J Phys* **46(1):** 23–31.

16. Harte J. (1988) *Consider a Spherical Cow: A Course in Environmental Problem Solving*. University Science Books, Mill Valley, Calif.

17. Fetter AL, Walecka JD. (1971) *Quantum Theory of Many-Particle Systems*. International Series in Pure and Applied Physics, McGraw-Hill, San Francisco.

18. Pines D. (1964) *Elementary Excitations in Solids: Lectures on Phonons, Electrons, and Plasmons.* W.A. Benjamin, New York.
19. Westfall RS. (1980) *Never at Rest: A Biography of Isaac Newton.* Cambridge University Press, Cambridge (England); New York.
20. French AP. (1971) *Newtonian Mechanics.* W.W. Norton, New York.
21. Chandrasekhar S. (1995) *Newton's Principia for the Common Reader.* Clarendon Press; Oxford University Press, Oxford [England]; New York.
22. Newton I, Cajori F, Crawford RT, Motte A. (1934) *Sir Isaac Newton's Mathematical Principles of Natural Philosophy and his System of the World.* University of California Press, Berkeley, Calif.
23. Keller EF. (1983) *A Feeling for the Organism: The Life and Work of Barbara Mc-Clintock.* W.H. Freeman, San Francisco.
24. Pedersen S, Bloch PL, Reeh S, Neidhardt FC. (1978) Patterns of protein synthesis in *E. coli*: A catalog of the amount of 140 individual proteins at different growth rates. *Cell* **14(1):** 179–190.
25. Zimmerman SB, Trach SO. (1991) Estimation of macromolecule concentrations and excluded volume effects for the cytoplasm of *Escherichia coli. J Mol Biol* **222(3):** 599–620.
26. Wu JQ, Pollard TD. (2005) Counting cytokinesis proteins globally and locally in fission yeast. *Science* **310(5746):** 310–314.
27. Goodsell DS. (1991) Inside a living cell. *Trends Biochem Sci* **16(6):** 203–206.
28. Minton AP. (2001) The influence of macromolecular crowding and macromolecular confinement on biochemical reactions in physiological media. *J Biol Chem* **276(14):** 10577–10580.
29. Luby-Phelps K. (2000) Cytoarchitecture and physical properties of cytoplasm: Volume, viscosity, diffusion, intracellular surface area. *Int Rev Cytol* **192:** 189–221.
30. Zimmerman SB, Minton AP. (1993) Macromolecular crowding: Biochemical, biophysical, and physiological consequences. *Annu Rev Biophys Biomol Struct* **22:** 27–65.
31. Neidhardt FC, Ingraham JL, Schaechter M. (1990) *Physiology of the Bacterial Cell: A Molecular Approach.* Sinauer Associates, Sunderland, Mass.

32. Goodsell DS. (1996) *Our Molecular Nature/The Body's Motors, Machines, and Messages.* Copernicus, New York.

33. Mitra K, Ubarretxena-Belandia I, Taguchi T, Warren G, Engelman DM. (2004) Modulation of the bilayer thickness of exocytic pathway membranes by membrane proteins rather than cholesterol. *Proc Natl Acad Sci USA* **101:** 4083–4088.

34. Sprong H, van der Sluijs P, van Meer G. (2001) How proteins move lipids and lipids move proteins. *Nat Rev Mol Cell Biol* **2(7):** 504–513.

35. Ursell T, Huang KC, Peterson E, Phillips R. (2007) Cooperative gating and spatial organization of membrane proteins through elastic interactions. *PLoS Comput Biol* **3(5):** e81.

36. Astbury WT. (1938) X-ray adventures among the proteins. *Trans Faraday Soc* **34:** 378–388.

37. Tanford C, Reynolds JA. (2001) *Nature's Robots: A History of Proteins.* Oxford University Press, Oxford; New York.

38. Hopfield JJ. (1974) Kinetic proofreading: A new mechanism for reducing errors in biosynthetic processes requiring high specificity. *Proc Natl Acad Sci USA* **71(10):** 4135–4139.

39. Ninio J. (1975) Kinetic amplification of enzyme discrimination. *Biochimie* **57(5):** 587–595.

40. Fischer EP, Lipson C. (1988) *Thinking about Science: Max Delbrück and the Origins of Molecular Biology,* 1st ed. Norton, New York.

41. Mangenot S, Hochrein M, Radler J, Letellier L. (2005) Real-time imaging of DNA ejection from single phage particles. *Curr Biol* **15(5):** 430–435.

42. Grayson P, Han L, Winther T, Phillips R. (2007) Real-time observations of single bacteriophage lambda DNA ejections *in vitro. Proc Natl Acad Sci USA* **104(37):** 14652–14657.

43. Inamdar MM, Gelbart WM, Phillips R. (2006) Dynamics of DNA ejection from bacteriophage. *Biophys J* **91(2):** 411–420.

44. Echols H, Gross C. (2001) *Operators and Promoters: The Story of Molecular Biology and Its Creators.* University of California Press, Berkeley.

45. Ackers GK, Johnson AD, Shea MA. (1982) Quantitative model for gene regulation by lambda phage repressor. *Proc Natl Acad Sci USA* **79(4):** 1129–1133.

46. Buchler NE, Gerland U, Hwa T. (2003) On schemes of combinatorial transcription logic. *Proc Natl Acad Sci USA* **100(9):** 5136–5141.
47. Bintu L, Buchler NE, Garcia HG, Gerland U, Hwa T, Kondev J, and Phillips R. (2005) Transcriptional regulation by the numbers: Models. *Curr Opin Genet Dev* **15(2):** 116–124.
48. Finzi L, Gelles J. (1995) Measurement of lactose repressor-mediated loop formation and breakdown in single DNA molecules. *Science* **267(5196):** 378–380.
49. Gillespie PG, Walker RG. (2001) Molecular basis of mechanosensory transduction. *Nature* **413(6852):** 194–202.
50. Perozo E, Cortes DM, Sompornpisut P, Kloda A, Martinac B. (2002) Open channel structure of mscl and the gating mechanism of mechanosensitive channels. *Nature* **418(6901):** 942–948.
51. Sukharev S, Betanzos M, Chiang CS, Guy HR. (2001) The gating mechanism of the large mechanosensitive channel mscl. *Nature* **409(6821):** 720–724.
52. Perozo E, Rees DC. (2003) Structure and mechanism in prokaryotic mechanosensitive channels. *Curr Opin Struct Biol* **13(4):** 432–442.
53. Sukharev S, Anishkin A. (2004) Mechanosensitive channels: What can we learn from 'simple' model systems? *Trends Neurosci* **27(6):** 345–351.
54. Nielsen C, Goulian M, Andersen OS. (1998) Energetics of inclusion-induced bilayer deformations. *Biophys J* **74(4):** 1966–1983.
55. Wiggins P, Phillips R. (2004) Analytic models for mechanotransduction: Gating a mechanosensitive channel. *Proc Natl Acad Sci USA* **101(12):** 4071–4076.
56. Wiggins P, Phillips R. (2005) Membrane-protein interactions in mechanosensitive channels. *Biophys J* **88(2):** 880–902.
57. Markin VS, Sachs F. (2004) Thermodynamics of mechanosensitivity. *Phys Biol* **1(1–2):** 110–124.
58. Perozo E, Kloda A, Cortes D, Martinac B. (2002) Physical principles underlying the transduction of bilayer deformation forces during mechanosensitive channel gating. *Nat Struct Biol* **9:** 696–703.
59. Faraday M. (1965) *Experimental Researches in Electricity.* Dover Publications, New York.

60. Maxwell JC, Thomson JJ. (1954) *A Treatise on Electricity and Magnetism*, unabridged 3rd ed. Dover Publications, New York.

61. Brush SG. (1983) *Statistical Physics and the Atomic Theory of Matter: From Boyle and Newton to Landau and Onsager*. Princeton Series in Physics, Princeton University Press, Princeton, NJ.

62. Pais A. (1982) *Subtle is the Lord: The Science and the Life of Albert Einstein*. Oxford University Press, Oxford (Oxfordshire); New York.

63. Wooten F. (1972) *Optical Properties of Solids*. Academic Press, New York.

64. Ashcroft NW, Mermin ND. (1976) *Solid State Physics*. Saunders College, Philadelphia.

65. Valenstein ES. (2005) *The War of the Soups and the Sparks: The Discovery of Neurotransmitters and the Dispute over How Nerves Communicate*. Columbia University Press, New York.

66. Crick F. (1988) *What Mad Pursuit: A Personal View of Scientific Discovery*. Alfred P. Sloan Foundation Series, Basic Books, New York.

67. Jaynes ET, Rosenkrantz RD. (1983) *E. T. Jaynes: Papers on Probability, Statistics and Statistical Physics*. Synthese library (D. Reidel; Sold and distributed in the USA and Canada by Kluwer Boston, Dordrecht; Boston Hingham, MA.)

68. Rickgauer JP, Fuller DN, Grimes S, Jardine PJ, Anderson DL, Smith DE. (2008) Portal motor velocity and internal force resisting viral DNA packaging in bacteriophage ϕ29. *Biophys J.*

69. Cohen JE. (2004) Mathematics is biology's next microscope, only better; biology is mathematics' next physics, only better. *PLoS Biol* **2(12):** e439.

70. N. R. C. U. C. on Undergraduate Biology Education to Prepare Research Scientists for the 21st Century. (2003) Bio 2010: Transforming undergraduate education for future research biologists.

chapter 11

Eppur si muove

Michele Parrinello[*]

Simulating conformational changes in proteins is of great importance and molecular dynamics simulations are expected to play a major role. However, such simulations are hampered by the very short time they can run. Here we present a complex strategy for alleviating this problem. We characterize the structure of proteins by a number of collective variables or order parameters and calculate the associated free energy surface. In particular, we present two efficient methods for exploring this surface, one being well-tempered metadynamics and the other a path-based method. With the help of these methods, very long time scale phenomena can be simulated.

1. Introduction

Understanding the structure, dynamics and interaction between proteins is one of the great scientific challenges of our time and progress in this area will underpin our ability to cure many diseases. However, the study of biological molecules poses severe problems to the experimentalists and theoreticians alike. Much of what we know about their macroscopic structure comes from X-ray, neutron scattering and NMR studies. The great relevance of solving a protein structure is demonstrated by the number of Nobel prizes that have been assigned for the determination of the structure of important biological molecules. However, these techniques provide only a static picture of biological molecules, and the way they work and interact with the environment has to be inferred indirectly.[1]

[*]Department of Chemistry and Applied Biosciences, ETH Zurich, USI Campus, Via Giuseppe Buffi 13, 6900 Lugano, Switzerland, e-mail: parrinello@phys.chem.ethz.ch.

While revolutionary new methodologies[2] are emerging, we are still a long way away from being able to study protein motion in the necessary detail. This situation leaves ample room for simulations to fill the void and complement the experimental data. However, while much progress can be reported and invaluable insight has been gained from atomistic simulations, much still remains to be done before the goal of simulating complex biological systems in an accurate and reliable way is achieved.

A detailed review of the history of atomistic simulation in biology would be beyond the scope of this chapter. However, in our view the first important step in this direction was taken by Stillinger and Rahman[3] when they successfully simulated water, whose relevance to biological systems hardly needs to be underlined. This work extended forever the scope of simulations from simple or model systems to as complex a liquid as water. This result provided the impetus for the many simulations of biologically relevant molecules that followed.[4] Nowadays atomistic modeling is a standard tool used by non-specialists, experimentalists and theoreticians alike. At the heart of this success is the painstaking construction of force fields able to represent intramolecular and intermolecular interactions. While this is still work in progress, the construction of relatively accurate force fields can be considered as one of the main achievements of modern-day theory.[5]

It is nonetheless important to recognize the limits of this approach. The force fields do not quite have the required accuracy; the atom motion is described by classical mechanics, the formation and breaking of chemical bonds is not allowed, electrons are implicitly confined to and remain in the ground state, and size and time scale are severely restricted. In each of these areas much progress has been made; particularly worthy of note is the coming of age of *ab initio* molecular dynamics, which is finding increasing application in the study of how enzymes work and of all the other biological processes involving chemical changes.[6] Furthermore, the advent of ever more powerful computers and the development of scalable codes that can optimally exploit parallelism are constantly pushing back the barriers with respect to accuracy, the size of the system studied and the time scale explored. Of these limitations the most challenging and the most relevant for the present discussion, which is focused on large-scale protein changes, is the limited time scale of simulations.

In order to understand the problem, let us recall that in an atomistic simulation most of the computer time is spent in the calculation of the forces. These in turn are used to update the position and velocity of the particles following a dynamics which is usually Newtonian. On the computer, the time is updated by discrete amounts. For the integration of the equation of motions to be accurate, the integration time step Δt has to be smaller than the fastest motions in the system. This implies that there is an intrinsic upper bound to the choice of Δt, which is of the order of fs. Combining this with the cost of calculating the forces, one finds that a reasonably sized system can be simulated for no more than ~ 100 ns if classical force fields are used or ~ 100 ps if the more costly *ab initio* approach is followed. This should be contrasted with the real-life time scale of protein motions, which easily reaches milliseconds and beyond.

The gulf between simulation time scales and actual large-scale protein motions is so wide that the solution cannot come any time soon from progress in computer technology. This is particularly relevant if we take into account the fact that future progress in computer power will come largely from ever more massive parallelism and not from any major increase in processor speed. In a parallel computer the simulation can be speeded up by partitioning the calculation of the forces among the different processors.[7] In order for this partition to be advantageous, the computational load that is taken over by a single processor has to last more than the time needed to communicate the outcome to the other processors. This poses a lower limit to the amount of work that each process can do, a limit that is determined by the speed of communication and the speed of a single cpu. None of these parameters is expected to increase very significantly in the foreseeable future. In terms of wall clock time this implies that through parallelism we cannot go below this intrinsic limit. What can be done, and this is certainly major progress, is to increase the size of the systems studied. Thus we can simulate much larger systems but we cannot push forward the time scale of simulations to any great degree.

This calls for new methods to be developed, not only to solve this technical issue but also for a new formulation of the problem. In fact, even if we had the technical ability to simulate proteins for much longer times, these trajectories would be so complex that understanding the underlying physics would be very difficult. Thus if we wish to really understand the process, we still need to construct a theoretical framework and a method of analysis such that we do not merely compute but also understand the results. In this chapter we shall describe the attempts made in our group to combine into the same procedure these two aspects of the problem: while solving the time scale problem, we at the same time construct the tools for understanding it in one integrated procedure. Emphasis will be given here to concepts and ideas and we shall refer to the existing literature for the technical details. Furthermore, this is not intended to be a review article and reference to the very fast-growing literature on the subject will be partial and incomplete.

2. The Time Scale Problem

The time scale problem has long been recognized as an important issue and a vast quantity of literature is available. In order to understand one of the roots of this problem, let us consider the classical case of a particle in a two-well potential energy surface (PES) (see Fig. 1). This is characterized by two metastable states A and B, separated by an energy barrier Δ much larger than the temperature $k_B T$. This is a classical example in theoretical chemistry, for which theories of increasing complexity have been proposed.[8,9] The prototype theoretical scheme for understanding this situation is transition state theory (TST), which predicts that the rate of transition between A and B can be estimated as $\kappa \approx \tau_{mol}^{-1} e^{-\Delta/k_B T}$ where $\tau_{mol} \approx 1\,\mathrm{ps}$ is a typical molecular time. It is clear that the value of κ is largely determined by its exponential dependence on $\Delta/k_B T$. For large barriers κ can be very small

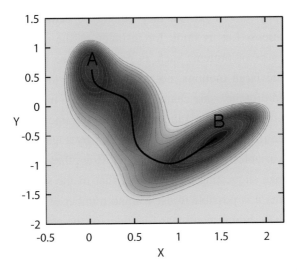

Fig. 1. Model two-dimensional potential energy surface exhibiting two main metastable states. The units are arbitrary. The lowest energy path (black line) is also shown.

and if one were to attempt a simulation that, say, started from A, the time κ^{-1} required to translocate to B would be too long for the transition $A \rightarrow B$ to be directly observable. On the other hand, if we know the location of the transition state we can use existing theories such as TST to evaluate κ. From the mathematical point of view, transition states are nothing other than saddle points on the PES of the systems, and a rather efficient procedure has been devised to identify such points. Another important concept to have emerged in the classical theory of chemical reactions is that of the intrinsic reaction path or reaction coordinate, which is identified by the set of points connecting A and B along the direction of smallest gradient. This is meant to identify the path that the system takes in going from one minimum to the other.

From this very short summary, it would appear that we have all the necessary tools to address this problem. Unfortunately, in the case of proteins, this theoretical armory is of limited help. The PES is very rugged, with an exponentially large number of saddle points. The structure of proteins is floppy and in general they cannot be defined solely in terms of atomic positions. Let us think for instance of the ubiquitous presence of floppy loops or of proteins such as prions, whose tail is unstructured and can execute large motions, not to mention the fact that even in the most structured proteins, side chains are very mobile in solution. However, proteins are not structurally random and we assume that these sets of structures, even if fluxional, could be identified by a number of descriptors or collective variables. This assumption is strengthened by the growing body of evidence suggesting, for instance, that in the study of the folding of small proteins it is possible to use a small number of such variables. We expect these to be highly nonlinear functions of the atomic coordinates R. These could be the average coordinates of a selected number of atoms, the contact map, the exposed hydrophobic surface, and so on. The use of these collective coordinates is not new, either in physics or in chemistry. Suffice it to recall here the introduction by Landau of order parameters in the theory of phase transitions or the use of the solvent coordinate in the Marcus theory of electron transfer. Identifying these coordinates is clearly a great step towards solving the problem. Let us suppose that we have identified M such coordinates, $S(R)$. It is then necessary to count the number of microscopic states that correspond to a given set of descriptors, thus considering the entropy of the system either explicitly, as we have here, or implicitly, as in more coarse-grained models. This implies shifting our attention from the PES to the free energy surface (FES). Given the probability distribution:

$$P(S) = \int dR \delta(S - S(R)) e^{-\frac{1}{k_{\mathrm{B}}T}U(R)} \bigg/ \int dRe^{-\frac{1}{k_{\mathrm{B}}T}U(R)} \tag{1}$$

where $U(R)$ is the system PES, the $F(S)$ can then be written as:

$$F(S) = -k_{\mathrm{B}}T \ln(P(S)). \tag{2}$$

This maps the original $3N$ dimensional R space into a much smaller M dimensional S space. This average focuses our attention on the relevant degrees of freedom and generates a low dimensional surface $F(S)$ that is much smoother than the original one.

Having laid down a possible description of the system's static properties in which the PES is replaced by the FES, we must now provide a formalism that replaces Newton's equations in the description of the S dynamics. Formally[10] we can use the Mori–Zwanzig projection technique and integrate out the $3N-M$ degrees of freedom not included in the S, thus arriving at a generalized Langevin equation for the time evolution of S. This procedure is not practical but shows that the exact equations for S are stochastic and the intricacies of the microscopic dynamics are hidden within a memory function that is non-local both in time and in the S coordinates. Calculating exactly the memory functions is possibly even more difficult than solving the microscopic dynamics. However, we can model the long time dynamics of proteins by assuming that the equations for the dynamics in the S space are well approximated by a high-friction ordinary Langevin equation:

$$\dot{S} = -\frac{D}{k_{\mathrm{B}}T}\frac{\partial F(S)}{\partial S} + \sqrt{2D}\,\xi(t) \tag{3}$$

which assumes that in proteins large-scale motions have a diffusive character. Here $\xi(t)$ is a white noise and we have assumed that the memory function is local in time and S space. In addition, for the sake of simplicity we take a constant diffusion coefficient D. This equation is the simplest paradigm for the S dynamics, which in the long time scale replaces the standard Newton equation governing the microscopic R evolution. It should be noted that the associated Fokker–Planck equation for the evolution of the probability distribution $P(S,t)$, takes the form[11]:

$$\frac{\partial P(S,t)}{\partial t} = \frac{D}{k_{\mathrm{B}}T}\frac{\partial}{\partial S}\left(P(S,t)\frac{\partial F(S)}{\partial S} + k_{\mathrm{B}}T\frac{\partial P(S,t)}{\partial S} \right) \tag{4}$$

which for $t \to \infty$ leads to the equilibrium distribution:

$$P(S) \propto \exp-\frac{F(S)}{k_{\mathrm{B}}T}.$$

These considerations outline a possible strategy for studying long time scale processes. First we explore $F(S)$, whose local minima identify metastable states. Later we calculate the diffusion coefficient D, or even better the memory function, of which D is an approximation. While in principle this program could be carried out for any choice of S, it becomes practical and useful only if the S are properly chosen. A possible way of estimating D can be found in Ref. 12.

3. Calculating the Free Energy Surface

Calculating free energies is one of the hardest tasks in computational statistical mechanics. The straightforward way would be to let the simulation run and evaluate the running histogram $N(S, t) = \int_0^t dt' \delta_{S,S(t')}$, which has the properties $N(S,0) = 0$ and $N(S, t) = \delta_{S,S(t')}$. Modulo an immaterial constant $F(S)$ can be estimated from $N(S,t)$ as:

$$F(S) = -k_B T \ln N(S, t).$$ (5)

Unfortunately such an estimate based on direct non-modified molecular dynamics suffers from the time scale problem and in the presence of high barriers only a fraction of the relevant S space will be explored. Other more efficient approaches need to be devised. Roughly speaking, attempts at solving this problem can be classified into two groups. The first focuses on direct evaluation of the reactive trajectories that bring the system from one minimum to the other. These methods include transition path sampling[8] and the super-symmetric Fokker–Planck equation.[13] Only later are the free energies reconstructed and the physics of the transition analyzed. The other group,[14-26] among which the present work can be classified, pushes the system out of its initial metastable state by adding a biasing potential that accelerates the exploration of other metastable states. In line with our previous discussion, we choose a history-dependent bias potential that depends on the microscopic variables through its dependence on S. This leads to a time-dependent potential $V(S(R), t)$, which modifies Newton's equation of motions as:

$$M\ddot{R} = -\frac{\partial U(R)}{\partial R} - \frac{\partial V(S(R), t)}{\partial R}. \tag{6}$$

Several suggestions have been made as to the choice of biasing potential. We describe here so-called well-tempered metadynamics,[27] which in our opinion has several advantages over previous efforts:

$$V(S,t) = k_{\mathrm{B}} \Delta T \ln\left(1 + \frac{1}{\tau} N(S,t)\right). \tag{7}$$

Here the parameters τ and ΔT have dimensions of time and temperature, respectively. With this choice Eq. (6) is supplemented with the following equation of motion for the biasing potential:

$$\dot{V}(S,t) = \frac{k_{\mathrm{B}} \Delta T}{\tau} e^{-\frac{V(S,t)}{k_{\mathrm{B}} \Delta T}} \delta_{S,S(t)}. \tag{8}$$

Equation (6) together with Eq. (8) defines well-tempered metadynamics.

The parameter τ controls the rate at which the biasing potential changes, and in practice is taken to be larger than the relaxation time of the degrees of freedom orthogonal to S. By using Eq. (7) it is seen that Eq. (8) is equivalent to:

$$\dot{V}(S,t) = \frac{k_{\mathrm{B}} \Delta T}{\tau} \frac{1}{1 + \frac{N(S,t)}{\tau}} \delta_{S,S(t)} \tag{9}$$

which implies that the rate of change of the bias potential at position S decreases with the time already spent there.

It can easily be shown that this dynamics converges to the distribution:

$$P(S) \propto \exp -\frac{F(S)}{k_{\mathrm{B}}(T + \Delta T)}. \tag{10}$$

As a consequence, in well-tempered metadynamics the sampling of the S space is performed at the enhanced temperature $T + \Delta T$. If ΔT is sufficiently large, the system can overcome the barriers in $F(S)$ several times during the simulation. As the simulation progresses the valleys in $F(S)$ are filled by a bias $V(S, t)$, whose local strength decreases with the time already spent there. Furthermore, the bias potential provides a practical way of estimating $F(S)$ using the relation:

$$F(S) = -\frac{T + \Delta T}{\Delta T} V(S, t) \tag{11}$$

which is valid at large times. In Fig. 2 it is shown how, with an appropriate ΔT, our enhanced dynamics works. It is seen that the $F(S)$ is not fully compensated by the bias potential, but rather scaled by a factor $T/(T + \Delta T)$.

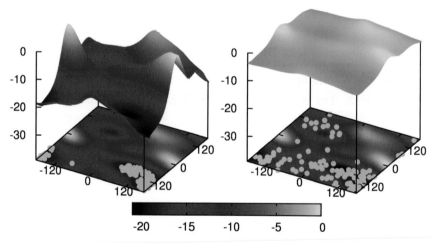

Fig. 2. Free energy surface of an alanine dipeptide in vacuum evaluated as a function of two dihedral angles (see Ref. 27). The original free energy is shown in a two-dimensional projection and as a surface in three dimensions (*left picture*). The turquoise dots represent the dihedral angles visited during an unbiased dynamics (*left*) and a well-tempered dynamics with $\Delta T = 1200$ K (*right*). On the vertical axis of the right-hand picture the sum of the free energy surface and the compensating bias potential is shown. The energies are measured in Kcal/mol.

With an appropriate choice of ΔT only the most significant parts of the $F(S)$ are explored.

In the limit of $\Delta T \to \infty$ which is known as metadynamics, the $V(S, t)$ fully compensates $F(S)$ and the system undergoes a fully diffusive motion in the S space. In the opposite limit of $\Delta T \to 0$, standard unbiased sampling is recovered.

4. Choosing the Collective Coordinates

The method described above permits an efficient calculation of $F(S)$ once an appropriate set of S has been chosen. A crucial question is how to choose the S. We have not yet come up with a method that can construct the appropriate S from the outset. While we do not rule out this possibility, it is likely that the result will be so complicated that penetrating its physical meaning might require a complex procedure, possibly as hard as, and certainly less rewarding than, constructing the S on the basis of physical intuition, as we have done so far. A saving grace is that it is relatively easy to tell whether a given set of S is appropriate. In fact all the slow modes have to be included, and finding out whether we have missed important slow modes is not difficult. The usual warning is a characteristic hysteretic behavior in which, once the system has moved from one basin to the other, it has difficulty in coming back to the initial metastable state. This is illustrated for a model system in Fig. 3, where only use of correct coordinates leads to correct sampling.

It must be added, however, that there is also much to be learned from failed attempts. A relevant case can be found in a recent study[28] of the inhibition of acetylcholinesterase by a small ligand (tetramethyl-ammonium). Initially only the distance from the target site was used as a collective variable. An analysis of the trajectory led to the conclusion that it was necessary to include π-cation interaction of the ligand with the inner aromatic pocket among the S. Once this was done the behavior was reversible and the problem was satisfactorily solved. Not only was it

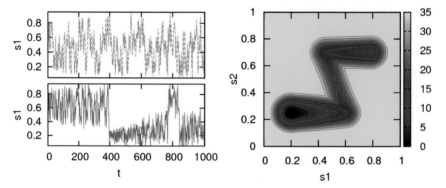

Fig. 3. The ability of well-tempered metadynamics to sample a model free energy surface (*right*) with two relevant degrees of freedom is illustrated. The green curve (*top left*), generated using $s1$ and $s2$ as collective variables, shows satisfactory ergodic behavior, with the system performing frequent transitions between one minimum and another. If instead only $s1$ is used (*bottom left*) the red curve shows infrequent and difficult transitions.

possible to calculate $F(S)$ but, more importantly, a full understanding of the underlying physics was gained. This illuminates once more the fact that combining simulation and physical intuition is generally the best way of solving complex problems.

A simple way of including all the slow modes is to enlarge the S space. Unfortunately this is not very practical, since the cost of reconstructing $F(S)$ grows exponentially with the dimension of S. Attempts have been made to alleviate this problem, which in many fields is referred to as a dimensional curse. Recently Piana and Laio[29] have proposed a method called bias switching, based on the original formulation of metadynamics, which has had great success in exploring multidimensional $F(S)$ but has been less effective in its quantitative reconstruction.

In another and different approach, Ensig *et al.*[30,31] first made a rough exploration of a multidimensional $F(S)$ and later identified within this space two metastable states and the lowest free energy path that joins one metastable state with the other. Later this path was improved and along this now one-dimensional manifold the free energy was accurately calculated.

In this way, we renounce the possibility of an accurate study of the global $F(S)$, but obtain all the information relevant to studying the particular transition, which is very often all that we are interested in. Still, the dimension of the initial S cannot be very large, as in this case even a rough exploration of $F(S)$ would be impossible.

5. Finding the Path in Free Energy Space

We now present a method for solving the dimensionality problem for the case in which we know the initial and final states (A and B) and simply wish to find the lowest free energy path that joins them.[32] This is still a very worthwhile aim, as in free energy space, this path plays a role similar to that played by the intrinsic reaction coordinate in the conventional study of chemical reactions. This intuitive notion has recently been put on to a more rigorous footing and it has been shown that, if Smoluchowsky dynamics applies, the lowest free energy path is the average over all the reactive trajectories and thus provides an excellent reaction coordinate.[33] Knowledge of this path can also be used to determine the transition rates. The question is how to obtain this path, and as we expect, in general, that several qualitatively different paths might join A to B, any search method must have a non-local capability. That is, it must be able to locate paths that are different in nature from the initial guess. We briefly present here such a method.

Let us now define in the space of S a reference path $S^0(t)$, $0 \le t \le 1$ joining the initial and final states S^A and S^B, parametrized such that $S^0(0) = S^A$ and $S^0(1) = S^A$. We now introduce two path-related coordinates:

$$z(R) \underset{\lim \lambda \to \infty}{=} -\lambda \ln \int_0^1 dte^{-\frac{1}{\lambda}\|S(R) - S^0(t)\|} \tag{12}$$

and:

$$s(R)\Big|_{\lim \lambda \to 0} = \frac{\int_0^1 dt\, e^{-\frac{1}{\lambda}\|S(R) - S^0(t)\|}}{\int_0^1 dt\, e^{-\frac{1}{\lambda}\|S(R) - S^0(t)\|}} \tag{13}$$

where through the symbol $\|\cdot\|$ we have introduced a metric that measures the distance between two points in the S space. As shown in Fig. 4, the coordinate $s(R)$ measures progress along the reactive coordinate, while $z(R)$ is its distance from the path. The set of these two coordinates can be used as a two-dimensional collective variable, and using, for instance, well-tempered metadynamics or any other free energy method we can reconstruct $F(s,z)$, which is the free energy surface as a function of s and z.

There are many advantages to this representation. While $F(s,z)$ is two-dimensional, the underlying space of the S can have very large M, thus removing the curse of dimensionality. Still the $F(s,z)$ carries a large amount of relevant information. Without any other analysis, much can be

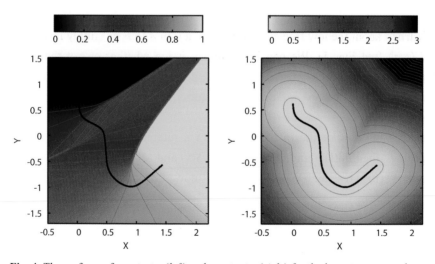

Fig. 4. The surfaces of constant s (*left*) and constant z (*right*) for the lowest energy path are shown. Note that close to the path the two surfaces are almost orthogonal and that s induces a foliation of the space.

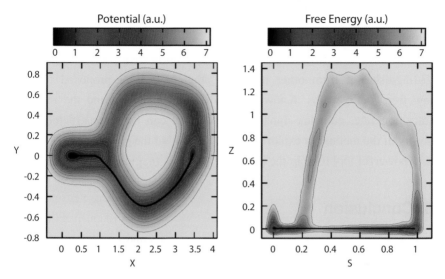

Fig. 5. Model potential energy surface with two competing pathways joining two metastable states (*left*). The lower path (drawn in the picture) is chosen as reference. On the right hand side $F(s,z)$ the imprint of the low free energy path at $z = 0$ and of the alternative upper path are clearly visible. This illustrates clearly the non-local sampling power of the method.

learned about the low free energy paths that join A to B by simply inspecting $F(s,z)$ (see Fig. 5). We can immediately judge whether or not the initial path is a good guess. In fact in the low z region, which is the closest to the path, $F(s,z)$ must be small because we expect the lowest free energy path $F(s,z=0)$ to be a line of local minima. Furthermore, any other low energy path will reveal itself as a low energy valley in $F(s,z)$ and can be identified and analyzed. We can go a step further and define the so-called string tension:

$$T\left[S^0(t)\right] = \int_0^1 dt F(s,0) \qquad (14)$$

which is a functional of the reference path, which is a local minimum when $S^0(t)$ coincides with a low free energy path. Thus one can vary $S^0(t)$ until T is minimized, and so find the lowest free energy path. Furthermore, by exploring the high z region of the (s, z) plane we can move away from the initial guess and discover qualitatively new transition pathways, which can be used as new guesses and similarly optimized. This latter ability distinguishes it from other methods[33] that have been proposed to find low free energy paths. Knowing the different pathways and the relevant transition rate, complex dynamical processes can be reconstructed. The ability of the method to explore large areas around the path can also be used as a powerful tool to map the most relevant $F(S)$ regions.

6. Conclusion

In this chapter, we have presented our strategy for solving the long time scale problems for proteins. From the above it should be clear that this approach is more general and could be applied to a large variety of problems. This strategy is the result of several years of efforts which are far from over. Along the way much has been learned and many problems resolved. We believe that when studying large-scale protein motions, the path method is preferable. It can deal with the multidimensional character of FES, is not affected by the dimensionality curse and can easily be parallelized. We regard as positive the fact that the method is not an automatic one; we would not expect it to be, nor would this be desirable. Finding the correct collective coordinates is "the" solution to the problem, and more than anything, it is the process that matters more than the final results themselves.

Acknowledgments

It is a pleasure to acknowledge the contribution of and heated discussions with past and present members of my group. In particular, most of the material presented here originates from my interaction with Alessandro Barducci, Massimo Bonomi, Davide Branduardi, Giovanni Bussi, Francesco Gervasio and Alessandro Laio. I should also like to thank Davide Branduardi, Giovanni Bussi and Francesco Gervasio for their help in preparing this chapter.

References

1. Henzler-Wildman K, Kern D. (2007) Dynamical personalities of proteins. *Nature* **450**: 964–972.
2. Park HS, Baskin JS, Kwon OH, *et al.* (2007) Atomic-scale imaging in real and energy space developed in ultrafast electron microscopy. *Nano Lett* **7(9)**: 2545–2551.
3. Stillinger FH, Rahman A. (1974) Improved simulation of liquid water by molecular dynamics. *J Chem Phys* **60**: 1545–1557.
4. McCammon JA, Gelin BR, Karplus M. (1977) Dynamics of folded proteins. *Nature* **267**: 585–590.
5. Dal Peraro M, Ruggerone P, Raugei S, Gervasio FL, Carloni P. (2007) Investigating biological systems using first principles Car-Parrinello molecular dynamics simulations. *Curr Opin Struct Biol* **17**: 149–156.
6. Carloni P, Roethlisberger U, Parrinello M. (2002) The role and perspective of *ab initio* molecular dynamics in the study of biological systems. *Acc Chem Res* **35(6)**: 445–464.
7. Phillips JC, Braun R, Wang W, Gumbart J, Tajkhorshid E, Villa E, Chipot C, Skeel RD, Kalé L, Schulten K. (2005) Scalable molecular dynamics with NAMD. *J Comp Chem* **26**: 1781–1802.
8. Bolhuis PG, Chandler D, Dellago C, Geissler PL. (2002) Transition path sampling: Throwing ropes over rough mountain passes, in the dark. *Annual Rev Phys Chem* **53**: 291.
9. Bennett CH (1977) Molecular dynamics and transition state theory: The simulation of infrequent events. In: Christoffersen RE (ed), *Algorithms for Chemical Computations*, p. 63. American Chemical Society, Washington DC.
10. Ottinger HC. (1988) General projection operator formalism for the dynamics and thermodynamics of complex fluids. *Phys Rev E* **57**: 1416–1998.
11. Gardiner CW. (2003) *Handbook of Stochastic Methods*. Springer-Verlag, Berlin Heidelberg.
12. Hummer G. (2005) Position-dependent diffusion coefficients and free energies from Bayesian analysis of equilibrium and replica molecular dynamics simulations. *New J Phys* **7**: Art. No. 34.

13. Mossa A, Clementi C. (2007) Supersymmetric Langevin equation to explore free-energy landscapes. *Phys Rev E* **75:** 046707.

14. Carter EA, Ciccotti G, Hynes JT, Kapral R. (1989) Constrained reaction coordinate dynamics for the simulation of rare events. *Chem Phys Lett* **156:** 472–477.

15. Sprik M, Ciccotti G. (1998) Free energy from constrained molecular dynamics. *J Chem Phys* **109:** 7737–7744.

16. Patey GN, Valleau JP. (1975) Monte-Carlo method for obtaining interionic potential of mean force in ionic solution. *J Chem Phys* **63:** 2334–2339.

17. Ferrenberg AM, Swendsen RH. (1988) New Monte-Carlo technique for studying phase-transitions. *Phys Rev Lett* **61:** 2635–2638.

18. Kumar S, Rosenberg JM, Bouzida D, Swendsen RH, Kollman PA. (1995) Multidimensional free-energy calculations using the weighted histogram analysis method. *J Comput Chem* **16:** 1339–1350.

19. Roux B. (1995) The calculation of the potential of mean force using computer-simulations. *Comput Phys Comm* **91:** 275–282.

20. Jarzynski C. (1997) Nonequilibrium equality for free energy differences. *Phys Rev Lett* **78:** 2690–2693.

21. Darve E, Pohorille A. (2001) Calculating free energies using average force. *J Chem Phys* **115:** 9169–9183.

22. Grubmuller H. (1995) Predicting slow structural transitions in macro-molecular systems — conformational flooding. *Phys Rev E* **52:** 2893–2906.

23. Voter AF. (1997) Hyperdynamics: Accelerated molecular dynamics of infrequent events. *Phys Rev Lett* **78:** 3908–3911.

24. Cvijovic D, Klinowski J. (1995) Taboo search — an approach to the multiple minima problem. *Science* **267:** 664–666.

25. Huber T, Torda AE, van Gunsteren WF (1994) Local elevation — a method for improving the searching properties of molecular-dynamics simulation *J Comput Aid Mol Des* **8:** 695–708.

26. Laio A, Parrinello M. (2002) Escaping free-energy minima. *Proc Natl Acad Sci USA* **20:** 12562–12566.

27. Ensing B, Laio A, Parrinello M, Klein ML (2005) A recipe for the computation of the free energy barrier and the lowest free energy path of concerted reactions. *J Phys Chem B* **109:** 6676–6687.

28. Ensing B, Klein ML. (2005) Perspective on the reactions between F- and CH3CH2F: The free energy landscape of the E2 and S(N)2 reaction channels. *Proc Natl Acad Sci USA* **102:** 6755–6759.

29. Barducci A, Bussi G, Parrinello M. (2008) Well-tempered meta-dynamics: A smoothly converging and tunable free-energy method. *Phys Rev Lett* **100:** Art. No. 020603.

30. Branduardi D, Gervasio FL, Cavalli A, Recanatini M, Parrinello M. (2005) The role of the peripheral anionic site and cation-pi interactions in the ligand penetration of the human AChE gorge. *J Am Chem Soc* **127:** 9147.

31. Piana S, Laio A. (2007) A bias-exchange approach to protein folding. *J Phys Chem B* **111(17):** 4553–4559.

32. Branduardi D, Gervasio FL, Parrinello M. (2007) From A to B in free energy space. *J Chem Phys* **126(5):** Art. No. 054103.

33. E WN, Ren WQ, Vanden-Eijnden E. (2007) Simplified and improved string method for computing the minimum energy paths in barrier-crossing events. *J Chem Phys* **126(16):** Art. No. 164103, APR 28 2007.

chapter 12

Protein Folding and Beyond
Energy Landscapes and the Organization of Living Matter in Time and Space

Peter G. Wolynes[*]

Proteins are highly dynamical entities. The folding, function and higher order assembly of biomacromolecules are key elements of the organization of living matter. Energy landscape theory is illuminating many of these issues of biomolecular dynamics. Folding is dominated by a funnel-like energy landscape with minimal ruggedness. Quantifying this notion using statistical mechanics has led to algorithms for learning about protein energetics using bioinformatics, designing novel proteins and predicting protein tertiary structure from sequence. The funneling of the landscape allows robust predictions to be made for folding kinetics and mechanism. The functional landscape is intimately connected with folding and residual frustration. Binding sites are often frustrated and can be located using algorithms based on energy landscape theory. The higher order assembly of cells is, however, far from equilibrium and requires new concepts beyond current energy landscape theories.

1. Introduction

We think something might be "alive" when we observe a disciplined organization of its movements. On this score, Earth's weather systems, human culture, computers and the legendary Rube Goldberg contraptions might all qualify as alive, with some controversy. Most controversy about classifying something as alive disappears when we use the term for humans, other animals, plants or bacteria. Ultimately these things can clearly be called alive

[*]Department of Chemistry and Biochemistry, Department of Physics, Center for Theoretical Biological Physics, University of California, San Diego, 9500 Gilman Drive, La Jolla, CA 92093, USA, e-mail: pwolynes@ucsd.edu.

because of the extraordinarily high level of dynamical organization found in those systems, which extends down to thermal motion at the molecular level, and is usually ruled by chaos. Although in the quantitative terms provided by the concept of entropy, the organization level of living matter is extremely high, relevant functional movement still occurs. Both movement and organization are needed for us to perceive the presence of life — no one thinks of diamonds as living! How organized structures can form but nevertheless retain the ability to move through a constrained diversity of states is the subject of the present chapter. I will argue that protein folding, the simplest biochemical act of organization, is now becoming understood, at least on the smaller length scales.[1] Evolution has led to macromolecules that spontaneously self-organize. The requisite robustness of folding that these molecules have had to exhibit in order to allow the organisms using them to have survived, leads also to many regularities in their behavior *in vitro*. Owing ultimately to evolutionary robustness, an especially simple theory for folding based on a funnel-like energy landscape that describes the relative stabilities of the myriad configurations of the biomolecules, allows protein folding kinetics to be quite predictable.[2,3] At the same time, using energy landscape ideas and structural data has made it possible to understand and quantitatively parameterize some of the key chemical physics of the solvent mediated forces which are the physical bases of this funneled energy landscape.[4–7] Again, the robustness of the funnel landscape allows even our current, somewhat incomplete, understanding of these forces to be practically useful. Energy landscape theory has, therefore, already led to algorithms for designing protein-like molecules *de novo*[8,9] and for predicting the tertiary structure of the smaller proteins from their sequence, at least to moderate resolution.[6,7,10–12] It is also becoming clear that although most of its random thermal motions must be disciplined for a protein to achieve specificity of biomolecular interactions, some thermally allowed motion must remain in a functioning biomolecule. Understanding this residual motion is important for understanding how proteins can behave as molecular machines, with a complexity less than many of Rube Goldberg's.

I will argue that the remaining motions of natural proteins often represent local "unfolding" problems whose unraveling may be as conceptually important to understanding function as resolving the folding problem has been to specificity and structure. Very recent ideas about how to pinpoint a local "anti-folding code" that locally quantifies frustration[13] are leading us to appreciate an intriguing relation between the energy landscapes of proteins required for their folding and those energy landscape features that are required for dynamical function.[14]

2. The Minimal Frustration Principle and the Funneled Energy Landscape

The huge number of interacting proteins in the cell requires that proteins must interact with a relatively high degree of specificity, otherwise most of cellular matter would be tied up in a mess of inappropriate interactions — instead of a cell we would have a scrambled egg (cell!). This functional specificity is achieved by the evolution of sequences that also lead to little scrambling even within each individual protein molecule. Many years ago Joe Bryngelson and I noticed that, in contrast to this specific structure seeking behavior, the natural thermodynamic states of a random heteropolymer would be either those of a highly fluctuating ensemble of structures having very few permanent interactions or a compact state with very slow dynamics characteristic of a gel, much like a single molecule version of a scrambled egg. The latter scrambled state resembles a glass or at least a highly viscous supercooled liquid. The very simplest theory of a random heteropolymer based on the Random Energy Model leads to the prediction of such a glass-like state.[15,16] This glass state emerges from the adventitious formation of random but nevertheless stabilizing contacts. The temperature at which glassy dynamics emerges depends on the statistical variation among energies of the conformations sampled in the compact state and also on the entropy of the set of all the conformations the molecule can take on without violating the rigid laws for stereochemistry of amino acid polymers (mostly set down by Pauling!). In the lowest random energy model approximation, which treats the landscape as uncorrelated, the ideal glass temperature T_G is given by a simple formula: $k_B T_G \cong \Delta E / \sqrt{S_C / k_b}$.

Here ΔE is the standard deviation of the energies of the ensemble compact configurations allowed by stereochemistry and excluded volume. S_C, correspondingly, is the configurational entropy of those states. At higher orders of approximation, this glass transition occurs within a compact "molten globule" state and the entropies and variances must be renormalized self-consistently to reflect the partial order of that liquid crystalline state.[17,18]

When correlations in the energy landscape are taken into account, also, one recognizes the occurrence of a characteristic higher temperature T_A above which the trapping in the minima generated by the randomly chosen interaction can be nearly completely neglected.[19,20]

These features obtained using the random energy model arguments are also confirmed by more complete statistical mechanical treatments using replica based theories.[21] While at a temperature below its T_G a random heteropolymer would fold eventually to a low free energy conformation, the kinetics of finding the lowest free energy state would be slow because there are many other candidate structures that are nearly as stable as the true ground state but entirely different in structure. These structurally very different alternate structures could also become more stable when the environmental conditions are changed: the folding of a random heteropolymer, intrinsically, is not robust. The absence of robust folding would make both the function within a simple organism and evolution between generations of organisms difficult.

The reason structurally disparate alternate structures can thermally compete with each other is that the low energy structures (for most random sequences) are actually stabilized by numerous small interactions. These interactions, by virtue of the random sequence, tend to still conflict with each other in most of the candidate low energy structures. This conflict between the various different constituent local interactions goes by the technical name of "frustration," a term first used in physics to describe magnets with conflicting local interactions between electronic spins,[22,23] like the triangular Ising antiferromagnet.

The energy of the alternative near-ground state structures can also be estimated using the scale of the statistical energy fluctuations and the configurational entropy. This estimate based again on the uncorrelated energy landscape gives $E_G = \Delta E \cdot \sqrt{S_C/k_b}$ where ΔE^2 is the variance of energies characterizing the ensemble of compact structures. Robust folding, as seen in natural proteins, is statistically rare according to this analysis based on the Random Energy Model. But while the average behavior is glassy, probabilistic analysis does show that robustness may nevertheless still be achievable for some sequences. Robustly foldable sequences can be found by choosing sequences such that the interacting pairs in the lowest energy structure are especially stable when they are individually satisfied. The large number of amino acid types makes achieving this unconflicted situation possible for some sequences. When this can be done, the ground state of the molecule will be much more stable than the estimate given by E_G. If the energy of the folded structure E_F is much below E_G, folding can occur at a temperature higher than T_G since the alternate structures typically having energy E_G will be entropically destabilized at the folding temperature T_F. Indeed if the sequence has a ground state that is sufficiently stable, folding may even occur at temperatures significantly higher than the onset temperature for activated dynamics T_A. In this case, the formation and dissolution of contacts not found in the ground state is such a rapid process that it can certainly be averaged over and sometimes this transient trapping can be nearly neglected, entering only as an internal friction effect.[16,24] For such a rare, consistently landscaped protein, alternative structures would be not only kinetically unimportant but would also become stable ground states only under strongly different environmental conditions. The condition for such robustness is $E_F \ll E_G$. Bryngelson and Wolynes noticed that the behaviors expected from systems lacking much frustration conformed much more to what was seen for evolved proteins than the statistical mechanical predictions for the very random energy landscape.

They, therefore, enunciated a "principle of minimal frustration" that can be quantitatively stated using energy landscape theory.[15] The minimal frustration principle holds that evolved proteins have a T_F that exceeds T_G or equivalently have a stable low energy structure with energy E_F that is lower than E_G by an amount that scales with the r.m.s. energy fluctuations. The Bryngelson–Wolynes analysis generalized and made more quantitative an earlier version of a "consistency principle" stated by Go[25] in more qualitative terms. Both these quantitative criteria (one based on temperature, the other energies) for minimal frustration derived from energy landscape theory can be written (again within the lowest level approximation) as $1 < T_F/T_G \cong (E - \bar{E})/(\Delta E \sqrt{S_C})$, where $\bar{E}$ is the average energy of the compact states (or "decoys").

We see then that the ratio of folding to glass transition temperature T_F/T_G can be said to depend on how "special" or statistically anomalous the ground state structure's energy is for that sequence. The statistical energy ratio without the entropy factor is sometimes called the energy Z-score.[26]

When the T_F/T_G ratio is large, there will be relatively few thermally occupiable local free energy minima associated with "mistakes" in folding. At the folding temperature the landscape, when T_F greatly exceeds T_G, is dominated by ensembles having structures in which a fraction of the native contacts form but only a few weak "non-native" interactions between the remaining parts of the molecule occur. Essentially, partially folded forms of the protein exist, but highly misfolded metastable structures with stabilized alternate structures do not. Because of this simplifying feature of minimally frustrated polymers, knowing the contributions made by a residue pair to native stability also tells us the energetic role of that pair in populated intermediate ensembles or in bottlenecks in the folding pathway, called transition state ensembles. The thermodynamic contribution of an interaction for one of these key ensembles will merely be scaled by the fraction of time that that pair contact is actually formed in the corresponding ensemble.[27] Rates will largely follow native state stability determined by the transition state ensemble, but only partially. The rates will follow what organic chemists call "linear free energy relations."[28] This regularity predicted by the funnel landscape provides the theoretical basis of the analysis of protein folding kinetics using the so-called ϕ values,[29] which can be determined through parallel kinetic studies on proteins made by site-directed mutagenesis. On a very rugged energy landscape the ϕ values become very difficult to interpret in structural terms because the kinetics of organization is determined by the stability of unknown (hard-to-predict) contacts, which are *not* present in the native structure. The presence of such difficult-to-predict contacts would determine the flow of structures towards the global free energy minimum, and knowledge of stability changes would give no clue as to their existence. For minimally frustrated proteins, in contrast, the knowledge of the protein structure along with measurement of kinetics of mutant proteins using a range of thermodynamic conditions that tune the global stability can give a clear picture of the key organizational steps in folding. Several good reviews of this approach exist and should be consulted for more details.[3,29]

The funnel-like nature of the folding landscape, in many cases, also allows the ϕ values to be predicted theoretically using knowledge of the native structure alone. Such predictions are accurate when the conformations belonging to the transition state ensemble have a large connected web of native contacts established, as is often, but not universally, the case.[2]

Predictions of kinetic rates based on topology alone break down for several distinct reasons — degeneracy of distinct folding routes is a common cause. Such degeneracy comes about rather often because a surprising number of proteins are structurally quite symmetric. The transition state ensemble can then break the overall symmetry seen in the final structure. In such a case, a small free energy difference favoring one of two symmetric routes over the other can easily change the kinetics.[3,30] When this symmetry breaking situation applies, the simpler form ϕ of value analysis can break down because mutations can shift the transition state ensemble rather dramatically: the mutations can no longer be treated as linear free energy perturbations but have nonlinear effects. Native topology models can also fail if the key ensembles have a sparse connectivity of contacts. This situation can occur for the smaller α-helical proteins. In some celebrated cases, the folding depends on just a handful of contacts. This is especially true for the controversial issue of the possible existence of downhill folding — a candidate for which is BBL. A sparse contact connectivity features in the folding of truncated BBL, a small residue protein fragment which has been studied extensively by the groups of Fersht[24,32] and Muñoz,[33] who have reached quite different conclusions as to its folding mechanism.

Simulations using a perfectly funneled landscape show that for BBL only a small number of tertiary contacts are actually involved in forming the bottleneck.[28] The crucial barrier determining contacts turn out to be different in the related but slightly different systems studied by the two different groups. The sensitivity of folding to this small cluster of contacts leads to the apparently discordant results of the experimentalists. For BBL, the free energy barrier is formed in such a delicate way that in fact the barrier appears to entirely disappear in the system studied by Muñoz, leading to "downhill folding." On the other hand, the barrier is predicted to *not* disappear for the version of the molecule studied by Fersht under his experimental conditions. This is consistent with the interpretation of BBL folding as two state with a barrier that the Fersht group has come to.

3. Minimal Frustration and the Design and Structure Prediction Problems

The minimal frustration principle ties into a trinity: sequence, structure and the energy function. If the energy function is accurate enough, minimal frustration provides a recipe for designing sequences to fold to a desired structure. Similarly, the knowledge of numerous structures with their corresponding sequences available through numerous laboratory structure determinations can be decoded into energy functions. The minimal frustration principle, therefore, even gives us a license to use bioinformatics to teach us some basic chemical physics!

Finally, again with a sufficiently refined energy function, the minimal frustration principle shows how to predict protein tertiary structure from sequence.[35,36] The mathematical formulation of the minimal frustration principle in terms of Z-scores is especially helpful in all of these tasks.

The value of landscape theory in structure prediction is that the minimum frustration principle lets us see how quantitatively accurate our energy function is by using statistics, without doing a completely exhaustive enumeration of conformations, that is, without searching for actual ground states. For a class of model energy functions with unknown parameters, the best parameter choices will lead to high T_F/T_G ratios for the known natural protein sequences. Since high T_F/T_G roughly translates into high $(E - \bar{E})/\Delta E$, the quality of an energy function can be rapidly assessed using statistics about the molten globule state. We can, therefore, search through a large family of energy functions rapidly by optimizing the Z-score of the known target structure's energy compared with minima generated by the same energy function.[35] Nevertheless, there is no guarantee that training a potential on one set of proteins will work for another set of proteins. Test predictions must be carried out. Indeed failure in such testing is expected if the model energy does not overlap with the true solvent-averaged, side-chain averaged free energy.

Optimization is now a common feature used in the development of many structure prediction codes.[6,7,10–12,37–41] Our group has successfully used optimized associative memory energy functions to predict targets in the CASP structure prediction exercises.[37] Low resolution successes such as these show that we are indeed beginning to learn how to minimize the frustration theorists encountered in previous decades of protein structure prediction.

Consistently made errors in predicting structures using optimized potentials can also signal to us that we are missing some basic, relevant physics. Recently, the study of binding problems has taught us that there is a class of solvent-mediated interactions in proteins of a type we had not expected. Such interactions were to some extent anticipated by thoughtful crystallographers,[43] but their quantitative magnitude was unclear. The major role of solvent water in folding has, of course, been well known for a long time. The biochemical role of an hydrophobic effect was understood very early on. The strongest signals in bioinformatically learned potentials are in fact such expected hydrophobic contributions. The best bioinformatic energy functions were learned by optimizing the landscapes for folding of small-to-moderate-size monomeric proteins. We therefore initially expected that the same potentials would also describe protein multimers, but using energy landscape analysis on dimers, it was discovered this was not always true. The experimental evidence in concert with funnel landscape based simulations, however, suggested that most dimers indeed do have funneled landscapes.[44,45] Therefore, when a fraction of the known dimer structures had anti-funneled binding landscapes according to Z-score, T_F/T_G criteria, we were perplexed at first.[4,5] However, a pattern for the proteins which seemed to have anti-funneling did emerge: the anti-funneled interfaces were extraordinarily hydrophilic. Unlike the protein interiors, these pairs of molecules were being held together at their interfaces by bridging solvent molecules. Now by recognizing the existence and nature of these interactions, we could design an energy potential with the right physics in which binding energies are modulated by the availability of water to bridge the sites. By employing Z-score optimization, the relative strengths of these water mediated interactions (often hydrophilic) compared to the usually dominant contact (hydrophobic) interactions could be determined. Introducing these solvent mediated interactions into structure-prediction algorithms significantly improves their performance.[6,7] The other side of the coin from structure prediction for structure-sequence relationships is protein design. Exactly parallel to its use in structure prediction, Z-score optimization provides a route for design of artificial foldable proteins. Such design may actually be an easier task than the prediction of structure for natural proteins because overdesign can allow a margin of error that makes up for errors in the potential used for the design. Z-score optimization in concert with combinatorial strategies has yielded foldable proteins[46–48] and this design can be automated.[8] An automated algorithm based on self-consistent Z-score optimization leads to high T_F/T_G and foldability in the laboratory. These automated designs used a physically based energy function which had been reversed engineered also using the T_F/T_G ratio optimization strategy.[10] This elegant approach enabled the Takada group automatically to design a sequence, quite different from naturally occurring ones, that folded into a targeted three-helix bundle topology. It was most exciting to see that when made in the laboratory their sequence gave a protein whose NMR spectrum was well-dispersed, indicative of having achieved a properly folded native state.

The current successes in designing foldable amino acid heteropolymers, even when the sequences are far from natural, rely in many ways on the known robustness of natural protein folding. Natural foldable proteins provide an existence proof for the solvability of the T_F/T_G optimization problem that encodes the minimal frustration principle. The interactions used for design can also be inferred, as we have seen, bioinformatically. Nucleic acids are also known to be able to achieve well-structured, foldable structures in three dimensions.[49] Three-dimensional ordering of active sites is crucial for the existence of ribozymes.

Can we go beyond these natural systems? Can foldability be achieved for other chemically distinct types of heteropolymers. A key concept in that quest is the notion of "foldamers" — units having a common backbone with variable side chains.[50] Energy landscape theory gives some guidance in foldamer design.[51]

Landscape theory also makes clear why having a large variety of amino acids with sufficiently distinct interactions allows large Z-scores to be achieved. It is much more difficult to design proteins with two kinds of amino acids than it is with the full 20 (or more when you include post-translational modification such as phosphorylation).[52,53]

We have seen that in addition to the Z-score according to landscape theory the T_F/T_G ratio depends on the configurational entropy per monomer. When we move outside the protein and ribozyme realm to take up the foldamer quest, landscape theory attracts our attention to this configurational entropy, which apart from the interactions themselves is the determining feature of whether a sufficient T_F/T_G ratio can be achieved. Decreasing S_C raises both T_F and T_G, but T_F rises more rapidly being inversely proportional to S_C itself, not its square root. Thus landscape theory implies that having a low backbone entropy is desirable in a family of foldamers. The more successful foldamers indeed have local rigidity designed into them (by intuition, as it were). The floppier foldamer candidates have not yet even led to predictably good secondary structure formers. Energy landscape theory also shows that foldability is enhanced when some rigidity can be achieved by pre-organization either locally (as in helix-formation for proteins[17]) or by exploiting the high specificity of Watson–Crick base pairing to induce rigidity and thus decrease the configurational entropy before tertiary contacts are made, as in ribozymes. For practical application in the real world, the ratio of T_F/T_G alone may not be sufficient to see whether self-assembly can occur, since the absolute environmental temperature is constrained. If dynamics is too slow, assembly suffers. If T_G is made too high, either by designed structural rigidity or pre-organization, the molecule will not rapidly assemble at room temperature. Interestingly, in ribozymes the tertiary contacts are rather sparser than they are in proteins (probably owing, in any event, to the stronger electrostatic repulsions in nucleic acids); this paucity of three-dimensional contacts acts to lower the Z-score of the interactions in ribozymes relative to that seen in three-dimensionally rich proteins. Ribozyme folding therefore seems to occur on a much rougher landscape than proteins.[49] The rugged landscape of RNA is apparently still navigable at room temperature by thermal Brownian motion.

4. Frustration and the Functional Landscape

As we mentioned, in living things, proteins must still move once they have folded. Limited local frustration can act to facilitate such motion and channel the thermal movements in functionally relevant ways.[13] Neither a protein's kinetic foldability nor its mutational robustness is completely compromised if a molecule has only localized frustration. Spatially localized frustration allows only specific protein dynamics to occur and not other undesirable motions; by appropriate evolution of frustrated regions only those motions that are needed for specific functions occur with large amplitude or rate. The alternate configurations caused by locally frustrating an otherwise largely unfrustrated structure will act as intermediates for only those specific thermal motions that are functionally necessary. Thus below its global unfolding temperature, the protein can function much like a macroscopic machine having only a few moving parts despite being a complex assemblage of many smaller subunits. Also, frustrated sites in monomers may become less frustrated in larger assemblies of proteins allowing specific association to realize frustration minimization as a physical act.[13] Quantitative methods of localizing frustration in proteins can thus give us insights into the functional protein energy landscape.

Ferreiro *et al.* have found that by using a sufficiently accurate energy function, the sites of frustration can be localized using a spatially local version of the global gap criterion formulation of the minimal frustration principle.[13] To see whether a specific interaction is frustrated, we can compare the contribution to the stabilization energy from a specific pair of residues to the statistical variations of the energies that would be found by replacing them with other amino acids in contact or by creating a different environment for the pair. When an individual native pair's contribution, normalized by the energy fluctuations (a local Z-score criterion in analogy to the global Z-score criterion for minimal frustration), is sufficiently large, that specific local interaction can be called minimally frustrated. The magnitude of the local threshold Z-score required for an interaction to be designated minimally frustrated depends on the configurational entropy lost in making that interaction. Global estimates of that entropy loss (used to describe folding thermodynamics) can be used, therefore, to set the threshold. On the other hand, if a native pair is sufficiently destabilizing compared with the other possibilities, we can call the pair interaction "highly frustrated." We would expect a high level of local frustration usually to be the result of an evolutionary constraint that somehow conflicts with robust folding. Because of the nonlinearity of entropy loss on assembly, not all of the individual pair interactions in a protein need to be minimally frustrated for the landscape of the protein as a whole to be funneled. Using the localized version of the landscape folding criterion derived in this way, along with the water-mediated potential inferred by us for structure prediction purposes, we have found that 40% of the interactions in natural proteins individually can be considered minimally frustrated, typically. Forty-five percent of the interactions in natural proteins by themselves would be classified as "neutral," but since the minimally frustrated pairs percolate through the structure the protein, as a whole, remains minimally frustrated. Only a small fraction of the interactions in proteins we have surveyed seem to be highly frustrated. These frustrated interactions are usually clustered. The location of these clusters seems to be highly correlated with the locations of binding sites or hinge regions involved in allosteric transitions.

A very nice example of the role of localized frustrated clusters in folding and function[54] has been analyzed for the protein Im7. Im7 binds to colicin. These binding partners have evolved in a biochemical arms race between bacteria. Colicin is a toxin secreted by some bacteria while Im7 acts defensively to neutralize it. The binding site in Im7 is frustrated in the monomer. A stable folding intermediate having non-native interactions arises from the residual frustration found in the native structure, which is localized precisely at the region involved in binding. The folding of Im7 by itself is anomalous.[55] The simple folding mechanism predicted by a fully funneled landscape does not apply. That frustration is, in fact, the source of the intermediate can be confirmed by quantitative local frustration analysis. This analysis predicts specific mutants that should reduce the level of frustration so as to make the landscape less rugged and more funneled. Simulations show that when one alleviates the frustration in that structure, that kinetic intermediate should disappear (according to a Hamiltonian based on water-mediated interactions model; the AMW model). On the other hand, the binding to colicin would probably be compromised.

While the frustration seen in Im7 seems to be connected simply to binding, frustration induced degeneracy also offers intriguing possibilities of allostery and regulation. Recently we have shown that phosphorylation of specific residues may sometimes act to reduce frustration locally, pinning a single functional structure in place. Our analysis of how the energy landscape of NFAT, a small regulatory protein, is changed by phosphorylation documents this regulatory strategy.[56]

5. The Organization of Living Matter on Larger Length Scales

A single degree of freedom still takes only $1k_BT$ to be organized, no matter how big it is in atomic terms. In a sense, that was the lesson Einstein imparted in his early studies of the Brownian motion of colloids. Nevertheless, it seems self-organization becomes harder as the object to be assembled becomes larger. Self-assembly by folding greatly reduces the entropy density, as we have seen, but large folded objects can have interactions scaling with the surface area of their interactions. This is why a protein crystal can be held together by a mere handful of, often adventitious and therefore far from robust, "crystal contacts." The problem of self-organization for large things is not stability but the kinetics of getting to a stable structure. Kinetic barriers can be avoided if sufficient entropy remains to fluidize the molecular motion until the last minute. Up to a point, robust self-assembly in the cell therefore often relies on folding upon binding — sort of a just-in-time policy for biomacromolecular economy.[57] Nevertheless, kinetic barriers for folding (even for folding upon binding) tend to scale with the size of the assembled system.[58] The larger barriers occur for frustrated systems but also may grow with size when the minimal frustration principle is respected, according to the capillarity picture. At some size scale the assembly rates generally become too slow and far-from-equilibrium assembly takes over in the cell.

This crossover already seems to occur when the largest proteins are folded *in vivo*. Rubisco, a very large protein, obligatorily requires chaperones in order to fold.[59] It is likely the chaperoned catalysis of proper folding is actually a kind of kinetic proofreading, in which misfolded protein molecules are ripped apart by unfolding machines and given a chance to try again.[60,61] The role of the ATP induced motions of a chaperone can be thought of as increasing the effective temperature for a limited fraction of the degrees of freedom.

There is life above the scale of single proteins, obviously. Everyone agrees the cell itself is not just a formless bag of macromolecules. Evidence exists that organization persists at the cellular scale and changes its nature at this supramolecular level. We are only beginning to probe the atomically weak but colloidally strong forces that can act on this larger scale length. It is also clear that in living things, motion on this size scale is not just equilibrium Brownian motion. Instead, chemical energy is flowing through the system: various molecular motors carry parts around in the cell in a hurly-burly reminiscent of a large city.[62,63] This motorization probably allows the problem of large barriers and high glass transition temperatures to be overcome much as in kinetic proofreading by chaperones. By comparing responses to fluctuations on the colloidal scale,[64] an effective temperature of the cytoskeleton has been measured to be 10,000 K. Obviously, this high temperature applies to only a minute fraction of the degrees of freedom in the cell. It becomes obvious that studying the organization of living matter on this longer length scale will require new tools, both experimental and theoretical. We can anticipate, however, that the strong foundation of energy landscape theory learned in the study of folding, will be of great help in finding the required conceptual framework for this far-from-equilibrium regime.

Acknowledgments

This work was supported by grants from the National Institutes of Health and the National Science Foundation.

References

1. Wolynes PG. (2005) Energy landscapes and solved protein folding problems. *Phil Trans R Soc* **A363**: 453–464.
2. Onuchic JN, Wolynes PG. (2004) Theory of protein folding. *Curr Op Struct Biol* **14**: 70–75.
3. Oliveberg M, Wolynes PG. (2005) The experimental survey of protein folding energy landscapes. *Q Rev Biophys* **38**: 245–288.
4. Papoian GA, Ulander J, Wolynes PG. (2003) The role of water mediated interactions in protein–protein recognition landscapes. *J Am Chem Soc* **125**: 9170–9178.
5. Papoian GA, Wolynes PG. (2003) The physics and bioinformatics of binding and folding — an energy landscape perspective. *Biopolymers* **68**: 333–349.
6. Papoian GA, Ulander J, Eastwood MP, Luthey-Schulten Z, Wolynes PG. (2004) Water in protein structure prediction. *Proc Natl Acad Sci USA* **101**: 3352–3357.
7. Zong C, Papoian GA, Ulander J, Wolynes PG. (2006) The role of topology, nonadditivity and water mediated interactions in predicting the structures of α/β proteins. *J Am Chem Soc* **128**: 5168–5176.
8. Jin WZ, Kambara O, Sasakawa H, Tamura A, Takada S. (2003) *De novo* design of foldable protein with smooth folding funnel: Automated negative design and experimental verification. *Structure* **11**: 581–590.
9. Park S, Xi Y, Saven J. (2004) Advances in computational protein design. *Curr Op Struct Biol* **14**: 487–494.
10. Fujitsuka Y, Takada S, Luthey-Schulten ZA, Wolynes PG. (2004) Optimizing physical energy functions for protein folding. *Proteins: Structure, Function and Genetics* **54**: 88–103.

11. Chikenji G, Fujitsaka Y, Takada S. (2003) A reversible fragment assembly method for *de novo* protein structure prediction. *J Chem Phys* **119**: 6895–6903.

12. Hardin C, Eastwood MP, Luthey-Schulten Z, Wolynes PG. (2000) Associative memory Hamiltonians for structure prediction without homology: Alpha-helical proteins. *Proc Natl Acad Sci USA* **97**: 14235–14240.

13. Ferreiro DU, Hegler JA, Komives EA, Wolynes PG. (2007) Localizing frustration in native proteins and protein assemblies. *Proc Natl Acad Sci USA* **104**: 19818–19824.

14. Frauenfelder H, Sligar S, Wolynes PG. (1991) The energy landscapes and motions of proteins. *Science* **254**: 1598–1603.

15. Bryngelson JD, Wolynes PG. (1987) Spin glasses and the statistical mechanics of protein folding. *Proc Natl Acad Sci USA* **84**: 7524–7528.

16. Bryngelson JD, Wolynes PG. (1989) Intermediates and barrier crossing in a random energy model (with applications to protein folding). *J Phys Chem* **93**: 6902–6915.

17. Luthey-Schulten ZA, Ramirez BE, Wolynes PG. (1995) Helix-coil, liquid crystal and spin glass transitions of a collapsed heteropolymer. *J Phys Chem* **99**: 2177–2185.

18. Koretke K, Luthey-Schulten Z, Wolynes PG. (1998) Self-consistently optimized energy functions for protein structure prediction by molecular dynamics. *Proc Natl Acad Sci USA* **95**: 2932–2937.

19. Wang J, Plotkin S, Wolynes PG. (1997) Configurational diffusion on a locally connected correlated energy landscape: Application to finite, random heteropolymers. *J. Phys I France* **7**: 395–421.

20. Plotkin SS, Wang J, Wolynes PG. (1996) A correlated energy landscape model for finite, random heteropolymers. *Phys Rev E* **53**: 6271–6296.

21. Pande VS, Grosberg A Yu, Tanaka T. (2000) Heteropolymer freezing and design: Towards physical models of protein folding. *Rev Mod Phys* **72**: 259–314.

22. Anderson PW. (1978) Concept of frustration in spin glasses. *J Less-Common Met* **62**: 291–294.

23. Toulouse G. (1977) Theory of frustration effect in spin glasses 1. *Commun Phys* **2**: 115–119.

24. Takada S, Wolynes PG. (1997) Microscopic theory of critical folding nuclei and reconfiguration activation barriers in folding proteins. *J Chem Phys* **107**: 9585–9598.

25. Go N. (1983) Theoretical studies of protein folding. *Ann Rev Biophys Bioeng* **12**: 183–210.

26. Bowie JU, Luthy R, Eisenberg D. (1991) A method to identify protein sequences that fold into a known three-dimensional structure. *Science* **253**: 164–170.

27. Onuchic JN, Socci ND, Luthey-Schulten ZA, Wolynes PG. (1996) Protein folding funnels: The nature of the transition state ensemble. *Fold Des* **1**: 441–450.

28. Grunwald E. (1984) Linear free energy relationships — a historical perspective. *Chem Tech* **14**: 698–705.

29. Fersht AR. (1999) *Structure and Mechanism in Protein Science.* WH Freeman & Company, New York.

30. Oloffson M, Hansson S, Hedberg L, Logan DT, Oliveberg M. (2007) Folding of S6 structures with divergent amino acid composition. *J Mol Biol* 237–248.

31. Ferguson N, Shartau PJ, Sharpe TD, Sato S, Fersht AR. (2004) One-state downhill versus conventional protein folding. *J Mol Biol* **344**: 295–301.

32. Ferguson N, Sharpe TD, Johnson CM, Schartau PJ, Fersht AR. (2007) Structural biology: Analysis of 'downhill' protein folding. *Nature* **445**: E14–15.

33. Sadqi M, Fushman D, Muñoz V. (2006) Atom-by-atom analysis of global downhill protein folding. *Nature* **442**: 317–321.

34. Cho S, Weinkam P, Wolynes PG. (2008) Origins of barrierless folding in BBL. *Proc Natl Acad Sci USA* **105**: 118–123.

35. Goldstein R, Luthey-Schulten Z, Wolynes PG. (1992) Optimal protein-folding codes from spin-glass theory. *Proc Natl Acad Sci USA* **89**: 4918–4922.

36. Goldstein RA, Luthey-Schulten ZA, Wolynes PG. (1992) Protein tertiary structure recognition using optimized Hamiltonians with local interactions. *Proc Natl Acad Sci USA* **89**: 9029–9033.

37. Hardin C, Eastwood MP, Luthey-Schulten Z, Wolynes PG. (2000) Associative memory Hamiltonians for structure prediction without

homology: Alpha-helical proteins. *Proc Natl Acad Sci USA* **97**: 14235–14240.

38. Hardin C, Eastwood MP, Prentiss MC, Luthey-Schulten Z, Wolynes PG. (2003) Associative memory Hamiltonians for protein structure prediction without homology: Alpha/beta proteins. *Proc Natl Acad Sci USA* **100**: 1679–1684.

39. Lu H, Skolnick J. (2001) A distance dependant atomic knowledge based potential for improved protein structure prediction. *Proteins* **44**: 223–232.

40. Pillardy A *et al.* (2001) Recent improvements in prediction of protein structure by global optimization of a potential energy function. *Proc Natl Acad Sci USA* **98**: 2329–2333.

41. Fain B, Levitt M. (2003) Funnel sculpting for *in silico* assembly of secondary structure elements of protein. *Proc Natl Acad Sci USA* **100**: 10700–10705.

42. Aloy P, Stark A, Hadley C, Russell R. (2003) Predictions without templates: New folds secondary structure and contents in CASP V. *Proteins* **53**: 436–456.

43. Jones S, Thornton JN. (1997) Analysis of protein–protein interaction sites using surface patches. *J Mol Biol* **272**: 121–132.

44. Levy Y, Onuchic JN, Wolynes PG. (2004) Protein topology determines binding mechanism. *Proc Natl Acad Sci USA* **101**: 13786–13791.

45. Levy Y, Papoian GA, Onuchic JN, Wolynes PG. (2004) The energy landscape analysis of protein dimers. *Israel J Chem* **44**: 281–297.

46. Street AG, Datta D, Gordon DB, Mayo SL. (2000) Designing protein beta sheet surfaces by Z-score optimization. *Phys Rev Lett* **84**: 5010–5013.

47. Calhoun JR, Kono H, Lahr S, Wang W, deGrado WF, Saven J. (2003) Computational design and characterization of a monomeric helical dinuclear metalloprotein. *J Mol Biol* **334**: 1101–1115.

48. Pokala N, Handel TM. (2005) Energy functions for protein design: Adjustment with protein–protein complex affinities, models for the unfolded state and negative design of solubility and specificity. *J Mol Biol* **347**: 203–227.

49. Thirumalai D, Woodson SA. (1996) Kinetics of folding of proteins and RNA. *Acc Chem Res* **29**: 433–439.

50. Gellman SH. (1998) Foldamers: A manifesto. *Acc Chem Res* **31:** 173–180.

51. Hill DJ, Mio M, Prince MJ, Hughes RB, Hughes TS, Moore JS. (2001) A field guide to foldamers. *Chem Rev* **101:** 3893–4012.

52. Wolynes PG. (1997) As simple as can be? *Nat Struct Biol* **4:** 871–874.

53. Riddle DS, Santiago JV, Bray-Hall ST, Doshi N, Grantcharova VP, Yi Q, Baker D. (1997) Functional rapidly folding proteins from simplified amino acid sequences. *Nat Struct Biol* **4:** 805–809.

54. Sutto L, Lätzer J, Hegler JA, Ferreiro DU, Wolynes PG. (2007) Consequences of localized frustration for the folding mechanism of the IM7 protein. *Proc Natl Acad Sci USA* **104:** 19825–19830.

55. Capadli AP, Kleanthous C, Radford SE. (2002) Im7 folding mechanism: Misfolding on a path to the native state. *Nat Struct Biol* **9:** 209–216.

56. Shen T, Zong C, Hamelberg D, McCammon JA, Wolynes PG. (2005) The folding energy landscape and phosphorylation: Modeling the conformational switch of the NFAT regulatory domain. *FASEB J* **19:** 1389–1395.

57. Dyson HJ, Wright PE. (2005) Intrinsically unstructured proteins and their functions. *Nature Revs Mol Cell Biol* **6:** 197–208.

58. Wolynes PG. (1997) Folding funnels and energy landscapes of larger proteins within the capillarity approximation. *Proc Natl Acad Sci USA* **94:** 6170–6174.

59. Todd MJ, Viitanen PV, Lorimer GH. (1954) Dynamics of the chaperonin ATPase cycle-implications for facilitated protein folding. *Science* **265:** 659–666.

60. Gulukota K, Wolynes PG. (1994) Statistical mechanics of kinetic proofreading in protein folding *in vivo*. *Proc Natl Acad Sci USA* **91:** 9292–9296.

61. Todd MG, Lorimer GH, Thirumulai D. (1996) Chaperonin facilitated protein folding: Optimization of rate and yield by an iterative annealing mechanism. *Proc Natl Acad Sci USA* **93:** 4030–4035.

62. Shen T, Wolynes PG. (2004) Stability and dynamics of motorized crystals and glasses. *Proc Natl Acad Sci USA* **101:** 8547–8550.

63. Shen T, Wolynes PG. (2005) Nonequilibrium statistical mechanical models for cytoskeletal assembly: Towards understanding tensegrity in cells. *Phys Rev* **E72:** 041927-11.
64. Lau AW, Hoffman BD, Davies A, Crocker JC, Lubensky TC. (2003) Microrheology, stress fluctuations and active behavior of living cells. *Phys Rev Lett* **91:** 198101–198104.

chapter 13

Protein Folding and Misfolding
From Atoms to Organisms

Christopher M. Dobson[*]

The manner in which a newly synthesized chain of amino acids folds into the unique structure of a functional globular protein depends both on the intrinsic properties of the amino acid sequence and on multiple influences within the crowded cellular milieu. Moreover, if proteins misfold, or fail to remain correctly folded, a common consequence is aggregation and the formation of amyloid deposits, a phenomenon that is implicated in many highly debilitating and increasingly common medical disorders, including Alzheimer's disease and Type II diabetes. In this chapter, we describe how the concerted application of a wide range of the experimental and theoretical techniques of physics and chemistry under laboratory conditions has allowed the fundamental principles of protein folding, misfolding and aggregation to be understood at an atomic level. We then discuss how recent approaches, initially using cells and then model organisms such as the fruit fly, enable the links between these physical and chemical principles and the molecular events that occur in a living system to be explored in a quantitative manner. The results provide compelling evidence for a common underlying origin of the amyloid-associated diseases based on an inherent tendency of functional proteins to convert into the generic alternative and frequently pathogenic amyloid form of structure. They also suggest strategies for the prevention and treatment of these diseases, and provide insight into the way in which the need to fold correctly and to avoid aggregation under physiological conditions has influenced the evolution of protein sequences and the properties that they encode.

One of the essential characteristics of a living system is the ability of its component molecular structures to self-assemble into their functional forms.[1] The folding of proteins into their compact three-dimensional

[*]University of Cambridge, Department of Chemistry, Lensfield Road, Cambridge CB2 1EW, UK.

structures is the most fundamental example of biological self-assembly; understanding this process therefore provides unique insight into the way in which evolutionary selection has influenced the properties of a molecular system for functional advantage.[2] The wide variety of highly specific structures that result from protein folding, and which serve to bring key functional groups into close proximity, has enabled living systems to develop astonishing diversity and selectivity in their underlying chemical processes by using a common set of just 20 amino acid building blocks.

As well as simply generating biological activity, however, we now know that protein folding is coupled to many other biological processes, including the trafficking of molecules to specific cellular locations and the regulation of the growth and differentiation of cells. In addition, only correctly folded proteins have the ability to remain soluble in crowded biological environments and to interact selectively with their natural partners. It is not surprising, therefore, that the failure of proteins to fold correctly, or to remain correctly folded, is the origin of a wide variety of pathological conditions.[3] In this chapter, we explore the underlying mechanism of protein folding and of the nature and consequences of protein misfolding and its links with disease. In order to achieve these objectives, we show that it is possible to relate processes studied in atomic detail in the test tube to their quantitative effects in living systems through the use of model organisms such as the fruit fly.

1. The Intrinsic Mechanism of Protein Folding

The folded structures of the native states of a large number of proteins have been known for many years, and are widely believed to correspond to the structures that are most thermodynamically stable for given sequences under physiological conditions.[4] The manner by which a polypeptide chain folds to its specific three-dimensional protein structure has, however, only recently been understood at anything approaching the atomic level. One of the major challenges has been the recognition that the total number of possible conformations of any polypeptide molecule is so large that an exhaustive search for any particular structure during folding from an ensemble of highly unstructured species would take an astronomical length of time. It is now clear, however, that the folding process does not involve a systematic examination of all possible conformations, but can instead be considered a stochastic search of the complex multidimensional space encoded by the protein sequence.[4–7]

The crucial feature of this conclusion is that such a search process can be very efficient if the free energy surface or "landscape" for a protein has the right shape (see Fig. 1). As native-like interactions between residues will, on average, be more stable than non-native ones under folding conditions, the landscape can be biased in such a way that only a minute fraction of all the possible conformations is sampled by any given protein molecule during its transition from a random coil to a native structure.[4–7] As the landscape, describing the free energies of the different possible conformations of the protein, is determined by the amino acid sequence, natural selection has enabled proteins to evolve so that they are able to fold rapidly and efficiently. Such a description is often referred to as the "new view" of protein folding[8] and illustrates how the application of ideas from chemical physics and statistical mechanics has provided a robust and universal conceptual basis for understanding this complex biological process in molecular detail.[4–7]

A key question concerning the mechanism of protein folding is how the energy landscape that is unique to a specific protein is defined by its amino acid sequence. The structural transitions taking place during folding *in vitro* can be investigated in atomic or near-atomic detail by a variety of innovative chemical and physical techniques, ranging from optical methods to NMR spectroscopy,[4] including the remarkable developments that now enable the behavior of single molecules to be defined.[9] Studies of a series of small proteins, typically having 60–100 residues, have been crucial for investigating the most basic steps in folding, as these proteins convert from

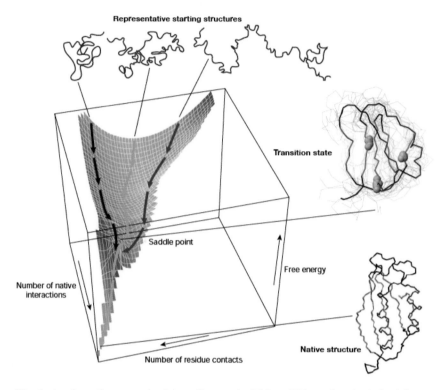

Representative starting structures

Transition state

Saddle point

Free energy

Number of native
interactions

Number of residue contacts

Native structure

Fig. 1. A schematic energy landscape for protein folding. This surface is derived from a computer simulation of the folding of a highly simplified model of a small protein. The surface serves to "funnel" the multitude of denatured conformations to the unique native structure. The critical region on a simple surface such as this one is the saddle point corresponding to the transition state, the barrier that all molecules must cross to be able to fold to the native state. Superimposed on this schematic surface is an ensemble of structures corresponding to the experimental transition state for the folding of a small protein; this ensemble was calculated by using computer simulations constrained by experimental data from mutational studies of the protein acylphosphatase.[13,14] The yellow spheres represent the three "key residues" in the structure; when these residues have formed their native-like contacts, the overall topology of the native fold is established. The structure of the native state is shown at the bottom of the surface, whilst at the top are indicated schematically some contributors to the distribution of unfolded states that represent the starting point for folding. Also indicated are highly simplified trajectories for the folding of individual molecules. From Ref. 2.

their unfolded states to their native states without the kinetic complications of highly populated intermediate states. In order to probe the folding of these small "two-state" proteins, a crucial technique has been to monitor the effects of specific mutations on the kinetics of folding and unfolding, as this approach enables the role of individual residues in the folding process to be analyzed.[10] Particular insight has come from the use of this strategy to analyze the nature of the transition states for folding, i.e., the critical regions of energy surfaces through which all molecules must pass to reach the native fold (see Fig. 1). The results of a large number of studies of these species suggest that the fundamental mechanism of protein folding involves the interaction of a relatively small number of key residues so as to generate a folding nucleus, about which the remainder of the structure rapidly condenses to form the native structure.[10]

Further details as to how such a mechanism is able to generate a unique fold have emerged from a range of theoretical studies, particularly involving molecular dynamics simulations.[11] Investigations that compare closely the simulation results with experimental observations have proved to be of particular value.[7,12] One approach that we have found to be particularly powerful incorporates experimental measurements directly into the simulations as restraints, and hence limits the regions of conformational space that are explored in each simulation to those that are compatible with the available experimental data[13]; this strategy reduces, dramatically, the time required to carry out a statistically significant set of simulations, and has enabled rather detailed ensembles of structures to be generated for the transition states of those proteins for which mutational data are available.[13,14]

These results suggest that, despite a high degree of conformational disorder, the large majority of members of the structural ensemble that describes the transition state have the same overall topology as the native fold. In essence, interactions involving the key residues force the chain to adopt a rudimentary native-like architecture. Although it is not yet clear exactly how the specific sequence encodes for such characteristics, it is likely to arise primarily from the pattern of hydrophobic and polar residues that drives preferential interactions of the side chains of specific residues as the structure becomes increasingly compact. Our results suggest that once the correct topology has been achieved, the native structure will then almost invariably be generated during the final stages of folding.[13,14] Conversely, if these key interactions are not formed, the protein cannot fold to a stable globular structure; this mechanism, therefore, acts also as a sequence-based "quality control" process by which misfolding can generally be avoided.

For proteins with more than about 100 residues, experiments show that the folding process is generally more complicated than that described above, as one or more partially structured intermediates is significantly populated during the folding process. One example of particular interest in this category is lysozyme, a protein that we and others have investigated in great detail, both because of the depth in which its folding is understood[7,15,16] and because of its link with a specific misfolding disease[3] that will be discussed later in this chapter. Of particular importance in terms of the mechanism of folding is that the structural properties of the intermediates observed in such studies provide important evidence about the manner in which the folding of a larger protein takes place.

A seminal finding of the detailed investigations of lysozyme and a number of other proteins in which intermediates have been characterized, is that folding takes place largely independently in different segments or domains of the protein.[7,17] In such cases, interactions involving key residues are likely both to establish the native-like fold within these domains, and to ensure that the latter then come together appropriately to form the correct overall structure.[17,18] The fully native structure is only acquired when all the native-like interactions are formed both within and between the domains; this happens in a final cooperative folding step when all the side-chains become locked in their unique close-packed arrangement and water is excluded from the interior of the protein.[19] Although details of the events of this type are understood at the molecular level for only a small number of proteins, this mechanism is attractive because it suggests that highly complex structures may be assembled in manageable pieces according to a set of universal principles that describe the folding of all proteins.

2. Protein Folding in the Cellular Environment

In a living system, proteins are synthesized on ribosomes, large "molecular machines" with molecular weights in excess of 2.5 MDa that self-assemble from more than 50 proteins and three large RNA molecules, from the genetic information encoded in the cellular DNA. Folding *in vivo* is thought in some cases to be co-translational, i.e., it is initiated during the process of protein synthesis that involves translation of the genetic information; in these circumstances, at least some elements of structure develop whilst the nascent chain is still attached to the ribosome.[20] Other proteins, however, appear to acquire their structures in the cellular environment after release from the ribosome, whilst yet others fold in specific compartments, such as mitochondria or the ER, following trafficking and translocation through the membranes that separate individual compartments of the cells of higher organisms.[21]

In order to begin to develop an understanding of protein folding in the cell in atomic detail, we have recently initiated NMR studies designed to probe the structures and properties of nascent chains attached to the ribosome; a major aim is to compare these characteristics with those derived from the *in vitro* studies discussed above where folding is typically initiated from chemically denatured states. The strategy that we have adopted to make such studies a possibility involves the use of biochemical techniques that stall protein biosynthesis at a particular point in the reaction, coupled with the use of methods that enable isotope labels detectable by NMR spectroscopy to be incorporated in the relatively mobile nascent chain but not in the 2.5 MDa ribosome particle itself.[22] Initial results appear extremely promising, and have already enabled us to observe the folding of a protein

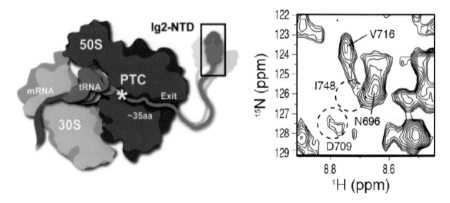

Fig. 2. Illustration of a ribosome (yellow and blue) with a nascent protein chain (grey) attached. A ribosome-bound nascent chain of a truncated tandem immunoglobulin domain has been studied by ultrafast NMR spectroscopic techniques (a section of a spectrum is shown on the right). In order to carry out the experiments, a strategy was developed in which isotopically-labeled amino acids are incorporated into polypeptide chains that are then stalled during protein synthesis on the ribosome. The spectra reveal that the domain of the protein whose sequence has emerged fully from the exit tunnel in the ribosome structure has folded to its fully native state, whilst the majority of the second domain that is still partially included in the tunnel is unfolded and highly dynamic; this result is indicated in the schematic illustration on the left. The region marked in red on this schematic picture has been found to interact weakly with the ribosome surface. Further details are given in Ref. 22, from which this figure is adapted.

domain on the ribosome and to obtain evidence for interactions between nascent chains and the ribosome surface (Fig. 2).

These results open the door to detailed studies of the various stages of folding as the polypeptide chain emerges from the ribosome tunnel, and of the interaction of cellular components with the nascent chain; it is even possible to imagine extending these studies to observe the folding processes in intact cells. Many details of the folding of a given protein will depend on the particular environment in which it takes place, although the fundamental principles of folding, discussed above, are undoubtedly universal. The opportunity to make such comparisons between folding in a range of specific environments, for example mimicking those in different regions of the cell, will be of enormous value in understanding the principles by which the folding of the various different types of protein takes place *in vivo*.

Despite the undoubted efficiency of folding in living systems, incompletely folded species populated during the folding process must inevitably expose at least some regions of structure that are buried in the native state; they are, therefore, prone to inappropriate interactions with each other and with other types of molecule within the crowded environment of a cell.[23] Living systems have therefore evolved a range of strategies to avoid such behavior, and of particular importance in this context are the large numbers of "molecular chaperones" now known to be present in all types of cells and cellular compartments.[20,21] Some chaperones have been found to interact with nascent chains as they emerge from the ribosome, whilst others are involved in guiding later stages of the folding process. Indeed, molecular chaperones, that, as the name implies, help to protect protein molecules as they acquire their mature structures, often work in conjunction with each other to ensure that the various stages in the folding of such systems are all completed efficiently.[20,21]

Many of the details of the actions of molecular chaperones have been determined from studies of their effects on folding *in vitro*, and the best characterized of the chaperones studied in this manner is the bacterial complex involving GroEL, a member of the family of "chaperonins," and its "co-chaperone" GroES. Many aspects of the sophisticated mechanism through which this coupled system functions are now well understood.[21] Of particular note is the fact that the chaperones in this family are multi-subunit complexes and contain a cavity in which incompletely folded polypeptide chains can enter and undergo the final steps in the formation of their native structures whilst sequestered and protected from the outside world. It is also becoming clear that only a subset of proteins requires interactions with GroEL in order to fold; interestingly, the proteins involved appear to be present in the cell at low levels and have high aggregation propensities, a finding consistent with the important role of chaperones as a protection against the development of aberrant interactions involving proteins.[24,25] Compelling evidence for an important role of molecular chaperones in preventing misfolding and its consequences comes, in addition, from the finding that the levels many of these species are substantially raised during cellular stress. Indeed, the designation of many chaperones as Hsps (heat shock proteins) reflects the fact that they were first identified in such a role.[21]

In eukaryotic systems such as ourselves, however, many of the proteins that are synthesized in a cell are destined for secretion to the extracellular environment. These proteins are translocated into the endoplasmic reticulum (ER), where folding takes place prior to secretion through the Golgi apparatus. The ER acts as the folding compartment for extracellular proteins and so contains a wide range of molecular chaperones and folding catalysts, and, in addition, the proteins that fold here must satisfy a "quality control" check prior to being exported (Fig. 3).[26,27] This quality control process involves a remarkable series of glycosylation and deglycosylation reactions that enables correctly folded proteins to be distinguished from misfolded ones.[26] Like the "heat shock response" in the cytoplasm, the "unfolded protein response" in the ER is also stimulated (upregulated) during stress and, as we shall see below, is strongly linked to the avoidance of misfolding diseases.[28] Such a process is particularly important as there appear to be few types of molecular chaperones located outside cells relative to the number found within cells where the intrinsic folding reactions take place, although the majority of misfolding diseases involves the appearance of extracellular protein aggregates. The best characterized extracellular molecular chaperone is clusterin, and studies of its interaction with species populated during the folding of a series of proteins, including studies we have carried out with lysozyme, suggest that it binds to partially unfolded intermediates and small aggregates and thereby prevents their conversion into larger and more intractable aggregates such as amyloid fibrils.[29,30]

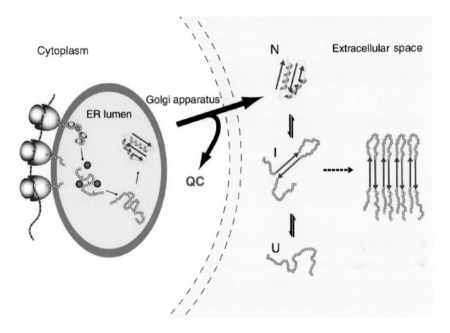

Fig. 3. Schematic representation of the possible mechanism of amyloid formation by a globular protein such as lysozyme. After synthesis on the ribosome, the protein folds in the endoplasmic reticulum (ER), aided by molecular chaperones that deter aggregation of incompletely folded species. The correctly folded protein is secreted from the cell and functions normally in its extracellular environment. Under some circumstances the protein unfolds at least partially, and becomes prone to aggregation. This can result in the formation of fibrils and other aggregates that can accumulate in tissue. Small oligomeric or pre-fibrillar aggregates as well as highly organized fibrils and plaques can give rise to pathological conditions in some disorders, notably the neurodegenerative diseases. N, I and U refer to native, partially unfolded (intermediate) and unfolded states of the protein, respectively. QC refers to the quality control mechanism that prevents incompletely folded proteins being secreted from the ER. From Ref. 38.

3. Protein Misfolding and its Biological Consequences

Folding and unfolding are the ultimate ways of generating and abolishing specific types of cellular activity. In addition, processes as apparently diverse as translocation across membranes, trafficking, secretion, the immune response and the regulation of cell growth are directly dependent on folding and unfolding events.[2] Failure to fold correctly, or to remain correctly folded, will, therefore, give rise to the malfunctioning of living systems and hence to disease.[31–33] Some of these diseases (e.g., cystic fibrosis[31] and various types of cancer[34]) result from the simple fact that if proteins do not fold correctly, they will not be present in sufficient quantities to exercise their proper function; many such disorders, normally called "loss of function" diseases, are familial because the probability of misfolding is often greater in mutational variants than in the wild type protein, because of the likelihood of their decreased stability and reduced cooperativity.

In other cases, proteins with a high propensity to misfold escape all the protective mechanisms and form aberrant intermolecular interactions with other such "misfolded" species to generate intractable aggregates within cells or (more commonly) in extracellular space. An increasing number of disorders (see Table 1), including Alzheimer's and Parkinson's diseases, the spongiform encephalopathies and type II diabetes, is known to be directly associated with the deposition of such aggregates in tissues, including the brain, heart and spleen.[32,33,35,36] Despite such differences in the location of the aggregates, and in the resulting pathological consequences, the underlying origins of these diseases are closely similar. In the next section, we shall look at these origins, initially through investigations of the nature of the "amyloid" aggregates that are associated with these disorders.

Each amyloid-associated disease involves predominantly the aggregation of a specific protein, although a range of other components, including additional proteins and carbohydrates is incorporated into the deposits when they form *in vivo*.[3] In the case of neurodegenerative diseases, the quantities of aggregates involved can sometimes be so small as to be almost undetectable, whilst in some systemic diseases — such as that associated with lysozyme discussed below — literally kilograms of protein can be found in one or more organs.[35] The characteristics of the soluble forms of the 20 or so proteins involved in the well-defined amyloid disorders are very varied — they range from intact globular proteins to largely unstructured peptide molecules — but the aggregated forms have many common characteristics.[37] Amyloid deposits, for example, show specific optical behavior (such as birefringence) on binding certain dye molecules such as Congo red.

The fibrillar structures typical of many of the aggregates have very similar morphologies (long, unbranched and often twisted structures a few nm in diameter) and a characteristic "cross-β" X-ray fiber diffraction pattern. The latter reveals that the organized core structure is composed of β-sheets whose strands run perpendicular to the fibril axis.[37] The ability of polypeptide chains to form such structures turns out, however, not to be restricted to the relatively small numbers of proteins associated with recognized clinical disorders, and, indeed, we have suggested that it could be a generic feature of polypeptide chains.[32,38] Compelling evidence for the latter statement is that fibrils can be formed *in vitro* by many peptides and proteins with no known disease association, including such well-known and highly studied molecules as myoglobin,[39] and also by homopolymers such as polyalanine, polythreonine or polylysine.[40]

The latter finding indicates that the ability to form the amyloid structure does not need to be encoded in the sequence of the protein; in essence, it is inherent in the intrinsic character of polypeptide chains, akin to analogous properties of many synthetic polymers, and this finding is reinforced by recent computer simulations of a simple model of a small homopolymeric peptide that self-assembles into a cross-β structure under a wide range of conditions (Fig. 4).[41] Of particular interest is the fact that a variety of different mechanisms of assembly is observed in the simulations, ranging from the direct assembly of single β-sheets to a process in which the peptides coalesce into a disorganized oligomer within which structural reorganization takes place to produce the cross-β structure; remarkably, the variety of assembly processes seen in an extended series of computer simulations has been observed experimentally in studies of a wide range of different systems.[3]

Table 1. A Selection of Some of the Major Human Diseases Associated with Misfolding and the Formation of Extracellular Amyloid Deposits or Intracellular Inclusions with Amyloid-like Characteristics. (Selected from Ref. 3 in which a more comprehensive list is given.)

Disease	Aggregating Protein or Peptide	Length of Protein or Peptide[a]	Structure of Protein or Peptide[b]
Neurodegenerative diseases			
Alzheimer's disease[c]	amyloid β peptide	40 or 42[f]	natively unfolded
Spongiform encephalopathies[c,e]	prion protein or fragments thereof	253	natively unfolded (1–120) and α-helical (121–230)
Parkinson's disease[c]	α-synuclein	140	natively unfolded
Amyotrophic lateral sclerosis[c]	superoxide dismutase 1	153	all-β, IG-like
Huntington's disease[d]	huntingtin with long polyQ stretches	3144[g]	largely natively unfolded
Familial amyloidotic polyneuropathy[d]	mutants of transthyretin	127	all-β, prealbumin-like
Non-neuropathic systemic amyloidoses			
AL amyloidosis[c]	immunoglobulin light chains or fragments thereof	ca. 90[f]	all-β, IG-like
AA amyloidosis[c]	fragments of serum amyloid A protein	76–104[f]	all-α, unknown fold
Senile systemic amyloidosis[c]	wild-type transthyretin	127	all-β, prealbumin-like
Hemodyalsis-related amyloidosis[c]	β2-microglobulin	99	all-β, IG-like
Finnish hereditary amyloidosis[d]	fragments of gelsolin mutants	71	natively unfolded
Lysozyme amyloidosis[d]	mutants of lysozyme	130	$\alpha + \beta$, lysozyme-fold

Table 1. (*Continued*)

Disease	Aggregating Protein or Peptide	Length of Protein or Peptide[a]	Structure of Protein or Peptide[b]
Non-neuropathic localised amyloidoses			
ApoAI amyloidosis[d]	fragments of apolipoprotein AI	80–93[f]	natively unfolded
Type II diabetes[c]	amylin	37	natively unfolded
Medullary carcinoma of the thyroid[c]	calcitonin	32	natively unfolded
Hereditary cerebral haemorrhage with amyloidosis[d]	mutants of amyloid β peptide	40 or 42[f]	natively unfolded
Injection-localized amyloidosis[e]	insulin	21 + 30[h]	all-α, insulin-like

[a] The data refer to the number of amino acid residues in the peptide or protein that is present in the aggregates and not to the number of residues in any precursor protein.

[b] This column reports the structural class and fold; both refer to the processed peptides or proteins that deposit into aggregates prior to aggregation and not to the precursor proteins.

[c] Predominantly sporadic although in some of these diseases hereditary forms associated with specific mutations are well documented.

[d] Predominantly hereditary although in some of these diseases sporadic cases are documented.

[e] 5% of cases are infectious (iatrogenic).

[f] Fragments of various lengths are generated and reported in *ex vivo* fibrils.

[g] Lengths refer to the normal sequences with non-pathogenic traits of polyQ.

[h] Human insulin consists of two chains (A and B with 21 and 30 residues, respectively) covalently bonded by disulphide bridges.

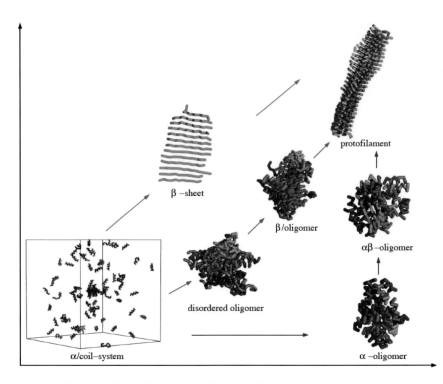

Fig. 4. Schematic illustration of a mechanism of aggregation derived from computer simulations. The simulations represent the aggregation of a 12-residue peptide composed of identical amino acids, and employ a simple "tube" model to describe the peptide structure.[42] The formation of the characteristic cross-β structure of amyloid fibrils is observed to emerge spontaneously, and can do so through a variety of apparently distinct processes that have been at the focus of intense experimental and theoretical studies.[3] These different processes appear as different manifestations of a common underlying process, called condensation ordering, and depend on the relative importance of hydrogen bonding and hydrophobic interactions. Highly hydrophobic polypeptide chains collapse first into disordered and highly dynamic oligomers and then rearrange into ordered assemblies, whilst more hydrophilic peptides assemble directly into arrays of β-strands. As well as enabling the various processes involved in aggregation to be identified, these simulations enable the nature of the nucleation process to be explored and provide insight into the origin of the toxicity of the oligomeric aggregates that appear in the intermediate stages of the process. From Ref. 41.

4. The Generic Nature of the Amyloid State

We have recently determined the atomic-level structure of a peptide molecule in amyloid fibrils by solid state NMR techniques, and the results show clearly the extended molecular conformation characteristic of β-strands and also the interesting fact that the side-chains are close-packed in remarkably specific orientations, at least within the central region of the structure.[42] Although no complete structure of an amyloid fibril has yet been determined in atomic detail, increasingly convincing models based on data from techniques such as X-ray fiber diffraction, cryo-EM and solid state NMR are now emerging[3]; one early example showing characteristic features that have been observed in general terms in a range of more recent studies of a variety of different systems, representing variations on a common theme, is shown in Fig. 5.[43] An important additional development is the successful crystallization of small peptides that show fibrillar like assemblies within three-dimensional crystals, enabling the nature of the interactions between specific residues in amyloid-like structures to be explored.[44]

More generally, we have been studying a range of different fibrils by means of experimental approaches originally developed within the rapidly developing field of nanotechnology, such as various forms of atomic force microscopy (AFM), in conjunction with computer simulation methods. Our findings reveal that amyloid fibrils represent a well-defined and closely related class of highly organized materials that can be compared and contrasted on the nanometer scale with well-established types of more conventional materials.[45] Remarkably, weight-for-weight amyloid fibrils are virtually as strong as steel and substantially more rigid than functional biological structures such as actin fibers in muscle and the microtubules that are associated with internal organization within cells. The origin of this extreme strength and rigidity appears to be that the core structure of the fibrils is primarily stabilized by strong intermolecular as well as intramolecular interactions, in particular arrays of hydrogen bonds involving the amide and carbonyl groups of the polypeptide main chain (Fig. 6).

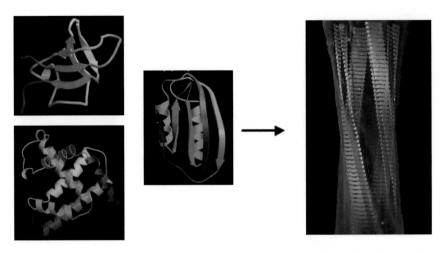

Fig. 5. Comparison of examples of native and amyloid structures of protein molecules. On the left are ribbon diagrams of the native structures of three small proteins: an SH3 domain (top), myoglobin (bottom) and acylphosphatase (middle). The native structures differ in their topologies and contents of α-helices and β-sheets resulting from the dominance of side-chain interactions within their highly evolved sequences. On the right is a molecular model of an amyloid fibril (image kindly provided by Helen Saibil, Birkbeck College, London from work reported in Ref. 43). The fibril was produced from the SH3 domain whose native structure is shown on the left, and consists of four "protofilaments" that twist around one another to form a hollow tube with a diameter of approximately 6 nm. The β-strands (flat arrows) are oriented perpendicular to the fibril axis and are linked together by hydrogen bonds involving main chain amide and carbonyl groups, many of which are intermolecular, to form a continuous structure in each protofilament. The protofilaments are held together by much weaker interactions involving primarily side-chain contacts. As the main chain is common to all polypeptides, the core protofilament structures of fibrils from different sequences have common features, differing only in detail as a result of differences in the non-dominant effects of side-chain packing. The arrow indicates that when the native states of globular proteins are destabilized, they tend to convert into the generic amyloid structure, as described in the text.

As the main chain is common to all polypeptides, this observation explains why fibrils formed from polypeptides of very different sequence have marked similarities, particularly in the fibril core structure, although differences in detail exist as a result of the need to accommodate the

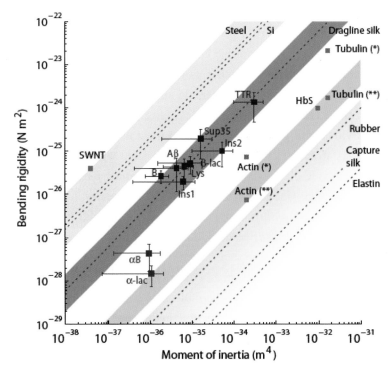

Fig. 6. Comparison of the mechanical properties among different classes of materials. The plot shows the correlation between the bending rigidity of a given material as a function of its cross sectional moment of inertia. A linear relationship within a specific type of material indicates that the forces stabilizing the differently sized samples of that material are identical. The blue band in the diagram encompasses the various examples of amyloid fibrils formed from different types of peptide or protein investigated in this study.[45] The close correlation of the rigidity and moment of inertia indicates similar interactions in each type of fibril, and analysis shows that the dominant contributions to the interactions are the main chain hydrogen bonds between the β-strands of the amyloid cross-β structure; further support for this conclusion comes from the fact that spider silk, the strength of which is also attributed to main chain hydrogen bonding, correlates closely with amyloid fibrils. The green band encompasses materials such as actin filaments that are held together by amphiphilic interactions characteristic of amino-acid side chains; the two examples of amyloid protofibrils examined in this study fall within this range, suggesting that the strong main chain interactions are not fully formed at this stage of the assembly process. Further details are given in Ref. 45, from which this figure is taken.

variable side-chains within the fibrillar structure.[37,38,45] In some cases, only a handful of the residues of a given protein may be involved in the core structure, with the remainder of the chain associated in some other manner with the fibrillar assembly; in other cases, almost the whole polypeptide chain appears to be involved. The generic amyloid structure, characteristic of the polymeric character of polypeptide chains, contrasts strongly with the highly individualistic globular structures of most natural proteins (Fig. 5); in these latter structures, the interactions associated with the very specific packing of the side chains can override, at least under carefully regulated conditions, the otherwise dominant main chain preferences.[38,39]

Even though the *ability* to aggregate to form amyloid fibrils appears to be generic, the *propensity* to do so under given circumstances can vary dramatically between different sequences.[3] It has proved possible to correlate the relative aggregation rates of a wide range of peptides and proteins with physicochemical features of the molecules such as charge, secondary structure propensities and hydrophobicity,[46] and indeed to predict the regions of a polypeptide chain that have the highest propensity to self-assemble and which are likely to be found in the fibril cores.[47] In a globular protein, for example, the polypeptide main chain and the hydrophobic side chains are largely buried within the folded structure. Only when they are exposed, for example when the protein is partially unfolded, e.g., at low pH or as the result of destabilizing mutations, or fragmented, e.g., by proteolysis, will conversion into amyloid fibrils be possible.

The propensities of such proteins to aggregate will, therefore, depend on the accessibility of such aggregation-prone species, a conclusion that is clearly demonstrated by detailed studies of the amyloidogenic mutational variants of lysozyme that we have found to decrease the stability and cooperativity of the native state.[48–50] Indeed, these experiments show that the effect of the disease-associated mutations is to decrease the energy difference between the native state and the intermediates populated in the normal folding of the protein, such that the latter are accessible to a much greater extent in the variants than in the wild-type protein.[48] The large mass of evidence now accumulated from studies of lysozyme that provides detailed insight into many aspects of the likely origin of systemic amyloid disease will not be described in detail here as it has recently been comprehensively reviewed.[49]

5. The Intrinsic Mechanism of Amyloid Formation

Experiments *in vitro* indicate that the formation of amyloid fibrils is generally characterized by a lag phase, followed by a period of rapid growth.[51,52] Such behavior is also typical of many well-understood non-biological processes, such as crystallization and gel formation and can be attributed to a nucleation mechanism, but is also characteristic of the assembly of fibrillar structures in which breakage occurs.[53,54] Each time a fibril breaks, it doubles the number of ends from which growth is able to occur and if the frequency of breakage is great enough relative to other steps in the growth process, it can have a profound effect on the kinetics of amyloid formation; indeed, it has been suggested that ready breakage of less stable fibrils could be an explanation for the transmissibility of some disease states, notably the prion disorders such as "mad cow disease" and Creutzfeldt-Jakob disease (CJD) (Table 1), but not of others, even though their underlying origins appear closely similar.[54]

One of the characteristics of processes of this type is that the lag phase, when present, can usually be eliminated by addition of pre-formed fibrils to fresh solutions, a process known as seeding. An interesting possibility is that seeding by chemically modified forms of proteins, resulting for example from degradation or oxidative stress, may be an important factor in triggering the aggregation process *in vivo* if these forms have a high aggregation propensity relative to the wild type sequence; in such cases, it could be an important factor stimulating the onset of disease.[55] It therefore becomes possible to rationalize even such complex biological characteristics, such as the transmissibility or otherwise of related diseases, in terms of well-established fundamental physicochemical principles.

It is clear from a wide variety of studies that there are striking similarities in the time-course of the aggregation of different peptides and proteins (Fig. 7), again supporting the view that the amyloid form of these molecules is a common or generic structure resulting from the inherent properties of polypeptide chains.[51,52] The first phase in amyloid formation

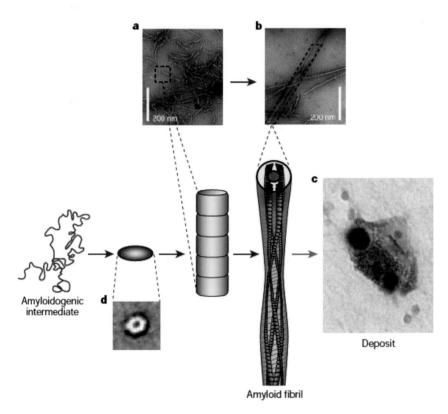

Fig. 7. A general mechanism of aggregation to form amyloid fibrils. Unfolded or partially folded proteins associate with each other to form small, soluble aggregates that undergo further assembly into protofibrils or protofilaments and then mature fibrils (top EM images). The latter sometimes accumulate in plaques or other structures such as the Lewy bodies associated with Parkinson's disease (microscope image on right). Some of the early aggregates appear to be amorphous or micellar in nature, although others form ring-shaped species (bottom EM image). From Ref. 2.

generally appears to involve the formation of soluble oligomers as a result of relatively non-specific interactions, although in some cases specific structural transitions, such as domain swapping, may be important.[56] The earliest species visible by electron or atomic force microscopy generally resemble small bead-like structures, sometimes linked together, and often described as amorphous aggregates or as micelles; these species could be analogous to those seen in the simulations illustrated in Fig. 4.

These early aggregates then appear to transform into species with more distinctive morphologies, often called "protofilaments" or "protofibrils," that are commonly short and curly, sometimes retaining bead-like elements visible in EM or AFM images; we have found that the stability of these types of aggregate is much less than mature amyloid fibrils, indicating that the characteristic main chain interactions have not yet developed fully.[45] These species must therefore undergo a reorganization step in order to transform into mature fibrils; it is possible that this process occurs within the structures, although it is also possible that in some cases at least the mature fibrils result from a distinct assembly process following dissociation of the component molecules from the protofibrils.[57] The various species that form early in the aggregation process are likely to be relatively disorganized structures that expose to the outside world a variety of segments of the protein that are normally buried in the globular state.[3,49] In some cases, however, these early aggregates appear to be quite distinctive structures, including well-defined annular species.[58]

6. Biological Regulation of the Different States of Proteins

Our findings strongly suggest that all proteins can, in principle, access a wide range of conformational states, and that the specific state of a protein that is adopted under specific conditions will depend on the relative thermodynamic stabilities of the various accessible conformations and on the kinetics of their interconversions (Fig. 8).[2,59] Amyloid fibrils are just one of the types

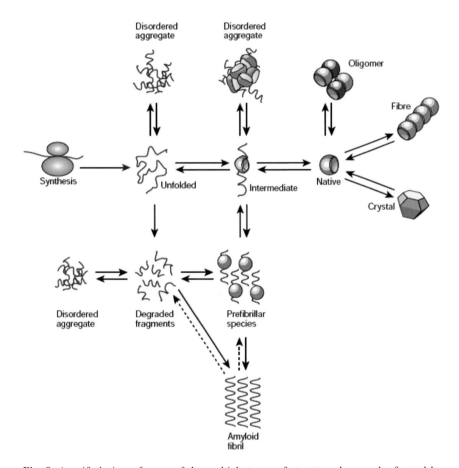

Fig. 8. A unified view of some of the multiple types of structure that can be formed by polypeptide chains. An unstructured chain, for example newly synthesized on a ribosome, may fold to a native structure, perhaps via one or more partially folded intermediates. It can, however, experience other fates such as degradation or aggregation. An amyloid fibril is just one form of aggregate, but it is unique in having a highly organized structure, as shown in Fig. 5. The populations and interconversions of the various states are determined by their relative thermodynamic and kinetic stabilities under any given conditions. In living systems, however, transitions between the different states are very highly regulated by control of the environment, and by the presence of molecular chaperones, proteolytic enzymes and other factors. Failure of such regulatory mechanisms is likely to be a major factor in the onset of misfolding diseases. From Ref. 2.

of aggregates that can be formed by proteins, a situation analogous to the multiple structural forms that are often found for non-biological polymeric systems such as polyethylene,[60] although a significant feature of the amyloid structure is that its highly organized hydrogen-bonded structure is likely to give it unique thermodynamic and kinetic stability.

It is interesting to note in this context that for amino acids and small peptides, as with other molecular species with relatively limited numbers of possible conformations, three-dimensional crystals can be formed[44] and represent the most highly organized of the different types of molecular assemblies; for larger unstructured polypeptides with vast numbers of different possible conformations, however, three-dimensional structures are likely to be inaccessible (unlike native states, the energy landscapes of these assemblies have not been sculpted during evolution) and the fibril form then represents the most highly organized structure that can, in practice, be adopted. Once formed, this latter type of stable aggregate may persist for long periods of time, allowing a progressive build up of deposits in tissue, and indeed enabling seeding of the subsequent conversion of additional quantities of the same protein into amyloid fibrils. It is, therefore, not surprising that biological systems have almost universally developed robust mechanisms to avoid the formation of such material. Nevertheless, there is increasing evidence that the remarkable material properties of amyloid structures[45] have been exploited by some species, including bacteria, fungi and even mammals, for specific (and carefully regulated) purposes.[3,61–63]

There is increasing evidence that evolutionary selection has acted to minimize the risk of inadvertently forming the generic amyloid form of peptides and proteins. Once example is that the occurrence of certain patterns of amino acids, such as alternating polar and hydrophobic residues that favor β-sheet structure of the type seen in amyloid fibrils, is lower than one would expect from random distributions of residue types.[64] Moreover, our recent studies suggest that the aggregation process that results in amyloid fibrils is nucleated in a similar manner to that of folding, but that the residues involved may well be located in very different regions of the sequence from those that nucleate folding.[65] Such "kinetic partitioning" means that mutations that occur during evolution could be selected for their ability to enhance folding at the expense of aggregation.

It is apparent, however, that biological systems have become robust not just by careful manipulation of the sequences of proteins but also by controlling, by means of molecular chaperones and degradation mechanisms, the particular conformational state adopted by a given polypeptide chain at a given time and under given conditions. This process can be thought of as being analogous to, and just as fundamental and important as, the way that biology regulates and controls the various chemical transformations that take place in the cell by means of enzymes. And just as the aberrant behavior of enzymes can cause metabolic disease, the aberrant behavior of the chaperone and other machinery regulating polypeptide conformations can contribute to misfolding and aggregation diseases.[66]

7. The Molecular Origins of Amyloid Disease

The ideas encapsulated in Fig. 8, therefore, serve as a physicochemical framework for understanding the fundamental events that underlie the control of protein conformational states and for the failures of such control mechanisms that give rise to misfolding diseases. For example, many of the mutations associated with the familial forms of deposition diseases, as discussed earlier for lysozyme, increase the population of partially unfolded states, and hence increase the propensity to aggregate, by decreasing the stability or cooperativity of the native structure.[36,50,67]

Other familial diseases are associated with the accumulation of amyloid deposits whose primary components are fragments rather than full length proteins; such fragments can be produced by aberrant processing or incomplete proteolysis, and are unable to fold into aggregation-resistant states. Yet other pathogenic mutations act by enhancing the propensities of such species to aggregate, for example by increasing their hydrophobicity or decreasing their charge.[46] And, in the case of the prion disorders such as Kuru or Creutzfeldt-Jakob disease, it appears that ingestion of pre-aggregated states of an identical protein, e.g., by voluntary or involuntary cannibalism or by means of contaminated pharmaceuticals or surgical instruments, can increase dramatically the inherent rate of aggregation through seeding, and perhaps ready breakage, and hence generate a mechanism for transmission.[51,68]

In some aggregation diseases, the large quantities of insoluble proteinaceous deposits may physically disrupt specific organs and hence cause pathological behavior.[35] But for neurodegenerative disorders, such as Alzheimer's disease, the primary symptoms almost certainly result from a "toxic gain of function" that is associated with aggregation.[69] The early prefibrillar aggregates of proteins associated with such diseases have been shown to be highly damaging to cells; by contrast, the mature fibrils appear relatively benign.[51,70] Moreover, we have recently found that similar aggregates of proteins that are not connected with any known disease can be equally toxic to cells, both when added to cell culture medium[71] and when microinjected into the brains of rats.[72] The generic nature of such aggregates and their effects on cells has recently been supported by the remarkable finding that antibodies raised against early aggregates of Aβ cross-react with early aggregates of a range of different peptides and proteins, and moreover inhibit their toxicity.[73,74]

It is possible that there are specific mechanisms that give rise to aggregate toxicity, for example as a result of annular species (Fig. 7) that resemble the toxins produced by bacteria that form pores in membranes and disrupt the ion balance in cells.[58] It is likely, however, that the relatively disorganized prefibrillar aggregates are toxic through a less specific and indeed generic mechanism, for example as a result of the exposure of non-native hydrophobic surfaces stimulating aberrant interactions with cell membranes or other cellular components.[75] In contrast to the exquisitely designed surfaces of the correctly structured molecules within the crowded cellular environment that have evolved to interact only with specific partners, therefore, the surfaces of any non-evolved polymeric aggregates that escape the various types of protective mechanisms, discussed below, are likely to interact inappropriately with many of the components of a biological system and hence will commonly cause malfunctions and potentially disease.

8. The Limitations of Molecular Evolution

Under normal circumstances, the actions of molecular chaperones and other "housekeeping" mechanisms are remarkably efficient in ensuring that such potentially toxic species are neutralized before they can do any damage.[21,75,76] Such neutralization could result simply from the efficient targetting of misfolded proteins for degradation, but it appears that molecular chaperones are also able to alter the partitioning between harmful and harmless forms of aggregates, as a result of changing the kinetic or thermodynamic stability of one or more of the multiple species accessible to a protein (Fig. 8).[77] If the efficiency of such protective mechanisms becomes impaired, however, the probability of pathogenic behavior must increase.[30,76] Such a scenario would explain why most of the amyloid diseases are associated with old age, where there is likely to be an increased tendency of proteins to become misfolded or damaged, ultimately at least coupled with a decreased efficiency of the protective chaperone and unfolded protein response.[78] It is ironic that through our success in increasing the life expectancy of the populations of the developed world, we are now seeing the limitations of our proteins and of the regulatory mechanisms that control their behavior.[75,79]

One of the characteristics of proteins that is implied in this explanation of misfolding diseases is that relatively small changes in their sequences as a result of mutation, or of their biological environment in old age are, at least in some cases, enough to cause a generic shift from normal (soluble) to abnormal (aggregation) behavior. This situation can be qualitatively rationalized by the argument that natural selection can only generate sequences that are good enough to allow the organism concerned to flourish relative to its potential competitors; in this context, the behavior of proteins in old age is unlikely to be of importance in such a selection process.[38,79]

Dramatic evidence for this supposition has recently emerged from an analysis of the relationship between experimental amyloid aggregation rates of human proteins and measurements of the levels of gene expression; the latter are likely to relate to the concentrations of the corresponding proteins in the organism itself. This analysis[80] shows that the correlation coefficient between the aggregation rates and expression levels of all the proteins for which both sets of data could be found, that includes proteins both associated and not-associated with amyloid disease, is an astonishing 0.97 (Fig. 9). In itself this result provides compelling evidence for the generic nature of the amyloid form of proteins and the mechanism by which it is formed.

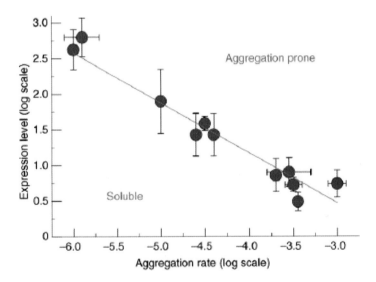

Fig. 9. Correlation between expression levels and the measured aggregation rates for a set of human proteins. The aggregation rates represent all the data obtained from a comprehensive search of the amyloid aggregation literature for studies carried out at pH values between 4.0 and 8.0. The expression levels are measured by the cellular mRNA concentrations and are taken from published databases. The standard deviations of the aggregation rates are reported only in four cases, as these values are generally not available or difficult to extract from the published data. Two proteins not involved in any known disease are shown by means of blue circles, whilst the others (shown in red) are all associated with amyloid diseases. From Ref. 80.

This very high degree of correlation is, however, exactly that predicted qualitatively by the reasoning given above concerning evolutionary selection. Specifically, it reflects the fact that a protein must be soluble enough to exist at the level that is optimal for the organism concerned, and this solubility is achieved by the selection during evolution of amino acid substitutions that reduce the propensity to aggregate. Most amino acid substitutions, however, increase the aggregation propensity of natural proteins,[47] so once evolutionary selection has achieved a sufficiently low aggregation propensity to allow the optimal level of the protein concerned to be achieved, random mutagenesis will, in general, prevent the aggregation propensity decreasing further; this combination of effects is likely to be the explanation for cytosolic proteins tending to fall very close to the line indicated in Fig. 9.

9. Testing the Generic View of Amyloid Disease

The conclusions and ideas of the molecular basis of amyloid disease that have been discussed so far, have been derived almost completely from experiments carried out in the test tube (*in vitro*) and in the computer (*in silico*). Despite the fact that there is strong circumstantial evidence to link them to events occurring in living systems (*in vivo*), including experiments with cells in culture, we wish to explore much more rigorously the way in which the myriad components of the intra- and extracellular environment affect the quantitative relationship between physicochemical properties, such as aggregation propensity, and its consequences in a living organism, and particularly to test the generic hypothesis for amyloid formation and its links with disease. To this end, we are using the fruit fly (*Drosophila meganister*) as a model organism to link the chemistry and physics of aggregation to its biological effects in higher organisms.[81] The advantage of this particular system for our purposes is that the short lifespan (typically about 30 days) and low unit cost relative to, for example transgenic rodent models, permits us to carry out a very large number of experiments in a reasonable timeframe to obtain data that are statistically highly significant.

The approach we have taken is to exploit the existence of transgenic fruit flies in which the 42-residue Aβ-peptide is expressed in the brain. Lines of flies had previously been generated in which deposits of the peptide can be seen to develop with time.[82,83] In addition, the flies develop locomotor defects, observed most easily in assays of their ability to climb up a glass surface, and have reduced lifespans. The nature of the deposits of the Aβ-peptide was found initially to occur within neurons and then to accumulate as extracellular deposits analogous to those seen in sufferers from Alzheimer's disease, as well as in transgenic mouse models designed to study this condition. It had also been found that flies expressing the Aβ-peptide having the E22G (Arctic) mutation, which results in a very early onset form of Alzheimer's disease in humans, have very much shorter lifespans than those expressing the wild-type peptide, and show a much earlier appearance of peptide-containing deposits within brain tissue and of locomotor defects.[82,83]

The overriding aim of our investigations is not to use the transgenic flies as a model for Alzheimer's disease as such, but to use them to explore the onset of the neuronal disfunction that is evident in the locomotor defects and reduced lifespan; indeed, it has been shown earlier that flies with such defects have, in addition, reduced learning abilities.[84] In particular, the strategy we have employed is to express a wide range of mutational variants of the Aβ-peptide with no known disease association but which are known to affect their *in vitro* aggregation behavior.[85] The particular objective of these studies is to investigate in this way the underlying nature of the process that gives rise to such disfunction and to see whether it is possible to identify the particular type of molecular species that is likely to be responsible for the observed neuronal damage. This particular line of transgenic flies expressing the Aβ-peptide itself is of particular value over other models that have been developed, in which the precursor APP protein is expressed and the Aβ-peptide appears as a result of the action of secretases; although this process is analogous to the origin of the amyloidogenic peptide in humans, expressing the peptide directly avoids the severe complications in analysis of the data that will result if the mutations affect the action of the secretases, and hence the length and quantity of the Aβ-peptide fragment that results for different variants.[83]

The conceptual basis that underlies these experiments is encapsulated in Fig. 9, and the accompanying explanation that indicates that at least many of our proteins are "on the edge" of aggregation, as they can have evolved only to be as robust as is necessary to allow the living system in which they are present to compete successfully for survival.[80] If we were to make mutations in the Aβ-peptide that increase or decrease its propensity for aggregation it should, on the arguments made earlier, increase or decrease, respectively, the severity of neuronal damage in the transgenic fly system. From our previous studies of aggregation *in vitro*, we can predict the changes in the intrinsic propensity to aggregate by using the algorithms based on physicochemical principles and derived from the experimental data.[46,47,85] We used this approach to design a series of some 20 single mutational variants of the 42-residue peptide in the first instance that we predict will give a spread of aggregation propensities.

The variation in intrinsic aggregation rates is predicted to be approximately an order of magnitude — rather modest in terms of the variations in the rates of different naturally occurring peptides and proteins that cover more than six orders of magnitude — and representative studies carried out *in vitro* have validated the accuracy of these predictions.[85] The results of this set of experiments are dramatic, and a flavor of their remarkable nature is illustrated in a snapshot of a climbing assay involving a sub-set of the mutated peptides (Fig. 10). This experiment illustrates the effect of introducing two different single residue mutations designed in each case to reduce the aggregation propensity of the wild-type peptide. It is immediately evident that the mutations have had a profound effect on the flies, showing the dramatic recovery of locomotor skills; similar experiments where mutations were designed to increase the aggregation propensity show equally striking decreases in such skills.[85] Taken together, these results support strongly the hypothesis that the origin of amyloid disease

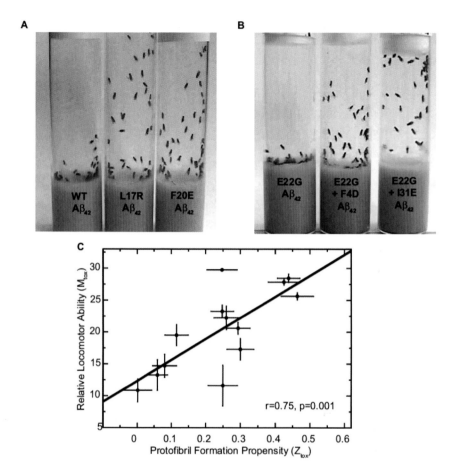

Fig. 10. The effect of mutations in the sequence of the 42-residue human Alzheimer $A\beta$-peptide on neuronal dysfunction in transgenic fruit flies. The upper left panel **(A)** illustrates a climbing assay of flies expressing the wild type sequence (left) and two mutational variants predicted to reduce the peptide's aggregation propensity; the more mobile the flies, the higher up the tube they can climb. The right hand upper panel **(B)** represents a similar experiment with flies expressing the $A\beta$-peptide containing the E22G "Arctic mutation" (left hand tube). The two left hand tubes are of peptides that contain mutations that decrease the propensity to form pre-fibrillar aggregates (protofibrils) in particular. The lower panel **(C)** shows the degree of correlation between the relative locomotor activity of a series of mutational variants against their predicted propensities to form protofibrils. Figure adapted from Ref. 90.

is fundamentally a consequence of the transition of a peptide or protein from its soluble and usually functional state to its generic and potentially pathogenic amyloid form.[38]

10. Defining the Molecular Origins of Neurodegeneration

The correlation between the effects of the mutations on the behavior of the fruit flies with the predictions of their effects on aggregation propensitities can be explored in a quantitative manner by defining a "toxicity" parameter based on locomotor ability and lifespan (Fig. 10); this procedure reveals that the correlation coefficient relating toxicity to the aggregation propensity of 17 mutational variants is an astonishing 0.85. We can conclude from this finding that, despite the vast machinery of biological processes associated with the regulation and management of peptide and protein expression and degradation, the onset of restricted movement and the lifespan of the flies is quantitatively dependent simply on the physicochemical properties of the aggregation-prone species. Moreover, the value of the correlation coefficient shows that the probability that neuronal dysfunction is not related directly to the aggregation of the $A\beta$-peptide, in this system at least, is less than 1 in 100,000.

Further experiments have been carried out to extend the scope of this study, and one set that is of particular interest relates to the introduction of a second mutation designed to "rescue" the flies expressing the $A\beta$-peptide containing the Arctic mutation (E22G) from the effects of its aggregation. Using the same procedure as described above, the predictive algorithm was used to design peptides with a second mutation that are expected to have a greatly reduced aggregation propensity. Again, the results are dramatic, and enabled the mobility and lifespan of the flies to be extended so as to be essentially the same as these properties for flies expressing the wild-type sequence.

Remarkably, however, a case has been found where flies expressing a peptide containing a double mutation (I31E/E22G), where the intrinsic aggregation propensity remains as high as that of the Arctic mutant itself, and yet the locomotor skills and lifespan have increased to levels comparable to those of the wild-type flies (Fig. 10). Moreover, histological analysis reveals large quantities of aggregated Aβ-peptide within the brains of the flies, in contrast to the situation with the rescue mutations that decrease the aggregation propensities, where no detectable deposits were evident at any stage of the lifespan of the flies. This finding indicates that the presence of deposits of aggregates does not in itself necessarily cause neurodegeneration, a finding consistent with observations of human autopsy data, where old people not suffering from Alzheimer's disease sometimes have extensive deposits of the Aβ-peptide in the form of plaques in the human brain,[86] and with similar observations made occasionally in mouse models of the disease.[87]

This apparently paradoxical result can be rationalized on the basis of arguments described above whereby toxicity in amyloid deposition is attributed to the early oligomeric aggregates rather than to the mature fibrils; if the population of the oligomers is low because, for example, they form relatively slowly but convert rapidly into fibrils, then the toxicity will also be low even if large deposits of the Aβ-peptide are present. Evidence that this situation is found in the present case comes from two sources.[85] First, an algorithm based on *in vitro* experimental data for the rate of conversion into prefibrillar aggregates rather than fibrillar ones, predicts that the toxicity of the I31E/E22G mutant should be as low as that resulting from expression of the wild-type Aβ-peptide. Second, EM analysis shows that the aggregation of I31E/E22G *in vitro* produces well-defined fibrillar species without the presence of significant quantities of the small prefibrillar aggregates, such as the bead-like structures seen with the wild-type protein and other mutants.

These experiments thus provide independent evidence for the proposition that oligomeric aggregates are responsible for cellular damage, and that they are the culprits in the onset of at least some of the diseases associated with the eventual appearance of amyloid fibrils. Moreover, studies of the effects of aggregation in another model organism, *C. elegans*, using gene knockout techniques, has provided evidence for the idea that the formation of relatively harmless large aggregates could have evolved to be a protective mechanism against neuronal damage.[88,89] We believe that the use of model organisms in the ways illustrated in these examples will play a major role in the quest to understand the general underlying links between physical and chemical principles and biological function, and specifically in the context of this chapter, the fundamental origins of the complex and increasingly common diseases that are associated with protein misfolding.[90]

11. The Prospects of Effective Therapies

In the specific context of protein misfolding and misassembly, events that will always have a finite probability of occurring given the complex and stochastic processes involved in normal folding and assembly, these studies have shown that under normal circumstances, molecular chaperones and other "housekeeping" mechanisms are remarkably efficient in ensuring that potentially toxic species such as oligomeric or prefibrillar amyloid aggregates are neutralized in living systems before they can cause significant damage.[21] Such neutralization can result from targeting them efficiently for degradation, from disrupting them to regenerate their soluble precursors, or from their conversion into less toxic aggregates such as fibrils and plaques.

The evidence discussed in this chapter suggests that the reason for the recent proliferation of aggregation diseases in the developed world in particular, is fundamentally due to the fact that at least some of our proteins are poised right at the boundary between solubility and insolubility.[80] In such a situation, relatively small changes in aggregation propensities (e.g., resulting from even a single mutation as in familial amyloid diseases such as that associated with lysozyme[49]), protein concentration (e.g., in dialysis-related amyloidosis[91]), or decreases in the efficiency of protective mechanisms or increases in the number of misfolded or damaged proteins (e.g., in old age[78]) can result in the initiation and slow accumulation of aggregates such as amyloid fibrils, that can in some cases result in the presence of toxic species such as fibril precursors.

These ideas, based initially on studies in "test tubes" or cells in culture, are being linked to the behavior of higher organisms through the use of model systems such as fruit flies as "living test tubes".[85] We see as a result of this type of approach, the way that the principles of chemistry and physics translate remarkably directly into the biological and physiological properties of living systems, reflecting the highly interdependent co-evolution of molecules and the biological environments in which they function. It is particularly satisfying in the light of the fact that living cells contain a remarkable concentration of molecular species, typically more than $300 \, g/L$,[23] to conclude that the importance of maintaining solubility and independence of these species is crucial in determining whether the behavior of a biological system is normal or aberrant.

This picture that our proteins, the most abundant and ubiquitous of all molecules in biology, are poised on the brink of an aggregation precipice may appear at first sight to be a very negative conclusion about the prospects for avoiding misfolding and deposition diseases in the future. There is, however, a very positive conclusion that can be drawn from these findings in that they indicate that only relatively small reductions in intrinsic physicochemical properties such as aggregation propensities, or in factors such as protein concentration or the efficiency of the various mechanisms, natural or otherwise, that serve to protect us from disease, can take us into the safety zone of solubility; such a situation is illustrated in the dramatic effects of the "rescue" mutations in the fly model of Alzheimer's disease.[85] Indeed, the vast increases in our understanding of the origins and means of progression of misfolding and aggregation diseases that have taken place in the last decade are beginning to allow the rational design of strategies to combat these debilitating disorders in different ways. The generic process of aggregation that has been outlined earlier indicates that there are several very specific steps in the process where directed therapeutic intervention looks highly promising.[92,93]

Ultimately, if one can achieve the ability to manipulate gene sequences in humans (e.g., by "gene therapy" or stem cell techniques) it should be possible to abolish conditions such as Alzheimer's disease, as we see in the case of the rescue mutations in transgenic fruit flies discussed above.[85] But until then certain classes of molecular therapeutics look particularly promising; as an example, a number of approaches based on antibodies or other specific binding agents are being explored, as such species can be targeted against a particular protein associated with a given disease, so as, for example, to stabilize the aggregation resistant native state or to reduce the concentration of aggregation prone species.[50,94,95] Moreover, the recent discovery that antibodies can be raised against different generic forms of aggregates, including oligomeric species, suggests that they can play a role analogous to natural chaperones.[73,74] In addition, the remarkable correlation between the events occurring *in vitro*, *in silico* and *in vivo*, not only represents a major breakthrough in showing the relevance of carefully designed studies in the test tube for understanding the equivalent processes in a living system, but also provides considerable insight into the relationships between chemistry, physics, biology and medicine.

12. Concluding Remarks

Application of the techniques and concepts of experimental and theoretical chemistry and physics over many years has provided great insight into the nature and properties of biological molecules at the atomic level, including the manner in which they undergo normal and aberrant self-assembly in laboratory environments; indeed, many of the fundamental principles have emerged at least in general terms from these studies.[3,4] Concurrently, the methods of biochemistry and cell biology have revealed much about how particular molecules are associated with specific functional processes in the cellular environment and the ways that such functions can be impaired.[3,75]

Further applications of these powerful approaches are likely to continue to increase the depth of our understanding of the fundamental events associated with the processes of protein folding, misfolding and aggregation.

The results discussed in this chapter also indicate that the use of model organisms such as the fruit fly can be of enormous value in testing new ideas as to the underlying molecular origins of the phenomena that give rise to disease in humans. In the case of the amyloid diseases, this approach has provided dramatic support for the generic hypothesis of amyloid-related diseases. It also represents a powerful means of exploring the genetic factors that influence such diseases, probing the effects of processes such as ageing, and of screening rapidly potential therapeutic compounds.[83,90] Most significantly, however, the substantial degree of progress that has already been made in recent years, provides grounds for great optimism that means will be found in the relatively near future to treat effectively or even to prevent at least the most common forms of this set of highly unpleasant and usually fatal disorders. Such progress is urgently needed in the light of the dramatic increase in the number of people who are suffering from, or vulnerable to, these conditions. Indeed, this class of diseases is rapidly moving to the top of the list of challenges to the healthcare and social support systems in an increasing number of countries around the world.

Acknowledgments

I should like to thank in particular the Wellcome Trust and Leverhulme Trust for generous funding over many years of the research activities described here, as well as the UK Research Councils, the European Commision, the Royal Society and numerous charitable organization for their crucial support, without which the work described in this article could not have been carried out. I should also like to thank very deeply all the students, research fellows and colleagues who have contributed to all aspects of this work, the names of many of whom appear in the references in this manuscript. This chapter is a substantially updated and extended version of an article (Ref. 2 in this manuscript) that was previously published in *Nature* **426:** 884–890 (2003).

References

1. Vendruscolo M, Zurdo J, MacPhee CE, Dobson CM. (2003) Protein folding and misfolding: A paradigm of self-assembly and regulation in complex biological systems. *Phil Trans R Soc Lond* **A361:** 1205–1222.
2. Dobson CM. (2003) Protein folding and misfolding. *Nature* **426:** 884–890.
3. Chiti F, Dobson CM. (2006) Protein misfolding, functional amyloid and human disease. *Annu Rev Biochem* **75:** 333–366.
4. Dobson CM, Sali A, Karplus M. (1998) Protein folding: A perspective from theory and experiment. *Angew Chem Int Ed Eng* **37:** 868–893.
5. Wolynes PG, Onuchic JN, Thirumalai D. (1995) Navigating the folding routes. *Science* **267:** 1619–1620.
6. Dill KA, Chan HS. (1997) From Levinthal to pathways to funnels. *Nature Struct Biol* **4:** 10–19.
7. Dinner AR, Sali A, Smith LJ, Dobson CM, Karplus M. (2000) Understanding protein folding via free energy surfaces from theory and experiment. *Trends Biochem Sci* **25:** 331–339.
8. Baldwin RL. (1994) Protein folding: Matching speed and stability. *Nature* **369:** 183–184.
9. Schuler B, Lipman EA, Eaton WA. (2002) Probing the free-energy surface for protein folding with single-molecule fluorescence spectroscopy. *Nature* **419:** 743–737.
10. Fersht AR. (1999) *Structure and Mechanism in Protein Science: A Guide to Enzyme Catalysis and Protein Folding.* W.H. Freeman, New York.
11. Shea JE, Brooks CL. (2001) From folding surfaces to folding proteins: A review and assessment of simulation studies of protein folding and unfolding. *Ann Rev Phys Chem* **52:** 499–535.
12. Fersht AR, Daggett V. (2002) Protein folding and unfolding at atomic resolution *Cell* **108:** 573–582.

13. Vendruscolo M, Paci E, Dobson CM, Karplus M. (2001) Three key residues form a critical contact network in a transition state for protein folding. *Nature* **409:** 641–646.

14. Paci E, Lindorff-Larsen K, Dobson CM, Karplus M, Vendruscolo M. (2005) Transition state contact orders correlate with protein folding rates. *J Mol Biol* **352:** 495–500.

15. Radford SE, Dobson CM, Evans PA. (1992) The folding of hen lysozyme involves partially structured intermediates and multiple pathways. *Nature* **358:** 302–307.

16. Hooke SD, Radford SE, Dobson CM. (1994) The refolding of human lysozyme: A comparison with the structurally homologous hen lysozyme. *Biochemistry* **33:** 5867–5876.

17. Salvatella X, Dobson CM, Fersht AR, Vendruscolo M. (2005) Determination of the folding transition states of barnase by using Φ_I-value restrained simulations validated by double mutant Φ_{IJ}-values. *Proc Natl Acad Sci USA* **102:** 12389–12394.

18. Vendruscolo M, Paci E, Karplus M, Dobson CM. (2003) Structure and relative free energies of partially folded states of proteins. *Proc Natl Acad Sci USA* **100:** 14817–14821.

19. Cheung MS, Garcia AE, Onuchic JN. (2002) Protein folding mediated by solvation: Water expulsion and formation of the hydrophobic core occur after the structural collapse. *Proc Natl Acad Sci USA* **99:** 685–690.

20. Hardesty B, Kramer G. (2001) Folding of a nascent peptide on the ribosome. *Prog Nucl Acid Res Mol Biol* **66:** 41–66.

21. Hartl FU, Hayer-Hartl M. (2002) Molecular chaperones in the cytosol: From nascent chain to folded protein. *Science* **295:** 1852–1858.

22. Hsu ST, Fucini P, Cabrita LD, Launay H, Dobson CM, Christodoulou J. (2007) Structure and dynamics of a ribosome-bound nascent chain by NMR spectroscopy. *Proc Natl Acad Sci USA* **104:** 16516–16521.

23. Ellis RJ. (2001) Macromolecular crowding: An important but neglected aspect of the intracellular environment. *Curr Opin Struct Biol* **11:** 114–119.

24. Kerner MJ, Naylor DJ, Ishihama Y, Maier T, Chang JC, Stines AP *et al.* (2005) Proteome-wide analysis of chaperonin-dependent protein folding in *Escherichia coli*. *Cell* **122:** 209–220.

25. Gaetano GG, Hartl FU, Dobson CM, Vendruscolo M. Unpublished data.

26. Hammon C, Helenius A. (1995) Quality control in the secretory pathway. *Curr Opin Cell Biol* **7**: 523.

27. Kaufman RJ, Scheuner D, Schröder M, Shen X, Lee K, Liu CY, Arnold SM. (2002) The unfolded protein response in nutrient sensing and differentiation. *Nature Rev Mol Cell Biol* **3**: 411–421.

28. Bence NF, Sampat RM, Kopito RR. (2001) Impairment of the ubiquitin-proteosome system by protein aggregation. *Science* **292**: 1552–1555.

29. Wilson MR, Easterbrook Smith SB. (2000) Clusterin is a secreted mammalian chaperone. *Trends Biochem Sci* **25**: 95–98.

30. Kumita JR, Poon S, Caddy GL, Hagan CL, Dumoulin M, Yerbury JJ, Stewart EM, Robinson CV, Wilson MR, Dobson CM. (2007) The extracellular chaperone clusterin potently inhibits human lysozyme amyloid formation by interacting with prefibrillar species. *J Mol Biol* **369**: 157–167.

31. Thomas PJ, Qu BH, Pedersen PL. (1995) Defective protein folding as a basis of human disease. *Trends Biochem Sci* **20**: 456–459.

32. Dobson CM. (2001) The structural basis of protein folding and its links with human disease. *Phil Trans R Soc Lond* **B356**: 133–145.

33. Horwich A. (2002) Protein aggregation in disease: A role for folding intermediates forming specific multimeric interactions. *J Clin Invest* **110**: 1221–1232.

34. Bullock AN, Fersht AR. (2001) Rescuing the functions of mutant p53. *Nature Rev Cancer* **1**: 68–76.

35. Tan SY, Pepys MB. (1994) Amyloidosis. *Histophathology* **25**: 403–414.

36. Kelly JW. (1998) Alternative conformation of amyloidogenic proteins and their multi-step assembly pathways. *Curr Opin Struct Biol* **8**: 101–106.

37. Sunde M, Blake CCF. (1997) The structure of amyloid fibrils by electron microscopy and X-ray diffraction. *Adv Protein Chem* **50**: 123–159.

38. Dobson CM. (1999) Protein misfolding, evolution and disease. *Trends Biochem Sci* **24**: 329–332.

39. Fändrich M, Fletcher MA, Dobson CM. (2001) Amyloid fibrils from muscle myoglobin. *Nature* **410**: 165–166.

40. Fändrich M, Dobson CM. (2002) The behaviour of polyamino acids reveals an inverse side-chain effect in amyloid structure formation. *EMBO J* **21:** 5682–5690.

41. Auer S, Dobson CM, Vendruscolo M. (2007) Revealing the nucleation barriers for protein aggregation and amyloid formation. *HFSP J* **1:** 137–146.

42. Jaroniec CP, MacPhee CE, Bajaj VS, McMahon MT, Dobson CM, Griffin RG. (2004) High resolution molecular structure of a peptide in an amyloid fibril determined by magic angle spinning NMR spectroscopy. *Proc Natl Acad Sci USA* **101:** 711–716.

43. Jiménez JL, Guijarro JI, Orlova E, Zurdo J, Dobson CM, Sunde M, Saibil HR. (1999) Cryo-electron microscopy structure of an SH3 amyloid fibril and model of the molecular packing. *EMBO J* **18:** 815–821.

44. Nelson R, Sawaya MR, Balbirnie M, Madsen AO, Riekel C, Grothe R, Eisenberg D. (2005) Structure of the cross-beta spine of amyloid-like fibrils. *Nature* **435:** 773–778.

45. Knowles TP, Fitzpatrick AW, Meehan S, Mott HR, Vendruscolo M, Dobson CM, Welland ME. (2007) Role of intermolecular forces in defining material properties of protein nanofibrils. *Science* **318:** 1900–1903.

46. Chiti F, Stefani M, Taddei N, Ramponi G, Dobson CM. (2003) Rationalisation of mutational effects on protein aggregation rates. *Nature* **424:** 805–808.

47. Pawar AP, DuBay KF, Zurdo J, Chiti F, Vendruscolo M, Dobson CM. (2005) Prediction of 'aggregation-prone' and 'aggregation-susceptible' regions in proteins associated with neurodegenerative diseases. *J Mol Biol* **350:** 379–392.

48. Booth DR, Sunde M, Bellotti V, Robinson CV, Hutchinson WL, Fraser PE, Hawkins PN, Dobson CM, Radford SE, Blake CCF, Pepys MB. (1997) Instability, unfolding and aggregation of human lysozyme variants underlying amyloid fibrillogenesis. *Nature* **385:** 787–793.

49. Dumoulin M, Kumita JR, Dobson CM. (2006) Normal and aberrant biological self-assembly: Insights from studies of human lysozyme and its amyloidogenic variants. *Accts Chem Res* **39:** 603–610.

50. Dumoulin M, Last AM, Desmyter A, Decanniere K, Canet D, Spencer A, Archer DB, Muyldermans S, Wyns L, Matagne A, Redfield C, Robinson CV, Dobson CM. (2003) A camelid antibody fragment inhibits amyloid fibril formation by human lysozyme. *Nature* **424:** 783–788.

51. Caughey B, Lansbury Jr, PT. (2003) Protofibrils, pores, fibrils, and neurodegeneration: Separating the responsible protein aggregates from the innocent bystanders. *Annu Rev Neurosci* **26:** 267–298.

52. Bitan G, Kirkitadze MD, Lomakin A, Vollers SS, Benedek GB, Teplow DB. (2003) Amyloid beta-protein (Abeta) assembly: Abeta 40 and Abeta 42 oligomerize through distinct pathways. *Proc Natl Acad Sci USA* **100:** 330–335.

53. Smith JF, Knowles TP, Dobson CM, MacPhee CE, Welland ME. (2006) Characterization of the nanoscale properties of individual amyloid fibrils. *Proc Natl Acad Sci USA* **103:** 15806–15811.

54. Tanaka M, Collins SR, Toyama BH, Weissman JS. (2006) The physical basis of how prion conformations determine strain phenotypes. *Nature* **442:** 585–589.

55. Nilsson MR, Driscoll M, Raleigh DP. (2002) Low levels of asparagine deamidation can have a dramatic effect on aggregation of amyloidogenic peptides: Implications for the study of amyloid formation. *Protein Sci* **11:** 342–349.

56. Schlunegger MP, Bennett MJ, Eisenberg D. (1997) Oligomer formation by 3D domain swapping: A model for protein assembly and misassembly. *Adv Protein Chem* **50:** 61–122.

57. Morozova-Roche LA, Zamotin V, Malisauskas M, Ohman A, Chertkova R, Lavrikova MA, Kostanyan IA, Dolgikh DA, Kirpichnikov MP. (2004) Fibrillation of carrier protein albebetia and its biologically active constructs: Multiple aligomeric intermediates and pathways. *Biochemistry* **43:** 9610–9619.

58. Lashuel HA, Hartley D, Petre BM, Walz T, Lansbury Jr, PT. (2002) Neurodegenerative disease: Amyloid pores from pathogenic mutations. *Nature* **418:** 291.

59. Dobson CM. (2003) Protein folding and disease: A view from the first horizon symposium. *Nature Rev Drug Disc* **2:** 154–160.

60. Krebs MR, MacPhee CE, Miller AF, Dunlop IE, Dobson CM, Donald AM. (2004) The formation of sperulites by amyloid fibrils of bovine insulin. *Proc Natl Acad Sci USA* **101**: 14420–14424.

61. True HL, Lindquist SL. (2000) A yeast prion provides a mechanism for genetic variation and phenotypic diversity. *Nature* **407**: 477–483.

62. Chapman MR, Robinson LS, Pinkner JS, Roth R, Heuser J, Hammar M, Normark S, Hultgren SJ. (2002) Role of *Escherichia coli* curli operons in directing amyloid fiber formation. *Science* **295**: 851–855.

63. Kelly JW, Balch WE. (2003) Amyloid as a natural product. *J Cell Biol* **161**: 461–462.

64. Broome BM, Hecht MH. (2000) Nature disfavours sequences of alternative polar and non-polar amino acids: Implications for amyloidogenesis. *J Mol Biol* **296**: 961–968.

65. Chiti F, Taddei N, Baroni F, Capanni C, Stefani M, Ramponi G, Dobson CM. (2002) Kinetic partitioning of protein folding and aggregation. *Nature Struct Biol* **9**: 137–143.

66. Macario AJL, Macario EC. (2002) Sick chaperones and ageing: A perspective. *Ageing Res Rev* **1**: 295–311.

67. Ramirez-Alvarado M, Merkel JS, Regan L. (2000) A systematic exploration of the influence of the protein stability on amyloid fibril formation *in vitro*. *Proc Natl Acad Sci USA* **97**: 8979–8984.

68. Prusiner SB. (1997) Prion diseases and the BSE crisis. *Science* **278**: 245–251.

69. Taylor JP, Hardy J, Fischbeck KH. (2002) Toxic proteins in neurodegenerative disease. *Science* **296**: 1991–1995.

70. Walsh DM, Klyubin I, Fadeeva JV, Cullen WK, Anwyl R, Wolfe MS, Rowan MJ, Selkoe DJ. (2002) Naturally secreted oligomers of amyloid beta protein potently inhibit hippocampal long-term potentiation *in vivo*. *Nature* **416**: 535–539.

71. Bucciantini M, Giannoni E, Chiti F, Baroni F, Formigli L, Zurdo J, Taddei N, Ramponi G, Dobson CM, Stefani M. (2002) Inherent cytotoxicity of aggregates implies a common origin for protein misfolding diseases. *Nature* **416**: 507–511.

72. Baglioni S, Casamenti F, Bucciantini M, Luheshi L, Taddei N, Chiti F, Dobson CM, Stefani M. (2006) Pre-fibrillar amyloid aggregates could be generic toxins in higher organisms. *J Neurosci* **26**: 8160–8167.

73. Kayed R, Head E, Thompson JL, McIntire TM, Milton SC, Cotman, CW, Glabe CG. (2003) Common structure of soluble amyloid oligomrs implies common mechanisms of pathogenesis. *Science* **300:** 486–489.

74. Kayed R, Glabe CG. (2006) Conformation-dependent anti-amyloid oligomer antibodies. *Methods Enzymol* **413:** 326–344.

75. Stefani M, Dobson CM. (2003) Protein aggregation and aggregate toxicity: New insigh   to protein folding, misfolding diseases and biological evolu   *l Med* **81:** 678–699.

76. Sherman MY, G   berg AL. (2001) Cellular defenses against unfolded proteins: A cell biologist thinks about neurodegenerative diseases, *Neuron* **29:** 15–32.

77. Muchowski PJ, Schaffar G, Sittler A, Wanker EE, Hayer-Hartl MK, Hartl FU. (2000) Hsp70 and Hsp40 chaperones can inhibit self-assembly of polyglutamine proteins into amyloid-like fibrils. *Proc Natl Acad Sci USA* **97:** 7841–7846.

78. Csermely P. (2001) Chaperone overload is a possible contributor to "civilization diseases". *Trends Gen* **17:** 701–704.

79. Dobson CM. (2002) Getting out of shape — protein misfolding diseases. *Nature* **418:** 729–730.

80. Tartaglia GG, Pechmann S, Dobson CM, Vendruscolo M. (2007) Life on the edge: A link between gene expression levels and aggregation rates of human proteins. *Trends Biochem Sci* **32:** 204–206.

81. Bilen J, Bonini NM. (2005) Drosophila as a model for human neurodegenerative disease. *Annu Rev Genet* **39:** 153–171.

82. Finelli A, Kelkar A, Song HJ, Yang H, Konsolaki M. (2004) A model for studying Alzheimer's Abeta42-induced toxicity in *Drosophila melanogaster. Mol Cell Neurosci* **26:** 365–375.

83. Crowther DC, Kinghorn KJ, Miranda E, Pase R, Curry JA, Duthie FA, Gubb DC, Lomar DA. (2005) Intraneuronal Abeta, non-amyloid aggregates and neurodegeneration in a Drosophila model of Alzheimer's disease. *Neuroscience* **132:** 123–135.

84. Iijima K, Liu HP, Chiang AS, Hearn SA, Konsolaki M, Zhong Y. (2004) Dissecting the pathological effects of human Abeta40 and Abeta42 in Drosophila: A potential model for Alzheimer's disease. *Proc Natl Acad Sci USA* **101:** 6623–6628.

85. Luheshi LM, Tartaglia GG, Brorsson AC, Pawar AP, Watson IE, Chiti F, Vendruscolo M, Lomas DA, Dobson CM, Crowther DC. (2007) Systematic *in vivo* analysis of the intrinsic determinants of amyloid-beta pathogenicity. *PLoS Biol* **5:** e290.

86. Dodart JC, Bales KR, Gannon KS, Greene SJ, DeMattos RB *et al.* (2002) Immunization reverses memory deficits without reducing brain Abeta burden in Alheimer's disease model. *Nature Neurosci* **5:** 452–457.

87. Westerman MA, Cooper-Blacketer D, Mariash A, Kotilinek L, Kawarabayashi T, Yourkin LH, Carlson GA, Yourkin SG, Ashe KH. (2002) The relationship between Abeta and memory in the Tg2576 mouse model of Alzheimer's disease. *J Neurosci* **22:** 1858–1867.

88. Cohen E, Bieschke J, Perciavalle RM, Kelly JW, Dillon A. (2006) Opposing activities protect against age-onset proteotoxicity. *Science* **313:** 1604–1610.

89. Lansbury PT. (1999) Evolution of amyloid: What normal protein folding may tell us about fibrillogenesis and disease. *Proc Natl Acad Sci USA* **96:** 3342–3344.

90. Luheshi LM, Crowther DC, Dobson CM. (2008) Protein misfolding and disease: From the test tube to the organism. *Curr Opin Chem Biol* **12:** 25–31.

91. Dobson CM. (2006) An accidental breach of a protein's natural defences. *Nature Struct Mol Biol* **13:** 295–297.

92. Cohen FE, Kelly JW. (2003) *Nature* **426:** 905–909.

93. Dobson CM. (2004) In the footsteps of alchemists. *Science* **304:** 1259–1262.

94. Schenck D. (2003) Amyloid-beta immunotherapy for Alzheimer's disease: The end of the beginning. *Nature Rev Neurosci* **4:** 49–60.

95. Dumoulin M, Dobson CM. (2005) Probing the origins, diagnosis and treatment of amyloid disease using antibodies. *Biochimie* **86:** 589–600.

chapter 14

A Systems Approach to Medicine Will Transform Healthcare

Leroy Hood*

A new approach to biology has emerged over the past ten years or so termed systems biology. This approach will not only transform how we approach biology, but how we understand and deal with disease as well. I will discuss why systems biology will transform biology in the 21st century, the emergence and essential principles of systems biology and then focus on its impact on medicine and how it will transform our current largely reactive medicine to a medicine that is predictive, personalized, preventive and participatory (P4). I will discuss this in a somewhat personal manner since I have been so intimately involved with the conceptualization and proof of principle for systems biology for the past 20 some years and for P4 medicine for the past five years.

1. Biology in the 21st Century

Biology will be a dominant science in the 21st century — just as chemistry was in the 19th century and physics was in the 20th century — and for a fascinating reason. The dominant challenge for all the scientific and engineering disciplines in the 21st century will be complexity — and biology is uniquely in a position now to solve the deep problems of its complexity and begin to apply this knowledge to the most challenging issues facing humankind. Biology will use systems approaches (holistic as contrasted with atomistic) and powerful new measurement and visualization technologies, as well as the new computational and mathematical tools that are emerging

*Institute for Systems Biology, 1441 North 34th St., Seattle, WA 98103-8904, USA, e-mail: hood@systemsbiology.org.

in the aftermath of the human genome project. Systems biology is a new approach to biology that attempts to understand biology in the context of its systems rather than studying genes and proteins one at a time, the *modus operenti* for biology for the past 40 years or so. Biology will use the fact that our examples of biological complexity can be tested by experimentation and analyzed computationally and mathematically. Biology will also use the emerging insight that biology can be viewed as an informational science.[1-5]

Solving the complexities of biology will enable scientists to achieve two important objectives: (1) biology will begin to solve some of humankind's most challenging problems, including healthcare for all, agriculture to provide food for all, nutrition and bio-energy; (2) biology will bring to the other scientific and engineering disciplines solutions to many of their most vexing problems — integrative strategies for computing to computer science; striking new chemistries to chemistry; molecular level machines for manipulating matter and measurements to engineering; new materials to material sciences; new ways of thinking about complexity to physics; and new ways of deducing relevant historical pasts to geology and archeology, etc. Thus 21st century biology will enrich most of the other scientific and engineering disciplines. This biological treasure trove of knowledge exists because biology has had 3 billion years of evolutionary trial and error to create, test and perfect novel scientific strategies and engineering solutions for most of the central problems of life, and these can obviously have broad applications in the other scientific and engineering disciplines.

2. How Might We Think About Systems Approaches to Complexity?

Suppose you were an engineer and wanted to understand how a radio converted radio waves into sound waves — how would you go about studying this phenomena? You might take a systems approach to this problem — first by taking the radio apart and attempting to understand the functions of its individual parts. This is generally what molecular and cellular biology has done for the past 40 years: studying individual genes and proteins. For the first time, the Human Genome Project in 2004 gave us a complete parts lists of all the genes, and by inference, all the proteins of organisms whose genomes had been sequenced. Second, you might assemble the parts into their electrical circuits and study what the individual circuits and integrated circuits can do in terms of converting radio into sound waves. In just this manner we can take a systems approach to human biology and medicine by studying not just the individual genes and proteins, but also the dynamical biological networks, proteins interacting with one another and with other biologically relevant molecules, that mediate the fundamental processes of life. The study of these networks lies at the heart of a new biological discipline termed systems biology.

3. Molecular Systems Biology — Initial Personal Thoughts and Conceptualization

Systems approaches to biology have been around for more than 100 years. In the late 1800s physiologists were interested in homeostasis in living organisms — certainly a systems problem. In the 20th century development biologists, immunologists and neurobiologists were taking systems approaches to their respective problems. However, with the advent of molecular biology and the human genome project, systems thinking could take a decidedly more molecular and more global approach.

For me, this molecular systems thinking began in the late 1970s and early 1980s. My labs at Caltech were next to those of Max Delbruck — a Nobel-Prize-winning physicist who pioneered quantitative biology. Max was never very impressed with immunology, my scientific topic of study at the time. He was skeptical about whether the deep problems of immunology, the immune response, tolerance and autoimmunity, could ever be solved by a one-gene or one-protein at a time approach. I argued that we had done very well in coming to understand the mechanisms of antibody diversity — and he said that diversity was an easy problem and not a deep problem of immunology. I gradually concluded that Max was at least in part correct, and started thinking about what it would take to successfully attack these big problems in immunology (which today still remain largely unsolved). I decided that one needed to study the systems of genes or proteins that mediate immune mechanisms, and not just individual genes or their protein counterparts. But we did not have the tools for identifying the components of biological systems and their interactions in the mid 1980s, nor did we have a complete parts list of all genes of humans or other model organisms, and by inference, all proteins. Clearly one needed to be able to do far more comprehensive analyses of the behaviors of mRNAs and proteins (e.g., changes in structure, expression levels, interactions and cellular localizations). This systems thinking also made the idea of the genome project attractive — because one could completely (more or less) define a parts list of genes and, by inference, mRNAs and proteins for any organism whose genome was sequenced — and thus hope to carry out comprehensive or global analyses (see below). I submitted a couple of systems-like biology grants to NSF that did not do very well because the reviewers were unclear as to what I was proposing (it was not clear that I completely understood what I had proposed either). In 1991 I wrote a chapter in the book *Code of Codes*[1] (about the human genome project that Dan Kevles and I co-edited) which contained a description of systems biology that could be used today (although I did not start using the term "systems biology" until a few years later). But our current view of systems biology emerged slowly over the late 1980s and early 1990s, fueled by the creation of an NSF Science and Technology Center initiated at Caltech in 1989 to integrate biology and technology, and then further matured by my establishing the cross-disciplinary department of Molecular Biotechnology at the University of Washington in 1992. Finally in 2000 I co-founded the independent Institute for Systems Biology (ISB) — perhaps the first organization focused entirely on systems biology.

4. Biology as an Informational Science and ISB

At ISB, we articulated the vision of transforming modern biology by pioneering the systems approaches to decipher biological complexity and by viewing biology as an informational science. We further argued that through the convergence of comprehensive systems approaches to biology (combining both holistic and reductionist approaches), the development of new technologies and the creation of powerful new computational/mathematical tools, the complexity of biology could be penetrated and understood. At ISB we wanted to pioneer and integrate approaches and tools for each of these areas. The view of biology as an informational science provides a powerful conceptual framework for dealing with complexity. Let me discuss some of the central concepts of the informational view of biology.

First, there are two fundamental types of biological information: the digital information of the genome and the outside or environmental information that impinges on and modifies the digital information. Indeed, it is the digital core of knowable information that distinguishes biology from all of the other scientific disciplines — for none of the others have this digitally encoded and hence readily decipherable core of information. Systems biology attempts to understand the integration of the digital and environmental information that mediates the three fundamental processes of life, evolution, development (in humans going from one cell in the fertilized egg to 10^{14} cells of many different types in the adult), and physiology (e.g., the immune response to an infection). Hence, the deep role of systems biology is to integrate the information from the digital genome and the environment to understand how life unfolds.

Second, biological information is captured, processed, integrated and transferred by biological networks — interacting sets of RNAs, proteins, the control regions of genes and small molecules — to the simple and complex biological molecular machines that actually execute the functions of life. Thus an understanding of the dynamical operation of biological networks in the context of evolution, development, physiology, or even disease is one of the central foci of ISB, as is understanding how molecular machines are constructed and function.

Third, biological information is encoded by a multiscale information hierarchy: DNA, RNA, proteins, interactions, biological networks, cells, tissues and organs, individuals, and finally, ecologies. The important point is that the environment impinges upon each of these levels of the information hierarchy and modulates the digital informational output from the genome. Hence, to understand how systems operate at a particular level — say understanding the 50 or so proteins that mediate the cell cycle — one should capture in a global manner each of the levels of information that lie between the phenotypic measurement (observable features of the cell cycle) and the core digital genome. And the information at each of these levels should be integrated in such a manner that the environmental modifications are identified, so as to understand how they impact the functioning of the systems.

Note that each of these levels of information pose chemical and technical challenges for their global (comprehensive) analysis, and this opens up the possibility of bringing in emerging technologies — microfluidics, nanotechnology, *in vivo* and *in vitro* molecular imaging, new chemistries for creating protein capture agents, and a host of others. To give one concrete example, we need to develop many new global chemistries to study proteins as they dynamically execute their functions: these measurements include their structure, their expression patterns, their chemical processing and modification, their half-lives, their interactions with other informational molecules and small molecules, their locations in the cell, their dynamically changing three-dimensional structures, and so on. It is obvious that many different aspects of chemistry will play a vital role in creating these new tools of systems biology. The capture, validation, storage, analysis, integration, visualization and graphical or mathematical modeling of data sets — dynamically captured and global in nature — pose a host of computational and mathematical challenges.

5. The Culture and Philosophy of Science at ISB

Some powerful new ideas that have begun to effectively integrate biology, medicine, technology and computation/mathematics are key to the development of ISB as an institution. The first three of these ideas are fundamentally scientific, the rest are strategic, and guide the implementation of the development of the institution.

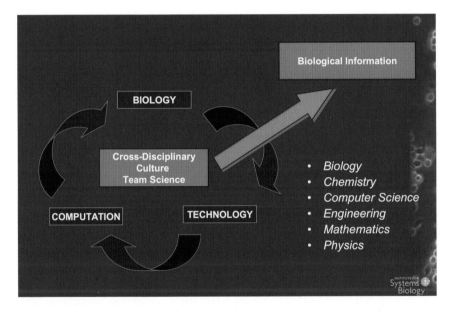

Fig. 1. A schematic diagram of the "virtuous cycle" of systems biology, namely, that the frontier problems of biology and medicine should drive the nature of the technologies developed and the measurements (or visualizations) arising from these new technologies should drive the nature of the computational and mathematical tools that are developed. Finally, the technologies and computational tools feed back on the biology and transform the data space that can be searched to understand the relevant biology. The realization of the virtuous cycle in an institution such as ISB requires a cross-disciplinary environment where biologists, chemists, computer scientists, engineers, mathematicians and physicists all learn one another's languages and work together in integrated teams to generate biological information — and hence new knowledge.

(1) The frontier problems of biology should dictate what technologies should be developed (making it possible to view new dimensions of biological data space). Likewise, the creation of new datasets from these technologies should drive the development of new computational and mathematical techniques for analyzing them (Fig. 1). New technologies and computational tools in turn enable the understanding of new levels of biological complexity. Thus the dynamics of emerging new technologies frame the rate at which new biological insights are generated.

(2) This integration of biology, technology and computation necessitates the creation of a cross-disciplinary environment bringing in close proximity biologists, chemists, computer scientists, engineers, mathematicians, physicians and physicists in an environment where they can learn one another's languages, work together in teams, and the non-biologists can learn relevant biology in a deep manner (Fig. 1). ISB is still struggling to learn how to perfect this difficult-to-achieve cross-disciplinary culture.

(3) Biology spans a spectrum of complexity, ranging from simpler model organisms — single-celled bacteria, yeast or bacteria — to more complex model organisms (mice), and ultimately humans. Most important is that biological experimentation is easier in simple species. Hence, ISB develops new measurement tools and computational approaches using simpler model organisms and then must learn how to apply these tools to higher model organisms — and ultimately to human complexity.

(4) ISB is deeply committed to an open-source philosophy: making our data and our tools readily available to the scientific community, and taking advantage of the collective input of this community to improve those tools.

(5) ISB is committed to remaining small, focused and highly interactive. To compensate for the limitations this philosophy imposes on faculty size, ISB needs to have an immediately available reservoir of expertise — a critical mass — in relevant biology, chemistry, computer science, engineering, mathematics and medicine residing within our senior scientists as well as our faculty members.

(6) One approach to attacking big scientific problems while remaining relatively small is to create strategic partners that bring to ISB the scientific, technological, computational and the medical expertise that we lack. We focus on partnering with the very best. ISB searches for these partners among academics, industry, and research institutions, as well as relevant institutions in foreign countries.

(7) ISB is determined to transfer its relevant knowledge to society — whether it is K-12 science education, spin-off companies employing ISB's novel biological or technology insights, or the education of society in science and technology.

(8) ISB believes in providing the scientific leadership that is necessary for catalyzing paradigm changes in how biology and medicine are carried out and in how biology and medicine are organized — and we have a long history of doing so.

6. What Are the Unique Features of Contemporary Systems Biology?

Systems approaches should be global, integrative and capture the dynamical nature of biological networks. Global or comprehensive measurements mean that the behaviors of all genes, all mRNAs, all proteins, all interactions, all small molecules, etc. be measured. Global measurements are possible with genomic measurements, but not yet with proteomic measurements. Integrative means that different hierarchical levels of biological information (e.g., DNA, RNA, protein, interactions, phenotypes, etc.) can be integrated together so as to capture the environmental contributions to each level of information — for only by understanding how the digital and environmental information are comprehensively integrated can one hope to understand how systems function. All biological networks change in a dynamical manner, and one of the grand challenges of systems biology is to be able to capture this dynamic behavior either graphically or mathematically. The systems approach to biology requires that the relevant components of the system be perturbed genetically and/or environmentally — and the response measured globally, dynamically and integratively. From this, one formulates models, either graphically or mathematically, whose validity can be further tested by additional perturbations, measurements and integrations. Thus this cycle is repeated until models can be formulated that are in accord with the experimental data. We note that data space is virtually infinite and hence one must carefully formulate relevant perturbations to explore just those regions of data space relevant to the system of interest. Conversely, not all global data sets have been collected in the right data space to draw correct inferences about the functioning of the systems of interest. Hence, care must be used in the integration of different data sets. These systems approaches can be applied to medicine (or disease) with very powerful results.

7. Systems Medicine Opens Up New Possibilities for Understanding Disease Mechanism and Providing New Approaches to Diagnosis, Therapy and Prevention

The central premise of systems medicine is that normal biological networks become disease perturbed by genetic and/or environmental factors — such a change alters the patterns of expression of the dynamically changing mRNAs and proteins they control during the course of disease progression. These altered patterns of information expression encode the dynamic pathophysiology of disease progression. Moreover, this view of disease opens up new possibilities for thinking about diagnosis, therapy and even prevention (Fig. 2).

A disease perturbation could be the result of specific, disease-causing DNA mutations, pathogenic organisms, or other pathological environmental factors such as toxins. Molecular fingerprints of pathological processes can take on many molecular forms, and can be identified by the analyses of proteins,[6,7] DNA,[8] RNA,[9] and metabolites,[10] as well as

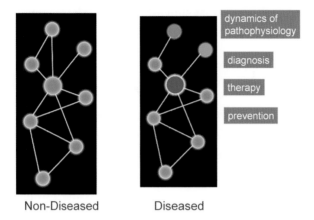

Non-Diseased   Diseased

Fig. 2. A schematic diagram of a normal biological network and a disease-perturbed biological network. The nodes represent mRNAs or proteins and the edges indication the nature of the interactions between nodes. The disease-perturbed network changes the levels of the information expressed as compared against the normal network (blue indicates normal levels and other colors indicate changed levels). Note that architecture of the disease-perturbed network has also changed in that the edge (relationship) between the green and red nodes has disappeared.

informative, post-translational modifications to these molecules such as protein phosphorylation.

Signals related to health and disease can be found in multiple sites. For instance, many bodily fluids such as the blood, urine, saliva, cerebral spinal fluid and so forth can be sampled to identify evidence of perturbed molecular fingerprints reflecting the altered expression patterns of genes and proteins in disease-perturbed biological networks. Of these, the blood is likely the most information rich organ (or fluid) in that it bathes all of the tissues in the body, and it is easily accessible for diagnostic procedures. In addition to the biomolecules secreted into the blood from cells and tissues throughout the body, the transcriptomes and proteomes of cells circulating in the blood (e.g., white blood cells) are also potentially an abundant source of biomedically important information.[11] Thus, the amount of information available in the blood about health and disease is enormous if we learn to read and interpret the molecular signals that reflect the operations of normal and disease-perturbed biological networks.

8. A Systems Approach to Prion Disease in Mice

As an example of an environmental disease perturbation, we have studied the dynamic onset of the infectious prion disease in several strains of mice at the level of mRNA in the brain (the affected organ), and have shown that a series of interlocking protein networks which surround the prion protein are significantly and dynamically perturbed across the 150-day span from disease initiation to death. Figure 3 shows differentially altered networks that were derived from comparisons of mRNA-expression patterns in normal and diseased brains at each of three time points after infection. They

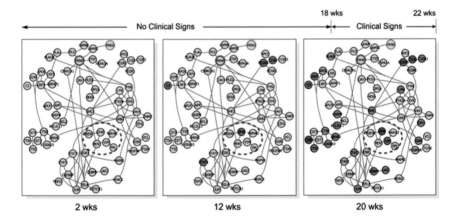

Fig. 3. A schematic model of a portion of the biological network for prion replication and accumulation. The nodes represent 67 different mRNA transcripts. The three panels represent the network at three different time points: 2 weeks, 12 weeks and 20 weeks (the animals die of prion infection at about 23 weeks). Yellow indicates that normal and diseased animals have identical transcript levels; red indicates that the disease transcripts are elevated in levels over their normal counterparts; and green indicates the disease transcripts are decreased in levels over their normal counterparts. These are unpublished data from Inyoul Lee, Daehee Huang, Hyuntae Yoo, George Carlson and Lee Hood.

involve 67 proteins in the prion replication and accumulation network. The initial network changes occur well before the clinical signs of the disease can be detected.

These dynamically changing, disease-perturbed networks lead to two important conclusions. First, some significant network nodal points change before the related clinical or histological changes are evident. Therefore, labeled biomarkers that are specific for the changing nodes or the biological processes they regulate could be used for *in vivo* imaging diagnostics even before symptoms arise, as has already been shown in patients.[6,7,12–14] Alternatively, if some of these altered nodes encode secreted proteins, they could provide readily accessible *in vitro* diagnostic blood markers for early disease detection. Second, many of the sub-networks of proteins that change during the onset of disease affect changes in phenotypic traits that are consistent with the pathology of the disease — that is, the network behavior of the disease-perturbed networks explains the pathophysiology of the disease. About 900 perturbed mRNAs appear to encode the core prion-disease response and only about 25% of these encode known pathology — the remainder encode new disease mechanisms previously unknown. During the progression of the prion-disease process, several hundred transcripts undergo significant gene expression changes well before the clinical signs of prion disease. Many of these potential early-disease "sentinels" are predicted to be secreted into the blood and therefore represent potential protein biomarkers for early disease diagnosis through blood protein analysis.

Another key issue that needs to be addressed is how best to decouple two primary dimensions of disease information: namely, disease stratification (e.g., which type of the several types of prostate cancer is present) and disease progression (e.g., stage of development of a particular prostate cancer stratification in time). Molecular signature changes can indicate the presence of different diseases (stratification) as well as their stage (progression). Thus, a key challenge in future will be to develop analysis tools that will enable us to distinguish the nature of a given individual's disease with regard to both stratification and progression states.

Let us now consider how a systems view of disease opens up new possibilities for the diagnosis of disease.

9. Blood — The Ideal Body Fluid for Protein Diagnostics

The blood is an ideal organ for identifying biomarkers because it interacts with virtually every organ and major tissue type in the body, and each secretes proteins into the blood. Moreover, the blood is easily accessible for diagnostic studies — and presumably the analysis of changes in protein levels or protein modifications that could serve as surrogates or reflections of health and disease. The blood is an enormously complex organ; it contains hundreds of thousands to millions of proteins whose concentrations range over 12 orders of magnitude. Since for proteins there is no equivalent to the polymerase chain reaction (PCR) amplification procedure, only the more abundant proteins can be seen by conventional protein analysis studies. It should also be pointed out that many different features of proteins ultimately must be characterized or quantified: these include protein identification, quantification, chemical modifications, alternative mRNA splicing products, cellular localizations, three-dimensional structures (and their dynamics), as well as their functions. We here are only concerned with the identification and quantification of proteins. Because diseases arise from disease-perturbed networks, proteins secreted by various organs into the blood can reflect the nature and location of disease in individuals.

There are two general current approaches to the analysis of protein mixtures. In one case protein-capture agents such as antibodies are used — and with appropriate controls these approaches can quantify the levels of proteins. These techniques include Western blot analyses, ELIZA assays, surface plasmon resonance and protein chips. A second approach is to use mass spectrometry. Since proteins have large masses that are difficult to analyze accurately by mass spectrometry, most proteins are analyzed after converting them into tryptic peptides (or other proteolytic fragments) whose masses are far smaller and hence more easily and accurately analyzed. Because blood is such a complex mixture, proteins of interest are often enriched by some type of fractional procedure (charge, hydrophobility, size and/or the presence of modifications such as glycosylation or phosphorylation) before (or after) their conversion into peptides and analysis by mass spectrometry. Isotopically labeled and synthesized peptides may be used to identify and quantify (relative or absolute) the same peptides from unknown samples (and hence one obtains a proxy for the quantification of their corresponding proteins) very effectively by mass spectrometry. Advances in mass spectrometry instrumentation and strategies for analysis have transformed the analysis of proteomes over the past 20 years and just in the last few years the progress has been particularly striking. These advances can be understood from a variety of recent reviews.[15,16]

The important point is that blood protein diagnostics requires two distinct technologies: discovery techniques to identify and validate useful blood biomarkers and typing techniques that can determine biomarker concentrations for perhaps thousands of different proteins in millions (or more) of different patients. The discovery techniques include the antibody- and mass spectrometry-based approaches discussed above. The typing techniques will require microfluidic and nanotechnology-based techniques discussed below.

10. A New Systems Approach to Diagnostics That Makes Blood a Window into Health and Disease — Organ-Specific Blood Protein Fingerprints

We discuss the organ-specific blood protein fingerprint approach to diagnostics because it lies at the very heart of the predictive medicine that will emerge over the next ten years. Our feeling is that this will be one of the central data gathering strategies for predictive and personalized medicine and, accordingly, we will discuss it in some detail. The idea is that disease arises from dynamically changing disease-perturbed networks, that the diseased organs will secrete proteins into the blood, and that if these proteins are encoded by disease-perturbed networks, their levels of expression in the blood will be altered in a manner that reflects the specific nature of the disease. Indeed, this idea is the basis for the very broad range of blood biomarker studies that are being carried out by many scientific centers.[17–21] The major difficulty with this simple view is that if you identify multiple blood proteins whose changes are specific for a particular disease state (as compared against normal controls) and then examine the same blood markers for that disease in bloods drawn from individuals, with say ten other pathologies, these marker proteins can change in unpredictable ways since multiple organs control the expression of most blood proteins and these organs respond differently to various environmental signals. The important point is that if a marker protein synthesized in five organs changes in the blood, we cannot be certain which organ(s) is(are) the origin of the change. While that marker may sample a biological network relevant to disease diagnosis, since its origin is not clear its use as a disease biomarker may raise more questions than it answers. The solution to this dilemma is clear: employ organ-specific blood protein biomarkers whose changes must therefore reflect changes only in the organ itself. If enough of these organ-specific blood proteins are sampled, they will represent a survey of many different biological networks in the organ of interest and will provide sufficient diagnostic information for any disease. Eventually, correlation of these organ specific biomarkers with more general biomarkers may prove to be the best long-term strategy for developing diagnostic fingerprints.

We have several lines of preliminary evidence that suggest this organ-specific blood protein approach will be effective. In prostate cancer, for example, there are disease-mediated altered patterns of mRNA and protein expression in the prostate. Some of these genes are expressed primarily in the prostate (organ-specific products) and some of these organ-specific proteins are secreted into the blood, where they collectively constitute a protein molecular fingerprint comprised of say 100 or more proteins whose relative concentration levels probably report the status of the biological networks in the prostate gland. We have demonstrated that changes in the blood concentrations of several of these prostate-specific blood proteins reflect the various stages of prostate cancer and, as discussed above, various brain-specific blood proteins also reflect the progression of prion disease in mice. We propose that the distinct expression levels of the individual proteins in each fingerprint represent a multi-parameter (and therefore potentially information rich) diagnostic indicator reflecting the dynamic behavior of, for example, the disease-perturbed networks from which they arise. The analysis of 50 or so organ-specific proteins should allow us to both stratify diseases in the organ as well as determine their stage of progression. We have identified 10s to 100s of organ-specific transcripts in each of the 40 or so organs that we have examined in mouse and human.

We can envision a time over the next 5–10 years when 50 or so organ-specific blood proteins will be identified for each of the 50 or so major organs and tissues in humans — so that computational analyses of the relative concentrations of the protein components in these organ-specific fingerprints will enable blood to become the primary window into health and disease. When we analyze data from these blood indicators, we also may be in a position to identify the dynamically changing disease-perturbed network(s). The analysis of these dynamic networks will allow us to study in detail the pathophysiology of the disease response and hence be in a position to think of new approaches to therapy and prevention. To generate the ability to assess simultaneously all 50 organ blood protein fingerprints in patients, we ultimately need to develop the microfluidic or nanotechnology tools for making perhaps 2500 rapid protein measurements (e.g., 50 proteins from each of 50 human organs) from a droplet of blood. Nanotechnology is necessary because only this severe miniaturization can allow the necessary thousands of measurements from a single drop of blood. In order to reach this stage we will also have to create the appropriate computational tools to capture, store, analyze, integrate, model and visualize the information arising from this approach.

11. Technologies That Will Transform P4 Medicine

There are several emerging technologies that will catalyze the emergence of P4 medicine — thus transforming the practice of healthcare. These technologies include the use of microfluidic and nanotechnology approaches to generate more sensitive protein measurement technologies, to increase the throughput of DNA sequencing, and to do single-cell analyses. These also include new *in vitro* and *in vivo* imaging technologies that I will not discuss here.

11.1. Blood Protein Measurements

P4 medicine will require that *in vitro* measurements of thousands of blood proteins be executed rapidly, automatically and inexpensively on small blood samples. This will require miniaturization, parallelization, integration and automation of tissue and blood purification and measurement technologies. In short, it will require making *in vitro* blood measurements inexpensively — perhaps for pennies or less per protein measured. This need for inexpensive *in vitro* measurements is driving the development of integrated microfluidics and nanotechnologies for *in vitro* diagnostics. We further argue that the complex changes in these protein levels must be analyzed computationally to search for the patterns that correlate with particular diseases — and indeed to stratify each disease as well as determine its stage of progression.

To meet the expanding requirements of systems medicine, *in vitro* measurements of proteins (and mRNAs, cells, etc.) whether for fundamental biological studies or for pathological analysis, must not just be inexpensive, but must also be sensitive, quantitative, rapid, and executed on very small quantities of tissues, cells, serum, etc. We are working towards developing chip platforms that can take a few microliters (a finger prick) of blood, separate the plasma from the serum, and then measure on the order of 1000 or more proteins and/or mRNAs quantitatively, with high sensitivity and specificity, and in a few minutes time. To illustrate how far we have come, let us discuss a new microfluidic approach to protein chip assays, termed DNA-encoded antibody libraries (DEAL) developed in the laboratory of my long-term collaborator, Jim Heath.[22]

The primary advantage of DEAL is that it uses a single, robust chemistry — that of DNA hybridization — to spatially localize and detect proteins, mRNAs, and cells, all in a multiplexed fashion (Fig. 4). Antibodies are typically too fragile to survive the fabrication procedures associated with assembling robust microfluidics chips, but DNA oligomers are significantly more robust. DEAL thus enables the detection of panels of protein biomarkers within a microfluidics environment and from very small quantities of biological material (100 nanoliters or so).[23] This amount of plasma can be readily separated from whole blood on-chip, thus allowing for the measurement of serum biomarkers from a fingerprick of whole blood. In addition, within the environment of flowing microfluidics, the rate-limiting

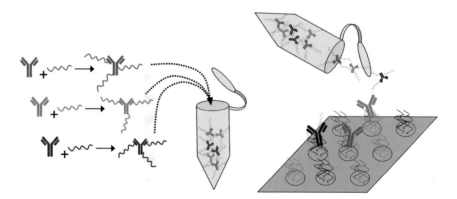

Fig. 4. A schematic diagram of the nature of the DEAL (DNA-encoded antibody libraries) protein chip. The protein capture agents (generally antibodies) are each labeled with a single strand DNA 20-mer and the complement of that 20-mer is covalently attached to a glass or silicon chip to address its respective capture agent. Then the covalently modified capture agents are poured over the protein chip and the individual capture agents are addressed by virtue of their interactions with their complementary 20-mer strands. Thus the red, green and blue protein capture agents can be specifically addressed to different positions on the protein chip.

step in performing a surface immunoassay is the kinetics of the binding of the analyte to the surface-bound capture agent.[24] Thus DEAL-based immunoassays can be executed very rapidly.

The versatility of DEAL also enables multiplexed cell sorting and localization, followed by few- or single-cell measurements of protein, RNA, and other biomolecules.[22] DEAL can be engineered into a highly sensitive and very rapid measurement technique, with a reported detection limit of 10 fM for the protein IL-2 — 150 times more sensitive than the corresponding commercial ELISA assay. This sensitivity can be applied to the isolation of rare cells based on combinations of cell-surface markers, enabling, for example, the isolation and addressing of individual cancer stem cells. DEAL can also be used to make single cell measurements of secreted proteins from each of these isolated single cells. Thus DEAL offers superb sensitivity and the ability to perform spatially multiplexed detection for characterization of rare cell types, such as circulating cancer cells or cancer stem cells. These advantages still face the inherent limitations of antibodies, so the development of new approaches to generating protein-capture agents is a critical part of future development of comprehensive blood diagnostics.

We can envision a time 5–10 years hence when small hand-held devices will be able to make these 2500 measurements from a fraction of a droplet of blood, send them via wireless to a server for analysis and consequently inform the patient and physician as to the status of the patient. While many fundamental scientific challenges remain to be solved before this goal is achieved, none of those challenges appear insurmountable at this point.

11.2. Blood Cell Diagnostics

One will also be able to use the blood cells as powerful diagnostic markers — either of infectious diseases or of genetic diseases. Microfluidic cell-sorting technologies for being able to sort blood cells into their ten or so individual types for analysis are now available. Even more important is the emergence of single-cell analytic tools where DNA, mRNAs or small RNAs, proteins and even metabolites can be analyzed rapidly from individual cells. The cells can also be perturbed with appropriate environmental stimuli to identify defects. It is possible that an appropriate analysis of immune cells (both innate and adaptive) from the blood will reveal important information about past antigenic history of the individual as well as the current state of immunological responsiveness. Thus blood cell diagnostics will be an important tool in the diagnosis of infectious diseases. Similarly, analysis of rare blood cells such as circulating cancer cells may also be utilized to guide therapies.

11.3. High Throughput DNA Sequencing of Individual Genomes

Another type of *in vitro* measurement will be a determination of the complete genome sequence of individuals using any one of a number of possible microfluidic and nanotechnology approaches. One attractive possibility is sequencing single strands of DNA with very little front-end preparation (e.g., no PCR amplification) on a massively parallel basis (billions of DNA strands simultaneously). A device like this will emerge over the next 5–10 years and will allow millions, if not billions, of individual human genome sequences to be determined rapidly, inexpensively and in a massively parallel manner. The 2500 organ-specific blood marker measurements and the complete individual human genome sequences will be the heart of the predictive and personalized medicine that will emerge over the next ten years or so.

12. Predictive, Personalized, Preventive and Participatory (P4) Medicine

The convergence of systems approaches to disease, new measurement and visualization technologies, and new computational and mathematical tools

- **Predictive:**
  - Probabilistic health history — DNA sequence
  - Biannual multi-parameter blood protein measurements
- **Personalized:**
  - Unique individual human genetic variation mandates individual treatment
  - Patient is his or her own control
- **Preventive:**
  - Strategies for re-engineering the behavior of disease perturbed networks with drugs
  - Vaccines
  - Focus on wellness
- **Participatory:**
  - Patient understands and participates in medical choices

Fig. 5. An outline of some important features of the individual components of P4 medicine.

suggests that our current largely reactive mode of medicine (wait until one is sick before responding) will over the next 5–20 years be transformed to P4 medicine[24–27] (Fig. 5). Two major techniques of *predictive* medicine will emerge over the next ten years. Individual genome sequences will be available, and increasingly we will be able to determine the likelihood of one's future health history (e.g., 50% probability of ovarian cancer by age 50). Hand-held devices to prick the finger and quantify 2500 organ-specific proteins from all human organs will send this information wirelessly to a server which in turn will analyze the information and email the patient and their physician a report — thus permitting an instantaneous assessment of current health status, perhaps twice a year. Predictive medicine will result in a *personalized* medicine that focuses on the illnesses of an individual and eventually their wellness, preventing disease rather than treating it. And we must remember that each of us differs on average by about 6 million nucleotides from our neighbors — hence, we are susceptible to differing combinations of diseases — and once again must be treated as individuals. Accordingly, from the assessment of genomes and environmental exposures will emerge initially a predictive and personalized medicine. Later, one will learn how to identify drugs to reengineer disease-perturbed networks — making them behave in a more normal fashion, or at least abrogating the most deleterious of their effects. In the future we will be able to design drugs to *prevent* networks from becoming disease perturbed. Hence, if you have an 80% chance of prostate cancer at 50, taking these preventive drugs beginning at 35 may reduce your disease probability to 2%. We will also take a systems approach to understanding the immune response and in doing so will develop general and powerful new methods for generating vaccines for infectious diseases and perhaps even for cancer. In any case vaccines will become a powerful tool of preventive medicine. Thus preventive medicine will enable a focus on wellness rather than on disease for each individual. Finally, if we can educate patients and their physicians as to the nature of predictive, personalized and preventive medicine with patient-oriented interpretations, this will also enable patients to understand more deeply and actively *participate* in personal choices about illness and wellness. Participatory medicine will necessitate the development of powerful new approaches to handling enormous amounts of personal information in a secure manner, and to a new form of medical education of individual patients as well as their physicians. Over the next 5–20 years medicine will become predictive, personalize, preventive and participatory (P4 medicine). The realization of P4 medicine is a major strategic goal of ISB. The vision of P4 medicine has emerged from ISB and has been delineated in a series of papers.[24,26–28]

13. Impact of P4 Medicine on the Practice of Medicine and Society

The systems view of medicine, will lead to five important revolutions in medicine.

First, in the long term, the organ-specific protein analyses of blood offers the potential to open up a new avenue for studying human biology —

when we learn to read and interpret the information inherent in secreted organ-specific protein patterns. The key patterns in these secreted proteins represent different network perturbations, and hence, different diseases. When secreted proteins enter the blood, they can provide a novel means to study biology in higher organisms and to identify drug targets through linking blood measurements to perturbations in underlying biological networks. Importantly, learning to study perturbations in underlying networks through secreted protein patterns (to whatever resolution is possible) will provide access to studying biological networks *in vivo* that are not amenable to direct experimentation. This may transform how we think about studying normal human development and physiology and the process of aging. Moreover, developing the capacity to identify *in vivo* network perturbations through secreted protein measurements in the blood will open up a new avenue to drug target identification and will provide a novel means to discover the perturbed sub-networks. Since taking blood is relatively non-invasive, it has less potential than biopsies or other invasive techniques of greatly distorting the system being studied. Thus, this approach has the potential to strongly complement existing approaches to study human biology. The effect of drugs, toxicity, human development and even aging may all be amenable to study through the blood if we can learn to read and interpret the signals in the proteins, which has the potential to be very significant in human biology.

Second, as the sensitivity of measurements increases (both *in vitro* and *in vivo*), we will achieve a digitalization of biology and medicine — that is, the ability to extract relevant information content from single individuals, single cells and ultimately single molecules — with its attendant economies of scale. Just as Moore's law led in time to the widespread digitalization of information technologies and communications, the exponentially increasing ability to extract quantized biological information from individual cells and molecules will transform biology and medicine in ways that we can only begin to imagine. For example, suppose that we could accurately measure the state and concentrations of 20,000 proteins in the blood rapidly

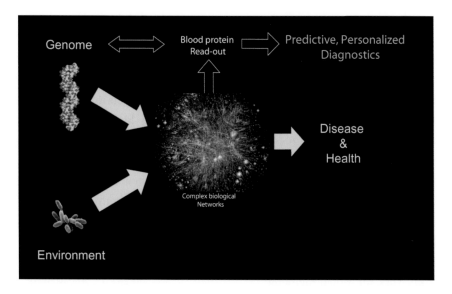

Fig. 6. This schematic diagram illustrates the interaction of the two types of biological information — the genome and the environment — mediate by biological network, and creating phenotypes such as the secretion of proteins into the blood. As is discussed in the text, the measurements of the concentrations of organ-specific blood proteins can provide diagnostic insight as to the state of health (or disease) of the individual. The diagnostic use of organ-specific blood proteins is one of the fundamental pillars of predictive and personalized medicine.

and inexpensively. What might one do with such information? Imagine the integration studies that one could carry out in analyzing how various environmental conditions perturb the basic genomic information (Fig. 6). And imagine what one could learn if we could reverse translate this massive phenotypic (protein) information into the dynamics of the corresponding biological networks. The digitalization of human health will have a far greater impact on society than the digitalization of information technologies and communications — for this digitalization is about improving human health.

Third, all of these changes — a systems approach to medicine with its focus on disease prevention and more efficient drug discovery, the introduction of increasingly inexpensive nanotechnology based diagnostics and *in vivo* measurement technologies, the highly accurate and specific molecular characterization of the systems biology of disease, and the digitalization of medicine — will start to reduce the inexorably increasing costs of healthcare. This can, in principle, enable us to afford to provide for the more than 45 million people in the United States who currently lack health insurance, to reduce the crushing costs of healthcare to society, and to export our digital predictive and preventive medical approach to the developing world. Just as the mobile phone has become a fundamental communication mode in developing countries, and has changed the lives of much of the world's population, so digital medicine of the 21st century can bring to the world's citizens a global and greatly improved state of human health and healthcare. In our view P4 medicine will become the foundational framework for global medicine.

Fourth, P4 medicine will, over the next ten years, transform the business plans of virtually every sector of the healthcare industry — pharmaceutical, biotech, medical instrumentation, diagnostics, IT healthcare, payers, providers, medical centers, medical schools, etc. For example, pharmaceutical companies are generally acknowledged as failing in their quest to produce effective and reasonably priced drugs. Systems medicine will bring presymptomatic diagnostics and the ability to stratify disease so that effective therapies can be successfully matched against specific diseases. It will provide powerful new approaches to assessing drug toxicity early in clinical trials. It will provide new approaches to assessing drug doses for individual patients, and for evaluating drug toxicities at a very early stage. A fascinating question is whether the pharmaceutical companies will be able to effectively employ these strategies of systems medicine? Another challenging example for P4 medicine centers about medical schools — medical schools are currently teaching physicians that will be practicing P4 medicine in the next 10–20 years — and these students are not getting the background and concepts that they will need for P4 medicine. Will medical schools be able to transform their teaching, research and eventually their patient care to encompass the P4 concepts? One might ask similar questions of every sector of the healthcare industry.

Finally, P4 medicine is a big problem that is amenable to powerful cross-disciplinary and systems-approach scientific attack. We have identified about 12 technical challenges, which if solved, would lead to P4 medicine. Likewise, we have identified about eight societal challenges arising from the impact of P4 medicine on society (ethical, legal, societal, regulatory, economic issues) — and they too must be dealt with at the same time the technical challenges are being solved if P4 medicine is to successfully emerge in the next 10–20 years. ISB is in the process of generating strategic partnerships to attack these two different categories of P4 challenges, technical and societal. ISB is exploring a series of fascinating possibilities for strategic partnerships — with individuals, with academic centers, with companies and even possibly with countries — that will not only bring critical scientific, engineering and clinical expertise, but will also open up striking new funding opportunities, and some of these opportunities are global in nature. Indeed, I see a globalization of science emerging over the next ten years or so — just as Tom Freidman described the globalization of the economy. And in a similar manner there are striking opportunities for those who will be at the leading edge of the globalization of the biological and medical sciences — for they can create powerful strategic partnerships leveraging leading-edge technical, biological and medical skills; they can gain access to novel medical material (patients, samples, etc.) and they will be able to recruit the most talented young scientists world-wide to work on a vision that will transform world health.

Acknowledgments

Portions of this paper were adapted from two articles that I recently wrote.[29,30] I would like to acknowledge many fascinating conversations with my close colleagues, David Galas and Jim Heath, on these topics. I also thank Shirley Meinecke for helping prepare the manuscript.

References

1. Kevles DJ, Hood L (eds). *The Code of Codes, Scientific and Social Issues in the Human Genome Project*, pp. 136–163. Harvard University Press, Cambridge, MA.

2. Ideker T, Galitski T, Hood L. (2001) A new approach to decoding life: Systems biology. *Annu Rev Genomics Hum Genet* **2**: 343–372.

3. Hood L, Galas DJ. (2003) The digital code of DNA. *Nature* **421**: 444–448.

4. Ideker T, Thorsson V, Ranish JA, Christmas R, Buhler J, Eng JK, Bumgarner R, Goodlett DR, Aebersold R, Hood L. (2001) Integrated genomic and proteomic analyses of a systematically perturbed metabolic network. *Science* **292(5518)**: 929–934.

5. Weston AD, Baliga NS, Bonneau R, Hood L. (2003) Systems approaches applied to the study of *Saccharomyces cerevisiae* and *Halobacterium sp. The Genome of Homo Sapiens, Volume LXVIII, Symposia on Quantitative Biology*, pp. 345–357. Cold Spring Harbor Press, New York.

6. Wang X, Yu J, Sreekumar A, Varambally S, Shen R, Giacherio D, Mehra R, Montie JE, Pienta KJ, Sanda MG, Kantoff PW, Rubin MA, Wei JT, Ghosh D, Chinnaiyan AM. (2005) Autoantibody signatures in prostate cancer. *N Engl J Med* **353**: 1224–1235.

7. Wang Y, Klijn JG, Zhang Y, Sieuwerts AM, Look MP, Yang F, Talantov D, Timmermans M, Meijer-van Gelder ME, Yu J, Jatkoe T, Berns EM, Atkins D, Foekens JA. (2005) Gene-expression profiles to predict distant metastasis of lymph-nodenegative primary breast cancer. *Lancet* **365**: 671–679.

8. Papadopoulou E, Davilas E, Sotiriou V, Georgakopoulos E, Georgakopoulou S, Koliopanos A, Aggelakis F, Dardoufas K, Agnanti NJ, Karydas I, Nasioulas G. (2006) Cell-free DNA and RNA in plasma as a new molecular marker for prostate and breast cancer. *Ann NY Acad Sci* **1075**: 235–243.

9. Scherzer CR, Eklund AC, Morse LJ, Liao Z, Locascio JJ, Fefer D, Schwarzschild MA, Schlossmacher MG, Hauser MA, Vance JM, Sudarsky LR, Standaert DG, Growdon JH, Jensen RV, Gullans SR.

(2007) Molecular markers of early Parkinson's disease based on gene expression in blood. *Proc Natl Acad Sci USA* **1043(3):** 955–960.

10. Solanky KS, Bailey NJ, Beckwith-Hall BM, Davis A, Bingham S, Holmes E, Nicholson JK, Cassidy A. (2003) Application of biofluid 1H nuclear magnetic resonance based metabonomic techniques for the analysis of the biochemical effects of dietary isoflavones on human plasma profile. *Anal Biochem* **323:** 197–204.

11. Buttner P, Mosig S, Lechtermann A, Funke H, Mooren FC. (2007) Exercise affects the gene expression profiles of human white blood cells. *J Appl Physiol* **102:** 26–36.

12. Golub TR, Slonim DK, Tamayo P, Huard C, Gaasenbeek M, Mesirov JP, Coller H, Loh ML, Downing JR, Caligiuri MA, Bloomfield CD, Lander ES. (1999) Molecular classification of cancer: Class discovery and class prediction by gene expression monitoring. *Science* **286:** 531–537.

13. Quackenbush J. (2006) Microarray analysis and tumor classification. *N Engl J Med* **354:** 2463–2472.

14. Ramaswamy S, Tamayo P, Rifkin R, Mukherjee S, Yeang CH, Angelo M, Ladd C, Reich M, Latulippe E, Mesirov JP, Poggio T, Gerald W, Loda M, Lander ES, Golub TR. (2001) Multiclass cancer diagnosis using tumor gene expression signatures. *Proc Natl Acad Sci USA* **98:** 15149–15154.

15. Aebersold R, Mann M. (2003) Mass spectrometry-based proteomics. *Nature* **422:** 198–207.

16. Domon B, Aebersold R. (2006) Mass spectrometry and protein analysis. *Science* **312:** 212–217.

17. Anderson NL, Anderson NG. (2002) The human plasma proteome: History, character, and diagnostic prospects. *Mol Cell Proteomics* **1:** 845–867.

18. Fujii K, Nakano T, Kawamura T, Usui F, Bando Y, Wang R, Nishimura T. (2004) Multidimensional protein profiling technology and its application to human plasma proteome. *J Proteome Res* **3:** 712–718.

19. Hood L, Heath JR, Phelps ME, Lin B. (2004) Systems biology and new technologies enable predictive and preventative medicine. *Science* **306:** 640–643.

20. Lathrop JT, Anderson NL, Anderson NG, Hammond DJ. (2003) Therapeutic potential of the plasma proteome. *Curr Opin Mol Ther* **5:** 250–257.

21. Lee HJ, Lee EY, Kwon MS, Paik YK. (2006) Biomarker discovery from the plasma proteome using multidimensional fractionation proteomics. *Curr Opin Chem Biol* **10:** 42–49.

22. Bailey RC, Kwong GA, Radu CG, Witte ON, Heath JR. (2007) DNA-encoded antibody libraries: A unified platform for multiplexed cell sorting and detection of genes and proteins. *J Am Chem Soc* **129(7):** 1959–1967. [Epub 2007 Jan 30].

23. Yang S, Undar A, Zahn JD. (2006) A microfluidic device for continuous, real time blood plasma separation. *Lab Chip* **6:** 871–880.

24. Zimmermann M, Delamarche E, Wolf M, Hunziker P. (2005) Modeling and optimization of high-sensitivity, low-volume microfluidic-based surface immunoassays. *Biomed Microdevices* **7:** 99–110.

25. Hood L, Heath JR, Phelps ME, Lin B. (2004) Systems biology and new technologies enable predictive and preventative medicine. *Science* **306(5696):** 640–643.

26. Heath JR, Phelps ME, Hood L. (2003) NanoSystems biology. *Molecul Imaging Biol* **5(5):** 312–325.

27. Weston AD, Hood L. (2004) Systems biology, proteomics, and the future of health care: Toward predictive, preventative, and personalized medicine. *J Proteome Res* **3:** 179–196.

28. Lin B, White JT, Lu W, Xie T, Utleg AG, Yan X, Yi EC, Shannon P, Khrebtukova I, Lange PH, Goodlett DR, Zhou D, Vasicek TJ, Hood L. (2005) Evidence for the presence of disease-perturbed networks in prostate cancer cells by genomic and proteomic analyses: A systems approach to disease. *Cancer Res* **65(8):** 3081–3091.

29. Price ND, Edelman LB, Lee I, Yoo H, Hwang D, Carlson G, Galas DJ, Heath JR, Hood L. (2007) Systems biology and the emergence of systems medicine. In: Ginsburg G & Willard H (eds), *Genomic and Personalized Medicine: From Principles to Practice*. Elsevier, in press

30. Hood L. (2008) A personal journey of discovery: Developing technology and changing biology. *Annu Rev Anal Chem* Vol. 1, Jul 2008, in press.

Chapter 15

The Neurobiology of Consciousness

*Christof Koch and Florian Mormann**

In recent years, the mystery of consciousness and its material basis has attracted scientific interest beyond the traditional field of philosophy. The neurobiological approach to consciousness aims at identifying its physico-chemical basis at the neuronal level of the brain. Electrophysiological, psychophysical, and functional imaging studies in humans and non-human animals have allowed brain scientists to narrow down on the neural substrates of consciousness and conscious perception. These findings, complemented by the development of a robust theoretical predictive framework, could eventually lead to a rational understanding of the phenomenon of consciousness.

1. The Neurobiological Approach to Consciousness

Consciousness is one of the most enigmatic features of the universe. People not only act but feel: they see, hear, smell, recall, plan for the future. These activities are associated with subjective, ineffable, immaterial feelings that are tied in some manner to the material brain. The exact nature of this relationship — the classical mind-body problem — remains elusive and the subject of heated debate. These first-hand, subjective experiences pose a daunting challenge to the scientific method that has, in many other areas, proven so immensely fruitful. Science can describe events microseconds following the Big Bang, offer an increasingly detailed account of matter and how to manipulate it, and uncover the biophysical and neurophysiological nuts and bolts of the brain and its pathologies. However, this same method

*Division of Biology, California Institute of Technology, 1200 E California Blvd, MS 216-76, Pasadena, CA 91125, USA, e-mail: koch@klab.caltech.edu.

has as yet failed to provide a satisfactory account of how first-hand, subjective experience fits into the objective, physical universe.

The brute fact of consciousness comes as a total surprise; it does not appear to follow from any phenomena in traditional physics or biology. Indeed, some modern philosophers even argue that consciousness is not logically supervenient to physics.[1] Supervenience is used to describe the relationship between higher-level and lower-level properties such that the property X supervenes on property Y if Y determines X. This implies, for example, that changing Y will, of necessity, change X. In that sense, biology is supervenient to physics. Put differently, two systems that are physically alike will also be biologically alike. Yet it is not at all clear whether two physically identical brains will have the same conscious state.

Note that it is not yet generally accepted that consciousness is a appropriate subject of scientific inquiry. A number of neuroscience textbooks provide extended details about brains over hundreds of pages yet leave out what it feels like to be the owner of such an awake brain, a remarkable omission.

People willingly concede that when it comes to nuclear physics or molecular biology, specialist knowledge is essential; but many assume that there are few relevant facts about consciousness and therefore everybody is entitled to their own theory. Nothing could be further from the truth.

There is an immense amount of relevant psychological, clinical and neuroscientific data and observations that need to be accounted for. Furthermore, the modern focus on the neuronal basis of consciousness in the brain — rather than on interminable philosophical debates — has given brain scientists tools to greatly increase our knowledge of the conscious mind.

Consciousness is a state-dependent property of certain types of complex, biological, adaptive, and highly interconnected systems. The best example of consciousness is found in a healthy and attentive human brain, e.g., the reader of this chapter. In deep sleep, consciousness ceases. Small lesions in the midbrain and thalamus can lead to a complete loss of consciousness, while destruction of circumscribed parts of the cerebral cortex can eliminate very specific aspects of consciousness, such as the ability to be aware of motion or to recognize faces, without a concomitant loss of vision in general.

Brain scientists are exploiting a number of empirical approaches that shed light on the neural basis of consciousness. This chapter reviews these approaches and summarizes what has been learnt.

2. What Phenomena Does Consciousness Encompass?

There are many definitions of consciousness.[2] A common philosophical one is "Consciousness is what it is like to be something," such as the experience of what it feels like to smell a rose or to be in love. This what-it-feels-like-from-within definition expresses the principal irreducible characteristic of the phenomenal aspect of consciousness: to experience something. "What it feels like to be me, to see red or to be angry" also emphasizes the subjective or first-person perspective of consciousness: it is a subject, an I, who is having the experiences and the experience is inevitably private.

What it feels like to have a particular experience is called the *quale* of that experience: the quale of red is what is common to such disparate conscious states as seeing a red sunset, the red flag of China, arterial blood, or a ruby gemstone. All four subjective states share "redness." There are countless qualia (the plural of quale): the ways things look, sound and smell, the way it feels to have a pain, the way it feels to have thoughts and desires, and so on. To have an experience means to have qualia, and the quale of an experience is what specifies it and makes it different from other experiences.

A science of consciousness must explain the exact relationship between phenomenal, mental states and brain states. This is the heart of the classical mind-body problem: What is the nature of the relationship between the immaterial, conscious mind and its physical basis in the electro-chemical interactions in the body? This problem can be divided into several sub-problems.

(i) Why is there any experience at all? Or, put differently, why does a brain state feel like anything? In philosophy, this is referred to by some as the Hard Problem (note the capitalization), or as the explanatory gap between the material, objective world and the subjective, phenomenal world.[1] Many scholars have argued that the exact nature of this relationship will remain a central puzzle of human existence, without an adequate reductionistic, scientific explanation. However, as similar sentiments have been expressed in the past for the problem of seeking to understand life or to determine what material the stars are made out of, it is best to put this question aside for the moment and not be taken in by defeatist arguments.

(ii) Why is the relationship among different experiences the way it is? For instance, red, yellow, green, cyan, blue, magenta are all colors that can be mapped onto the topology of a circle. Why? Furthermore, as a group, these color percepts share certain communalities that make them different from other percepts, such as seeing motion or smelling a rose.

(iii) Why are feelings private? As expressed by poets and novelists, we cannot communicate an experience to somebody else except by way of example.

(iv) How do feelings acquire meaning? Subjective states are not abstract states but have an immense amount of associated explicit and implicit feelings. Think of the unmistakable smell of dogs coming in from the rain or the crunchy texture of potato chips.

(v) Why are only some behaviors associated with conscious states? Much brain activity and associated behavior occur without any conscious sensation.

3. The Neuronal Correlates of Consciousness

Progress in addressing the mind-body problem has come from focusing on empirically accessible questions rather than on eristic philosophical arguments. Key is the search for the neuronal correlates — and ultimately the

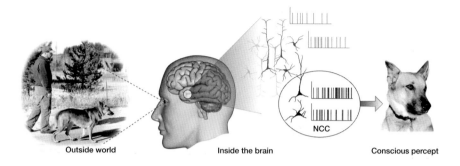

Outside world   Inside the brain   Conscious percept

Fig. 1. The Neuronal Correlates of Consciousness (NCC) are the minimal set of neural events and structures — here synchronized action potentials in neocortical pyramidal neurons — sufficient for a specific conscious percept or a conscious (explicit) memory. From Ref. 4.

causes — of consciousness. As defined by Crick and Koch,[3,68] the neuronal correlates of consciousness (NCC) are the *minimal neuronal mechanisms jointly sufficient for any one specific conscious percept* (Fig. 1).

This definition of NCC stresses the attribute *minimal* because the question of interest is which subcomponents of the brain are actually essential. For instance, it is likely that neural activity in the cerebellum does not underlie any conscious perception, and thus is not part of the NCC. That is, trains of spikes in Purkinje cells (or their absence) will not induce a sensory percept although they may ultimately affect some behaviors (such as eye movements).

On the other hand, the definition does not focus exclusively on the *necessary* conditions for consciousness because of the great redundancy and parallelism found in neurobiological networks. While activity in some population of neurons may underpin a percept in one case, a different population might mediate a related percept if the former population is lost or inactivated.

Every phenomenal, subjective state will have associated NCC: one for seeing a red patch, another one for seeing grandmother, yet a third one for hearing a siren, etc. Perturbing or inactivating the NCC for any one specific conscious experience will affect the percept or cause it to disappear. If the NCC could be induced artificially, for instance by cortical microstimulation in a prosthetic device or during neurosurgery, the subject would experience the associated percept.

What characterizes the NCC? What are the communalities between the NCC for seeing and for hearing? Will the NCC involve all pyramidal neurons in cortex at any given point in time? Or only a subset of long-range projection cells in frontal lobes that project to the sensory cortices in the back? Only layer 5 cortical cells? Neurons that fire in a rhythmic manner? Neurons that fire in a synchronous manner? These are some of the proposals that have been advanced over the years.[5]

It should be noted that discovering and characterizing the NCC in brains is not the same as a theory of consciousness (cf. Sec. 11). Only the latter can tell us why particular systems can experience anything, why they are conscious, and why other systems — such as the enteric nervous system or the immune system — are not. However, understanding the NCC is a necessary step toward such a theory.

4. The Neurobiology of Free Will

A further aspect of the mind-body problem is the question of free will, a vast topic. Answering this question goes to the heart of the way people think of themselves. The spectrum of views ranges from the traditional and deeply embedded belief that we are free, autonomous, and conscious actors to the view that we are biological machines driven by needs and desires beyond conscious access and without willful control.

Of great relevance are the classical findings by Libet and colleagues[6] of brain events that precede the conscious initiation of a voluntary action. In this elegant experiment, subjects were sitting in front of an oscilloscope, tracking a spot of light moving every 2.56 sec around a circle. Every now and then, "spontaneously," subjects had to carry out a specific voluntary action, here flexing their wrist. If this action is repeated sufficiently often while electrical activity around the vertex of the head is recorded, a *readiness potential* (*Bereitschaftspotential*) in the form of a sustained scalp negativity develops long before the muscle starts to move. Libet asked subjects to silently note the position of the spot of light when they first "felt the urge" to flex their wrist and to report this location afterwards. This temporal marker for the awareness of willing an action occurs on average 200 msec before initiation of muscular action (with a standard error of about 20 msec), in accordance with commonsense notions of the causal action of free will. However, the readiness potential can be detected at least 350 msec before awareness of the action. In other words, the subject's brain signals the action at least half a second before the subject feels that he or she has initiated it.

This simple result has been replicated but, because of its counterintuitive implication that conscious will has no causal role, continues to be vigorously debated.[7]

Psychological work in both normal individuals as well as in patients reveals further dissociations between the conscious perception of a willed action and its actual execution: subjects believe that they perform actions that they did not do while, under different circumstances, subjects feel that they are not responsible for actions that are, demonstrably, their own.[8]

Yet whether volition is illusory or is free in some libertarian sense does not answer the question of how subjective states relate to brain states. The perception of free will, what psychologists call the *feeling of agency* or *authorship* (e.g., "I decided to lift my finger"), is certainly a subjective state with an associated quale no different in kind from the quale of a toothache or seeing marine blue. So even if free will is a complete chimera, the subjective feeling of willing an action must have some neuronal correlate.

Direct electrical brain stimulation during neurosurgery[9] as well as fMRI experiments implicate medial pre-motor and anterior cingulated cortices in generating the subjective feeling of triggering an action.[10] In other words, the neural correlate for the feeling of apparent causation involves activity in these regions.

5. Quantum Mechanics and Consciousness

Neurobiologists and cognitive scientists implicitly assume that the relevant variables giving rise to consciousness are to be found at the neuronal level, among the synaptic release or the action potentials in population of neurons, rather than at the molecular, or even the sub-molecular, level — that the activity of neurons is responsible for all higher brain functions, including behavior, memory, emotion, and consciousness.

There is, however, also an alternative tradition that dates back to the early days of quantum mechanics[11] and to the supposedly critical role of the conscious observer in the measuring process. As is well known, it can be difficult to separate the process to be observed from the act of observing it. In more recent times, proposals for the critical role of quantum mechanics for consciousness have come from the physicist Roger Penrose[12] and others,[13] as well as from the contemplative-meditative Buddhist school that also emphasizes the illusory nature of the object-subject distinction.[14]

The role of quantum mechanics for the photons received by the eye and for the molecules of life is undisputed. But although brains obey the laws of quantum mechanics, they do not seem to exploit any of its special features. Of particular interest here is *quantum entanglement*, the observation that the quantum states of multiple objects, such as two coupled electrons, may be highly correlated even though they are spatially separated, violating our intuition about locality (entanglement is the key feature of quantum mechanics exploited in quantum computers).

Molecular machines, such as the light-amplifying components of photoreceptors, pre- and post-synaptic receptors and the voltage- and ligand-gated channel proteins that span cellular membranes and underpin neuronal excitability, are so large that they can be treated as classical objects. Two key biophysical operations underlie information processing in the brain: chemical transmission across the synaptic cleft, and the generation of action potentials. These both involve thousands of ions and neurotransmitter molecules, coupled by diffusion or by the membrane potential that extends across tens of micrometers. Both processes will destroy any coherent quantum states. It follows that a neuron either spikes at a particular point in time or does not, but it is not in a superposition of spike and non-spike states. Thus, spiking neurons can only receive and send classical, rather than quantum, information.

The question of whether quantum mechanics governs the information processing in the brain also relates to the question of whether or not the brain as a complex dynamical system can be regarded as purely deterministic, i.e., *predictable* in a physical sense. A deterministic system of arbitrary complexity is in principle predictable if its initial state can be measured with arbitrary precision. The crucial question is whether or not quantum-mechanical uncertainty intrinsically limits this precision. The issue of determinism and predictability, in turn, relates to the problem of the free will (see Sec. 4).

At the moment, there is no evidence that any components of the nervous system — a 37°C wet and warm tissue strongly coupled to its environment — display quantum entanglement. At the cellular level, the interaction of neurons seems to be governed by classical physics.[15,16]

Of course, it is very difficult to rigorously conclude that quantum mechanics is irrelevant to consciousness.[†] History is littered with pronouncements that certain things are not possible. The vertebrate brain is, after all, the product of an evolutionary selection process optimized over hundreds of millions of iterations.

[†]An interesting proposal is that by Bernroider[17] that the two K^+ ions in the selectivity filter of the closed voltage-dependent potassium channel — whose structure has been crystallized and characterized at the Angstrom level of resolution by MacKinnon and his colleagues[18] — are in a superposition of two quantum states (S1/S3 and S2/S4). Even if true for such heavy ions — of molecular mass 39 u — it is not known how such an entanglement within a single channel could lead to entanglement across two or more potassium channels, let alone across neurons. Yet it is worth watching these and other developments and to keep an open mind.

6. Consciousness in Other Species

Data about subjective states come not only from people who can talk about their subjective experiences but also from non-linguistic competent individuals — newborn babies or patients with complete paralysis of nearly all voluntary muscles (locked-in syndrome) — and, most importantly, from animals other than humans. There are three reasons to assume that many species, in particular those with complex behaviors such as mammals, share at least some aspects of consciousness with humans:

(i) *Similar neuronal architectures*: Except for size, there are no large-scale, dramatic differences between the cerebral cortex and thalamus of mice, monkeys, humans and whales. In particular, the macaque monkey is a powerful model organism to study visual perception because it shares with the human visual system three distinct cone photopigments, binocular stereoscopic vision, a foveated retina and similar eye movements.

(ii) *Similar behavior*: Almost all human behaviors have precursors in the animal literature. Take the case of pain. The behaviors seen in humans when they experience pain and distress — facial contortions, moaning, yelping or other forms of vocalization, motor activity such as writhing, avoidance behaviors at the prospect of a repetition of the painful stimulus — can be observed in all mammals and in many other species. Likewise for the physiological signals that attend pain: activation of the sympathetic autonomous nervous system resulting in change in blood pressure, dilated pupils, sweating, increased heart rate, release of stress hormones, and so on. The discovery of cortical pain responses in premature babies shows the fallacy of relying on language as the sole criterion for consciousness.[19]

(iii) *Evolutionary continuity*: The first true mammals appeared at the end of the Triassic period, about 220 million years ago, with primates proliferating following the Cretaceous-Tertiary extinction event, about 60 million years ago, while humans and macaque monkeys did not diverge until 30 million years ago.[20] *Homo sapiens* is part of an evolutionary continuum with its implied structural and behavioral continuity, rather than an independently developed organism.

While certain aspects of consciousness, in particular those relating to the recursive notion of self and to abstract, culturally transmitted knowledge, are not widespread in non-human animals, there is little reason to doubt that other mammals share conscious feelings — sentience — with humans. To believe that humans are special, are singled out by the gift of consciousness above all other species, is a remnant of humanity's atavistic, deeply held belief that *homo sapiens* occupies a privileged place in the universe, a belief with no empirical basis.

The extent to which non-mammalian vertebrates — such as tuna, cichlid and other fish, crows, ravens, magpies, parrots and other birds, or even invertebrates, such as squids, or bees, with complex, non-stereo-typed behaviors including delayed-matching, non-matching-to-sample and other forms of learning[21] — are conscious is difficult to answer at this point in time (but see Ref. 22). Without a sounder understanding of the neuronal architecture necessary to support consciousness, it is unclear where in the animal kingdom to draw the Rubicon that separates species with at least some conscious percepts from those that never experience anything and that are nothing but pure automata.[23]

7. Level of Arousal and Content of Consciousness

There are two common, but quite distinct, usages of the term consciousness, one revolving around *arousal* and *states of consciousness* and another one around the *content of consciousness* and *conscious states*.

7.1. States of Consciousness and Conscious States

To be conscious of anything, the brain must be in a relatively high state of arousal (sometimes also referred to as *vigilance*). This is as true of wakefulness as it is of REM sleep that is vividly, consciously experienced in dreams, although usually not remembered. The level of brain arousal, measured by electrical or metabolic brain activity, fluctuates in a circadian manner and is influenced by lack of sleep, drugs and alcohol, physical exertion, etc. in a predictable manner. High arousal states are usually associated with some conscious state — a percept, thought or memory — that has a specific content. We see a face, hear music, remember an incident, plan an experiment, or fantasize about sex. Indeed, it is unlikely that one can be awake without being conscious of something. Referring to such conscious states is conceptually quite distinct from referring to states of consciousness that fluctuate with different levels of arousal. Arousal can be measured behaviorally by the signal amplitude that triggers some criterion reaction (for instance, the sound level necessary to evoke an eye movement or a head turn toward the sound source). Clinicians use scoring systems such as the Glasgow Coma Scale to assess the level of arousal in patients.

Different levels or states of consciousness are associated with different kinds of conscious experiences. The awake state in a normal functioning individual is quite different from the dreaming state (for instance, the latter has little or no self-reflection) or from the state of deep sleep. In all three cases, the basic physiology of the brain is changed, affecting the space of possible conscious experiences. Physiology is also different in *altered states of consciousness*, for instance after taking psychedelic drugs when events often have a stronger emotional connotation than in normal life. Yet another state of consciousness has been reported to occur during certain meditative practices, when conscious perception and insight are said to be enhanced compared to the normal waking state.

In some obvious but difficult to rigorously define manner, the *richness of conscious experience* increases as an individual transitions from deep sleep to drowsiness to full wakefulness. This richness of possible conscious experience could be quantified using notions from complexity theory that incorporate both the dimensionality as well as the granularity of conscious experience (e.g., the integrated-information-theoretical account of consciousness[24]). Inactivating all of visual cortex in an otherwise normal individual would significantly reduce the dimensionality of conscious experience since no color, shape, motion, texture or depth could be perceived. As behavioral arousal increases, so does the range and complexity of behaviors that an individual is capable of. A singular exception to this progression is REM sleep where most motor activity is shut down in the *atonia* that is characteristic of this phase of sleep, and the person is difficult to wake up. Yet this low level of behavioral arousal goes, paradoxically, hand in hand with high metabolic and electrical brain activity and conscious, vivid states.

These observations suggest a two-dimensional graph (Fig. 2) in which the richness of conscious experience (its representational capacity) is plotted as a function of levels of behavioral arousal or responsiveness.

7.2. Global Disorders of Consciousness

Global disorders of consciousness can likewise be mapped onto this plane (Fig. 2). Clinicians talk about *impaired states of consciousness* as in "the *comatose state*," "the *persistent vegetative state*" (PVS), and "the *minimally conscious state*" (MCS). Here, state refers to different levels of consciousness, from a total absence in coma, PVS and general anesthesia, to a fluctuating and limited form of conscious sensation in MCS, sleep walking or during a complex partial epileptic seizure.[26]

The repertoire of distinct conscious states or experiences that are accessible to a patient in MCS is presumably minimal (mainly pain and discomfort, possibly sporadic sensory percepts), immeasurably smaller than the possible conscious states that can be experienced by a healthy brain. In the limit of brain death, the origin of this space has been reached with no experience at all (Fig. 2). A more desirable state is global anesthesia during which the patient should not experience anything — to avoid psychological trauma — but the level of arousal during the operation should be compatible with clinical exigencies. While anesthetics may, in principle, be useful for

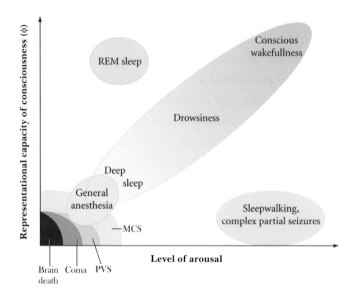

Fig. 2. Physiological and pathological brain states can be situated in a two-dimensional graph. Here increasing levels of behaviorally determined arousal are plotted on the *x*-axis and the "richness" or "representational capacity of consciousness"[24] is plotted on the *y*-axis. Increasing arousal can be measured by the threshold to obtain some specific behavior (for instance, spatial orientation to a sound). Healthy subjects cycle during a 24-hour period from deep sleep with low arousal and very little conscious experience to increasing levels of arousal and conscious sensation. In REM sleep, low levels of behavioral arousal go hand-in-hand with vivid consciousness. Conversely, various pathologies are associated with little or no conscious content. Modified from Ref. 25.

the study of consciousness, they have as yet failed to provide breakthrough insights about the Neural Correlates of Consciousness. This is mainly due to their diverse mechanisms of molecular action, targeting receptors throughout large parts of the brain. As our ability increases to differentiate subreceptor variation and target them with molecular tools by means of genetic *in vivo* studies,[27] this is likely to change.

Given the absence of any accepted theory for the minimal neuronal criteria necessary for consciousness, the distinction between a PVS patient — who shows regular sleep-wave transitions and who may be able to move their eyes or limbs or smile in a reflexive manner as in the case of Terri Schiavo in Florida — and an MCS patient — who can communicate (on occasion) in a meaningful manner (for instance, by differential eye movements) and who shows some signs of consciousness — is often difficult in a clinical setting. Functional brain imaging may prove immensely useful here.

Blood-oxygen-level-dependent functional magnetic resonance imaging (BOLD fMRI) recently demonstrated that a patient in a vegetative state following a severe traumatic brain injury showed the same pattern of brain activity as normals when asked to imagine playing tennis or visiting all the rooms in her house.[28] Differential brain imaging of patients with such global disturbances of consciousness (including akinetic mutism) reveal that dysfunction in a widespread cortical network including medial and lateral prefrontal and parietal associative areas is associated with a global loss of consciousness.[25] Impaired consciousness in epileptic seizures of the temporal lobe was likewise found to be accompanied by a decrease in cerebral blood flow in frontal and parietal association cortex and an increase in midline structures such as the mediodorsal thalamus.[29]

7.3. Localized Brain Lesions Affecting Consciousness

In contrast to diffuse cortical damage, relatively discrete bilateral injuries to midline (paramedian) subcortical structures can also cause a complete loss of consciousness. These structures are therefore part of the *enabling* factors that control the level of brain arousal (as determined by metabolic or electrical activity) and that are needed for any form of consciousness to occur. One such example is the heterogeneous collection of more than two dozen nuclei (on each side) in the upper brainstem (pons, midbrain and in the posterior hypothalamus) collectively referred to as the *reticular activating system* (RAS). These nuclei — three-dimensional collections of neurons with their own cytoarchitecture and neurochemical identity — release distinct neuromodulators such as acetylcholine, noradrenaline/norepinephrine, serotonin, histamine and orexin/hypocretin. Their axons project widely throughout the brain (Fig. 3). These neuromodulators control the excitability of thalamus and forebrain and mediate the alternation between wakefulness and sleep as well as the general level of both behavioral and brain arousal. Acute lesions of nuclei in the RAS can result in loss of consciousness and coma. However, eventually the excitability of thalamus and forebrain can recover and consciousness can return.[30] Another enabling factor for consciousness are the five or more intralaminar nuclei (ILN) of the thalamus. These receive input from many brainstem nuclei and project strongly to the basal ganglia and, in a more distributed manner, into layer I of much of neocortex. Comparatively small (1 cm^3 or less), bilateral lesions in the thalamic ILN completely knock out all awareness.[31]

In summary, a plethora of nuclei with distinct chemical signatures in the thalamus, midbrain and pons must function for a subject to be in a

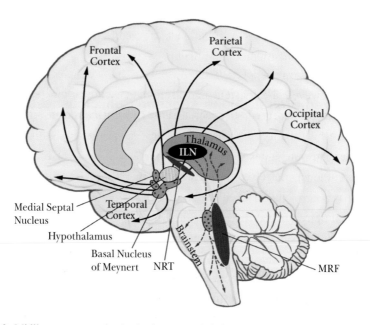

Fig. 3. Midline structures in the brainstem and thalamus necessary to regulate the level of brain arousal. Small, bilateral lesions in many of these nuclei cause a global loss of consciousness. From Ref. 4.

sufficient state of brain arousal to experience anything at all. These nuclei belong to the enabling factors for consciousness. Conversely, it is likely that the specific content of any one conscious sensation is mediated by neurons in cortex and their associated satellite structures, including the amygdala, thalamus, claustrum and the basal ganglia.

7.4. Split-Brain Studies

The brain has a remarkable degree of bilateral symmetry. The mind, however, has but a single stream of consciousness, not two. Under ordinary conditions, the two hundred million fibers making up the corpus callosum, together with the anterior commissure and other minor bundles, integrate neural activity in the two halves of the forebrain such that only a single, integrated percept arises.

In certain cases of intractable epileptic seizures, part or all of the *corpus callosum* are surgically cut. Remarkably, after recovery, these patients usually act, speak, and feel no different than before. They do not complain of a loss of half their visual field or of other dramatic deficits. Upon closer inspection, however, a persistent and profound disconnection (split-brain) syndrome can be observed. If specific sensory information is provided to one or the other hemisphere, the information is not shared with its twin. A split-brain patient with typical (left hemispheric) language dominance is unable to name an image of an object presented in the left visual hemifield, but can pick this object from a group of objects using his left hand. This procedure involves conscious perception followed by a targeted motor response with the neural correlates necessarily being constrained to one hemisphere.

The primary conclusion from split-brain patients, work for which Roger Sperry was awarded the Nobel prize in 1981, is that both hemispheres are independently capable of conscious experience.[32,33] Whatever the neuronal correlates of consciousness, these must exist independently in both cortical hemispheres.

Another example of a complex impairment of conscious perception, not to be confused with the split-brain syndrome, is the neglect syndrome, properly called visuo-spatial hemi-neglect. It can be found after extensive damage typically to the right brain hemisphere with affection of the inferior parietal cortex. Patients with a hemi-neglect syndrome show an impaired awareness of their left visual hemifield despite the visual pathway being completely intact.

8. The Neuronal Basis of Conscious Perception

The possibility of precisely manipulating visual percepts in time and space has made vision a preferred modality in the quest for the NCC. Psychologists have perfected a number of techniques — masking, binocular rivalry, continuous flash suppression, motion-induced blindness, change blindness, inattentional blindness — in which the seemingly simple and unambiguous relationship between a physical stimulus in the world and its associated percept in the privacy of the subject's mind is disrupted.[34] In particular, a stimulus can be perceptually suppressed for seconds or even minutes at a time: the image is projected into one of the observer's eyes but is invisible, not seen. In this manner the neural mechanisms that respond to the subjective percept rather than the physical stimulus can be isolated, permitting the footprints of visual consciousness to be tracked in the brain. In a *perceptual illusion*, the physical stimulus remains fixed while the percept fluctuates. The best known example is the *Necker cube* whose 12 lines can be perceived in one of two different ways in depth (Fig. 4).

A perceptual illusion that can be precisely controlled is *binocular rivalry*.[35] Here, a small image, e.g., a horizontal grating, is presented to the left eye, and another image, e.g., a vertical grating, is shown to the corresponding location in the right eye. In spite of the constant visual stimulus, observers consciously see the horizontal grating alternate every few seconds with the vertical one. The brain does not allow for the simultaneous perception of both images.

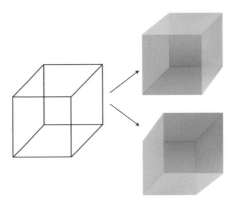

Fig. 4. The Necker Cube: The left line drawing can be perceived in one of two distinct depth configurations shown on the right. Without any other cue, the visual system flips back and forth between these two interpretations. From Ref. 4.

Macaque monkeys can be trained to report whether they see the left or the right image. The distribution of the switching times and the way in which changing the contrast in one eye affects these leaves little doubt that monkeys and humans experience the same basic phenomenon. In a series of elegant experiments, Logothetis and colleagues[36] recorded from a variety of visual cortical areas in the awake macaque monkey while the animal performed a binocular rivalry task. In primary visual cortex (V1), only a small fraction of cells weakly modulate their response as a function of the percept of the monkey. The majority of cells responded to one or the other retinal stimulus with little regard to what the animal perceived at the time. Conversely, in a high-level cortical area such as the inferior temporal (IT) cortex along the *ventral ("what?") pathway*, almost all neurons responded only to the perceptually dominant stimulus (in other words, a "face" cell only fired when the animal indicated by its performance that it saw the face and not the pattern presented to the other eye; see Fig. 5), implying that the NCC involves activity in neurons in inferior temporal cortex.

Does this imply that the NCC is local to IT? At this point, no definitive answer can be given. However, given known anatomical connections, it is

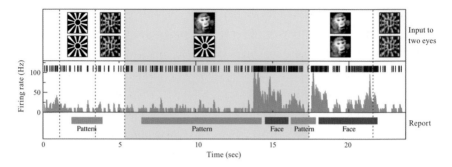

Fig. 5. A few seconds in the life of a typical IT cell while a monkey experiences binocular rivalry. The upper row indicates the visual input, with dotted vertical lines marking stimulus transitions. The second row shows the individual spikes, the third the smoothed firing rate, and the bottom row the monkey's behavior. The animal was taught to press a lever when it saw either one or the other image, but not both. The cell responded only weakly to either the sunburst pattern or to its optical superposition with the image of a monkey's face around 5 sec. During binocular rivalry (gray zone), the monkey's perception vacillated back and forth between seeing the face and seeing the bursting sun. Perception of the face was consistently accompanied (and preceded) by a strong increase in firing rate. From N. Logothetis (private communication) as modified by Ref. 4.

possible that specific reciprocal interactions between IT cells and neurons in parts of the prefrontal cortex are necessary for the NCC. This is compatible with the broadly accepted notion that the NCC must involve positive feedback to insure that neural activity is persistent and strong enough to exceed some threshold and be broadly distributed to multiple cognitive systems, including working memory, planning and language.

In a related perceptual phenomenon, *flash suppression*, the percept associated with an image projected into one eye is suppressed by flashing another image into the other eye (while the original image remains). Its methodological advantage over binocular rivalry is that the timing of the perceptual transition is determined by an external trigger rather than by an internal event. The majority of cells in IT cortex and in the superior temporal sulcus of monkeys trained to report their percept during flash suppression follow the animal's percept. That is, when the cell's preferred stimulus is perceived, the cell responds. If the picture is still present on the retina but is perceptually suppressed, the cell falls silent, even though legions of primary visual cortex neurons fire vigorously to this stimulus.[37,38] Single neuron recordings in the medial temporal lobe of epilepsy patients during flash suppression likewise demonstrate abolishment of their responses when their preferred stimulus is present but perceptually masked.[39]

In a powerful combination of binocular rivalry and flash suppression, a stationary image in one eye can be suppressed for minutes on end by continuously flashing different images into the other eye (continuous flash suppression).[40] This paradigm lends itself naturally to further investigate the relationship between neural activity — whether assayed at the single neuron or at the brain voxel level — and conscious perception.

A number of fMRI experiments have exploited binocular rivalry and related illusions to identify the hemodynamic activity underlying visual consciousness in humans. They demonstrate quite conclusively that BOLD activity in the upper stages of the ventral pathway (e.g., the fusiform face area and the parahippocampal place area) as well as in early regions, including V1 and the lateral geniculate nucleus (LGN), follow the percept and not simply the retinal stimulus.[41] Furthermore, a number of elegant fMRI experiments[42,43] support the hypothesis that V1 is necessary, but not sufficient for visual consciousness for normal seeing.[44] It remains an open question as to whether V1 is necessary to enjoy the vivid visual dreams that are consciously experienced and that occur most frequently during REM sleep when the eyes are closed.[45]

9. Other Perceptual Puzzles of Contemporary Interest

The attributes of even simple percepts seem to vary along a continuum. For instance, a patch of color has a brightness and a hue that are variable, just as a simple tone has an associated loudness and pitch. However, is it possible that each particular, consciously experienced, percept is all-or-none? Might a pure tone of a particular pitch and loudness be experienced as an atom of perception, either heard or not, rather than gradually emerging from the noisy background? The perception of the world around us would then be a superposition of many elementary, binary percepts.[46]

Is perception continuous, like a river, or does it consist of series of discontinuous batches, rather like the discrete frames in a movie[47,48]? In cinematographic vision,[49] a rare form of visual migraine, the subject sees the movement of objects as fractured in time, as a succession of different configurations and positions, without any movement in-between. The hypothesis that visual perception is quantized in discrete batches of variable duration, most often related to EEG rhythms in various frequency ranges (from theta to beta), is an old one. This idea is being revisited in light of the discrepancies of timing of perceptual events within and across different sensory modalities. For instance, even though a change in the color of an object occurs simultaneously with a change in its direction of motion, it may not be perceived that way.[50–52]

What is the relationship between endogenous, top-down attention and consciousness? Although these are frequently coextensive — subjects are usually conscious of what they attend to — there is a considerable tradition in psychology that argues that these are distinct neurobiological processes.[53] This question is receiving renewed "attention" due to the development of ever more refined and powerful visual masking techniques[34] that independently manipulate attention and consciousness. Indeed, it has been shown that attention can be allocated to a perceptually invisible stimulus[54] and that subjects can be conscious of a stimulus without attending to it. When exploring the neural basis of these processes, it is therefore critical to not confound attention with consciousness and *vice versa*.

10. Forward Versus Feedback Projections

Many actions in response to sensory inputs are rapid, transient, stereotyped, and unconscious.[55] They could be thought of as cortical reflexes and are characterized by rapid and somewhat stereotyped responses that can take the form of rather complex automated behavior as seen, e.g., in complex partial epileptic seizures. These automated responses, sometimes called *zombie behaviors*,[56] could be contrasted by a slower, all-purpose conscious mode that deals more slowly with broader, less stereotyped aspects of the sensory inputs (or a reflection of these, as in imagery) and takes time to decide on appropriate thoughts and responses. Without such a consciousness mode, a vast number of different zombie modes would be required to react to unusual events.

A feature that distinguishes humans from most animals is that we are not born with an extensive repertoire of behavioral programs that would enable us to survive on our own ("physiological prematurity"). To compensate for this, we have an unmatched ability to learn, i.e., to consciously acquire such programs by imitation or exploration. Once consciously acquired and sufficiently exercised, these programs can become automated to the extent that their execution happens beyond the realms of our awareness. Take as an example the incredible fine motor skills exerted in playing a Beethoven piano sonata or the sensorimotor coordination required to ride a motorcycle along a curvy mountain road. Such complex behaviors are possible only because a sufficient number of the subprograms involved can be executed with minimal or even suspended conscious control.

In fact, the conscious system may actually interfere somewhat with these automated programs[57]: focusing consciousness onto the smooth execution of a complex, rapid and highly trained sensory-motor task — dribbling a soccer ball, to give one example — can interfere with its smooth execution, something well known to athletes and their trainers.

From an evolutionary standpoint it makes sense to have both automated behavioral programs that can be executed rapidly in a stereotyped and automated manner, and a slightly slower system that allows time for thinking and planning more complex behavior. This latter aspect, planning, may be one of the principal functions of consciousness.

It seems possible that visual zombie modes in the cortex mainly use the *dorsal ("where?") pathway* in the parietal region.[55] However, parietal activity can affect consciousness by producing attentional effects on the ventral stream, at least under some circumstances. The basis of this inference are clinical case studies and fMRI experiments in normal subjects.[58] The conscious mode for vision depends largely on the early visual areas (beyond V1) and especially on the ventral stream.

Seemingly complex visual processing (such as detecting animals in natural, cluttered scenes) can be accomplished by the human cortex within 130–150 msec,[48,59] way too fast for eye movements and conscious perception to occur. Furthermore, reflexes such as the oculovestibular reflex take place at even more rapid time scales. It is quite plausible that such behaviors are mediated by a purely feed-forward moving wave of spiking activity that passes from the retina through V1, into V4, IT and prefrontal cortex, until it affects motor-neurons in the spinal cord that control the finger press (as in a typical laboratory experiment). The hypothesis that the basic processing of information is feed-forward is supported most directly by the short times (approx. 100 msec) required for a selective response to appear in IT cells.

Conversely, conscious perception is believed to require more sustained, reverberatory neural activity, most likely via global feedback from frontal regions of neocortex back to sensory cortical areas.[44] These feedback loops would explain why in backward masking a second stimulus, flashed 80–100 msec after onset of a first image, can still interfere (mask) with the percept of the first image. The reverberatory activity builds up over time until it exceeds a critical threshold. At this point, the sustained neural activity rapidly propagates to parietal, prefrontal and anterior cingulate cortical regions, thalamus, claustrum and related structures that support short-term memory, multi-modality integration, planning, speech and other processes intimately related to consciousness. Competition prevents more than one or a very small number of percepts to be simultaneously and actively represented. This is the core hypothesis of the *global workspace* model of consciousness.[60,61] Sending visual information to more frontal structures would allow the associated visual events to be decoded and placed into context (for instance, by accessing various memory banks) and to have this interpretation feed back to the sensory representation in visual cortex.[62]

In brief, while rapid but transient neural activity in the thalamo-cortical system can mediate complex behavior without conscious sensation, it is surmised that consciousness requires sustained but well-organized neural activity dependent on long-range cortico-cortical feedback.

11. An Information-Theoretical Theory of Consciousness

At present, it is not known to what extent animals whose nervous systems have an architecture considerably different from the mammalian neocortex are conscious (see Sec. 6). Furthermore, whether artificial systems, such as computers, robots or the World Wide Web as a whole, which behave with considerable intelligence, are or can become conscious (as widely assumed in science fiction, e.g., the paranoid computer *HAL* in the film *2001*), remains completely speculative. What is needed is a theory of consciousness, which explains in quantitative terms what type of systems, with what architecture, can possess conscious states.

While discovering and characterizing the NCC is a necessary step in understanding consciousness, such an opportunistic, data-driven approach cannot explain why certain structures and processes have a privileged relationship with subjective experience. For example, why is it that neurons in corticothalamic circuits are essential for conscious experience, whereas cerebellar neurons, despite their huge numbers, are most likely not? And what is wrong with cortical zombie systems that makes them unsuitable for yielding subjective experience? Or why is it that consciousness wanes during slow-wave sleep early in the night, despite levels of neural firing in the thalamocortical system that are comparable to the levels of firing in wakefulness?

Information theory may be such a theoretical approach that establishes at the fundamental level what consciousness is, how it can be measured, and what requisites a physical system must satisfy in order to generate it.[1,63]

The most promising candidate for such a theoretical framework is the *information integration theory of consciousness.*[24] It posits that the most important property of consciousness is that it is extraordinarily *informative.* Any one particular conscious state rules out a huge number of alternative experiences. Classically, the reduction of uncertainty among a number of alternatives constitutes information. For example, when a subject consciously experiences reading this particular phrase, a huge number of other possible experiences are ruled out (consider all possible written phrases that could have been written in this space, in all possible fonts, ink colors, and sizes, think of the same phrases spoken aloud, or read and spoken, and so on). Thus, every experience represents one particular conscious state out of a huge repertoire of possible conscious states.

Furthermore, information associated with the occurrence of a conscious state is *integrated* information. An experience of a particular conscious state is an integrated whole. It cannot be subdivided into components that are experienced independently.[63] For example, the conscious experience of this particular phrase cannot be experienced as subdivided into, say, the conscious experience of how the words look independently of the conscious experience of how they sound in the reader's mind. Similarly, visual shapes cannot be experienced independently of their color, nor can the left half of the visual field of view be experienced independently of the right half.

Based on these and other considerations, the theory claims that *a physical system can generate consciousness to the extent that it can integrate information.* This idea requires that the system has a large repertoire of available states (information) yet cannot be decomposed into a collection of causally independent subsystems (integration).

Importantly, the theory introduces a measure of a system's capacity to integrate information. This measure, called ϕ, is obtained by determining the minimum repertoire of different states that can be produced in one part of the system by perturbations of its other parts.[24] ϕ can loosely be thought of as the representational capacity of the system (as in Fig. 2). Although ϕ is not easy to calculate exactly for realistic systems, it can be estimated. Thus, by using simple computer simulations, it is possible to show that ϕ is high for neural architectures that conjoin functional specialization with functional integration, like the mammalian thalamocortical system. Conversely, ϕ is low for systems that are made up of small, quasi-independent modules, like the cerebellum, or for networks of randomly or uniformly connected units.[24]

The notion that consciousness has to do with the brain's ability to integrate information has been tested directly by transcranial magnetic stimulation (TMS). In TMS a coil is placed above the skull and a brief and intense magnetic field generates a weak electrical current in the underlying grey matter in a noninvasive manner. Massimini *et al.*[64] compared multichannel EEG of awake and conscious subjects in response to TMS pulses to the EEG when the same subjects were deeply asleep early in the night — a time during which consciousness is much reduced. During quiet wakefulness, an initial response at the stimulation site was followed by a sequence of waves that moved to connected cortical areas several centimeters away. During slow wave sleep, by contrast, the initial response was stronger but was rapidly extinguished and did not propagate beyond the stimulation site. Thus, the fading of consciousness during certain stages of sleep may be related, as predicted by the theory, to the breakdown of information integration among specialized thalamocortical modules.

12. Conclusion

Ever since the Greeks first formulated the mind-body problem more than two millennia ago, it has been the domain of armchair speculations and esoteric debates with no apparent resolution. Yet many aspects of this ancient set of questions now fall squarely within the domain of science.

Progress in the study of the NCC on the one hand, and of the neural correlates of non-conscious behaviors on the other, will hopefully lead to a better understanding of what distinguishes neural structures or processes that are associated with consciousness from those that are not.

The growing ability of neuroscientists to manipulate in a reversible, transient, deliberate and delicate manner identified populations of neurons using methods from molecular biology[65] in combination with optical tools[66] opens up the possibility of moving from correlation — observing that a particular conscious state is associated with some neural or hemodynamic activity — to causation. Exploiting these increasingly powerful tools depends on the simultaneous development of appropriate behavioral assays and model organisms amenable to large-scale genomic analysis and manipulation.[67]

It is the combination of such fine-grained neuronal analysis in animals with ever more sensitive psychophysical and brain imaging techniques in humans, complemented by the development of a robust theoretical predictive framework, that will hopefully lead to a rational understanding of consciousness, one of the central mysteries of life.

Acknowledgments

This work was supported by the NIMH, NSF, DARPA, the Marie Curie Foundation of the European Community, and the Mathers Foundation.

References

1. Chalmers DJ. (1996) *The Conscious Mind: In Search of a Fundamental Theory.* Oxford University Press, New York.
2. Koch C, Tononi G. (2007) Consciousness. In: *New Encyclopedia of Neuroscience.* Elsevier, in press.
3. Crick F, Koch C. (1990) Towards a neurobiological theory of consciousness. *Semin Neurosci* **2:** 263–275.
4. Koch C. (2004) *The Quest for Consciousness: A Neurobiological Approach.* Roberts, Denver, CO.

5. Chalmers DJ. (2000) What is a neural correlate of consciousness? In: Metzinger T (ed), *Neural Correlates of Consciousness: Empirical and Conceptual Questions*, pp. 17–40. MIT Press, Cambridge.

6. Libet B, Gleason CA, Wright EW, Pearl DK. (1983) Time of conscious intention to act in relation to onset of cerebral activity (readiness-potential): The unconscious initiation of a freely voluntary act. *Brain* **106**: 623–642.

7. Haggard P, Eimer M. (1999) On the relation between brain potentials and conscious awareness. *Exp Brain Res* **126**: 128–133.

8. Wegner DM. (2002) *The Illusion of Conscious Will*. MIT Press, Cambridge, MA.

9. Fried I, Katz A, McCarthy G, Sass KJ, Williamson P, Spencer SS, Spencer DD. (1991) Functional organization of human supplementary motor cortex studied by electrical stimulation. *J Neurosci* **11**: 3656–3666.

10. Lau HC, Rogers RD, Haggard P, Passingham RE. (2004) Attention to intention. *Science* **303**: 1208–1210.

11. von Neumann J. (1932) *Mathematische Grundlagen der Quantenmechanik*. Springer, Berlin.

12. Penrose R. (1989) *The Emperor's New Mind*. Oxford University Press, Oxford.

13. Hameroff S. (1998) The Penrose–Hameroff "Orch OR" model of consciousness. *Philos Trans R Soc Lond A* **356**: 1869–1896.

14. Wallace A. (2007) *Hidden Dimensions — The Unification of Physics and Consciousness*. Columbia University Press, New York.

15. Koch C, Hepp K. (2006) Quantum mechanics and higher brain functions: Lessons from quantum computation and neurobiology. *Nature* **440**: 611–612.

16. Koch C, Hepp K. (2008) The relation between quantum mechanics and higher brain functions: Lessons from quantum computation and neurobiology. In: Chiao RY, Cohen LM, Leggett AJ, Phillips WD, Harper Jr. CL (eds), *Amazing Light: Visions for Discovery: New Light on Physics, Cosmology and Consciousness*. Cambridge University Press, in press.

17. Bernroider G, Roy S. (2005) Quantum entanglement of K ions, multiple channel states and the role of noise in the brain. *SPIE* **5841**: 205–214.

18. Doyle DA, Cabral JM, Pfuetzner RA, Kuo A, Gulbis JM, Cohen SL, Chait BT, MacKinnon R (1998) The structure of the potassium channel: Molecular basis of K^+ conduction and selectivity. *Science* **280:** 69–77.

19. Slater R, Cantarella A, Gallella S, Worley A, Boyd S, Meek J, Fitzgerald M. (2006) Cortical pain responses in human infants. *J Neurosci* **26:** 3662–3666.

20. Allman JM. (2000) *Evolving Brains.* Scientific American Library, New York.

21. Giurfa M, Zhang S, Jenett A, Menzel R, Srinivasan MV. (2001) The concepts of 'sameness' and 'difference' in an insect. *Nature* **410:** 930–933.

22. Edelman DB, Baars JB, Seth AK. (2005) Identifying hallmarks of consciousness in non-mammalian species. *Conscious Cogn* **14:** 169–187.

23. Griffin DR. (2001) *Animal Minds: Beyond Cognition to Consciousness.* University of Chicago Press, Chicago, IL.

24. Tononi G. (2004) An information integration theory of consciousness. *BMC Neurosci* **5:** 42–72.

25. Laureys S. (2005) The neural correlate of (un)awareness: Lessons from the vegetative state. *Trends Cogn Sci* **9:** 556–559.

26. Schiff ND. (2004) The neurology of impaired consciousness: Challenges for cognitive neuroscience. In: Gazzaniga MS (ed), *The Cognitive Neurosciences III*, pp. 1121–1132. MIT Press, Cambridge, MA.

27. Rudolph U, Antkowiak B. (2004) Molecular and neuronal substrates for general anaesthetics. *Nat Rev Neurosci* **5:** 709–720.

28. Owen AM, Cleman MR, Boly M, Davis MH, Laureys S, Pickard JD. (2006) Detecting awareness in the vegetative state. *Science* **313:** 1402.

29. Blumenfeld H, McNally KA, Vanderhill SD, Paige AL, Chung R, Davis K, Norden AD, Stokking R, Studholme C, Novotny EJ Jr, Zubal IG, Spencer SS. (2004) Positive and negative network correlations in temporal lobe epilepsy. *Cereb Cort* **14:** 892–902.

30. Villablanca JR. (2004) Counterpointing the functional role of the forebrain and of the brainstem in the control of the sleep-waking system. *J Sleep Res* **13:** 179–208.

31. Bogen JE. (1995) On the neurophysiology of consciousness: I. An Overview. *Conscious Cogn* **4:** 52–62.

32. Bogen JE. (1993) The callosal syndromes. In: Heilman KM, Valenstein E (eds), *Clinical Neurosychology*, pp. 337–407. Oxford University Press, New York.

33. Gazzaniga MS. (1995) Principles of human brain organization derived from split-brain studies. *Neuron* **14**: 217–228.

34. Kim C-Y, Blake R. (2004) Psychophysical magic: Rendering the visible 'invisible'. *Trends Cogn Sci* **9**: 381–388.

35. Blake R, Logothetis NK. (2002) Visual competition. *Nat Rev Neurosci* **3**: 13–21.

36. Logothetis N. (1998) Single units and conscious vision. *Philos Trans R Soc Lond B* **353**: 1801–1818.

37. Leopold DA, Logothetis NK. (1996) Activity changes in early visual cortex reflect monkeys' percepts during binocular rivalry. *Nature* **379**: 549–553.

38. Sheinberg DL, Logothetis NK. (1997) The role of temporal cortical areas in perceptual organization. *Proc Natl Acad Sci USA* **94**: 3408–3413.

39. Kreiman G, Fried I, Koch C. (2002) Single-neuron correlates of subjective vision in the human medial temporal lobe. *Proc Natl Acad Sci USA* **99**: 8378–8383.

40. Tsuchiya N, Koch C. (2005) Continuous flash suppression reduces negative afterimages. *Nat Neurosci* **8**: 1096–1101.

41. Rees G, Frith C. (2007) Methodologies for identifying the neural correlates of consciousness. In: Velmans M, Schneider S (eds), *The Blackwell Companion to Consciousness*, pp. 553–566. Blackwell, Oxford, UK.

42. Haynes JD, Rees G. (2005) Predicting the orientation of invisible stimuli from activity in human primary visual cortex. *Nat Neurosci* **8**: 686–691.

43. Lee SH, Blake R, Heeger DJ. (2007) Hierarchy of cortical responses underlying binocular rivalry. *Nat Neurosci* **10**: 1048–1054.

44. Crick FC, Koch C. (1995) Are we aware of neural activity in primary visual cortex? *Nature* **375**: 121–123.

45. Braun AR, Balkin TJ, Wesensten NJ, Gwadry F, Carson RE, Varga M, Baldwin P, Belenky G, Herscovitch P. (1998) Dissociated pattern of

activity in visual cortices and their projections during human rapid eye movement sleep. *Science* **279**: 91–95.

46. Sergent C, Dehaene S. (2004) Is consciousness a gradual phenomenon? Evidence for an all-or-none bifurcation during the attentional blink. *Psychol Sci* **15**: 720–728.

47. Purves D, Paydarfar JA, Andrews TJ. (1996) The wagon wheel illusion in movies and reality. *Proc Natl Acad Sci USA* **93**: 3693–3697.

48. VanRullen R, Koch C. (2003) Visual selective behavior can be triggered by a feed-forward process. *J Cogn Neurosci* **15**: 209–217.

49. Sacks O. (2004) In the river of consciousness. *New York Rev Books* **51**: 41–44.

50. Zeki S. (1998) Parallel processing, asynchronous perception, and a distributed system of consciousness in vision. *Neuroscientist* **4**: 365–372.

51. Bartels A, Zeki S. (2006) The temporal order of binding visual attributes. *Vision Res* **46**: 2280–2286.

52. Stetson C, Cui X, Montague PR, Eagleman DM. (2006) Motor-sensory recalibration leads to reversal of action and sensation. *Neuron* **51**: 651–659.

53. Koch C, Tsuchiya N. (2007) Attention and consciousness: Two distinct brain processes. *Trends Cogn Sci* **11**: 16–22.

54. Naccache L, Blandin E, Dehaene S. (2002) Unconscious masked priming depends on temporal attention. *Psychol Sci* **13**: 416–424.

55. Milner AD, Goodale MA. (1995) *The Visual Brain in Action.* Oxford University Press, Oxford, UK.

56. Koch C, Crick FC. (2001) On the zombie within. *Nature* **411**: 893.

57. Beilock SL, Carr TH, MacMahon C, Starkes JL. (2002) When paying attention becomes counterproductive: Impact of divided versus skill-focused attention on novice and experienced performance of sensorimotor skills. *J Exp Psychol Appl* **8**: 6–16.

58. Corbetta M, Shulman GL. (2002) Control of goal-directed and stimulus-driven attention in the brain. *Nat Rev Neurosci* **3**: 201–215.

59. Thorpe S, Fize D, Marlot C. (1996) Speed of processing in the human visual system. *Nature* **381**: 520–522.

60. Baars BJ. (1988) *A Cognitive Theory of Consciousness.* Cambridge University Press, New York, NY.

61. Dehaene S, Sergent C, Changeux JP. (2003) A neuronal network model linking subjective reports and objective physiological data during conscious perception. *Proc Natl Acad Sci USA* **100:** 8520–8525.

62. Jazayeri M, Movshon JA. (2007) A new perceptual illusion reveals mechanisms of sensory decoding. *Nature* **446:** 912–915.

63. Tononi G, Edelman GM. (1998) Consciousness and complexity. *Science* **282:** 1846–1851.

64. Massimini M, Ferrarelli F, Huber R, Esser SK, Singh H, Tononi G. (2005) Breakdown of cortical effective connectivity during sleep. *Science* **309:** 2228–2232.

65. Han X, Boyden ES. (2007) Multiple-color optical activation, silencing, and desynchronization of neural activity, with single-spike temporal resolution. *PLoS ONE* **2:** e299.

66. Adamantidis AR, Zhang F, Aravanis AM, Deisseroth K, de Lecea L. (2007) Neural substrates of awakening probed with optogenetic control of hypocretin neurons. *Nature* **450:** 420–424.

67. Lein ES, Hawrylycz MJ, Ao N, *et al.* (2007) Genome-wide atlas of gene expression in the adult mouse brain. *Nature* **445:** 168–176.

68. Crick FC, Koch C. (2003) A framework for consciousness. *Nat Neurosci* **6:** 119–127.

chapter 16

Computer-Aided Drug Discovery
Physics-based Simulations from the
Molecular to the Cellular Level

*J. Andrew McCammon**

Advances in chemical theory, structural biology, and computer technology continue to increase the power of physics-based methods for biomolecular modeling. Computer simulations at the molecular level have recently played a key role in the discovery of the first in a new class of antiviral drugs for HIV/AIDS. Simulations at the supramolecular and cellular level will play an increasing role in drug discovery in the coming years.

1. Introduction

Computation has long served as a helpful tool in modern drug discovery. One of the original approaches in computer-aided drug discovery, the Hansch method, accounts for the physical properties of substituents in the spirit of the Hammett correlations in physical organic chemistry.[1] Other computational approaches have subsequently proven helpful in drug discovery. Structure-based computational methods are particularly notable.[2–4] The present account outlines the progress that has been made with physics-based simulations in drug discovery. At the most detailed level, such methods make use of quantum mechanics or molecular mechanics to describe the atomic interactions among prospective drugs or other ligands

*Howard Hughes Medical Institute, NSF Center for Theoretical Biological Physics, Department of Chemistry and Biochemistry, Department of Pharmacology, University of California at San Diego, La Jolla, CA 92093-0365, USA, e-mail: jmccammon@ucsd.edu.

and the macromolecular targets to which these compounds may bind. The earliest studies of this kind were essentially static in nature, with the molecules treated as rigid or nearly rigid entities. Subsequently, dynamical simulations have proven very effective in guiding drug discovery by allowing consideration of the conformational adaptation that is typically associated with ligand binding. Current work includes the extension of physics-based approaches to the supramolecular and subcellular levels. Such models are expected to be valuable in considering the concerted action of one or more drugs at multiple molecular targets.

In the most quantitative approaches to computer-aided drug discovery, one attempts to calculate the strength of interaction of ligands with the molecules to which they might bind. The latter may include not only the desired target macromolecule, but also other macromolecules in studies of binding selectivity. Such calculations were attempted, for example, by the Scheraga group as early as 1972.[5] In this work, the binding energies of substrate molecules to an enzyme were estimated by use of molecular mechanics methods that had been introduced 26 years earlier by Dostrovsky *et al.*,[6] by Hill[7] and by Westheimer and Mayer.[8] Limitations in computer power did not allow the sampling of configurations required to include solvation or entropic effects in this work.

In 1986, Wong and McCammon combined the statistical mechanical theory of free energy with atomistic simulations of solvent and solutes to calculate, for the first time, the relative free energy of binding of different small inhibitor molecules to an enzyme.[9] The necessary statistical mechanical theory had been available for many years, but to capture the solvation, entropic, and other temperature-dependent factors essential for a complete thermodynamic treatment, several other advances were required. One was the introduction of molecular dynamics simulations into biochemistry in 1977.[10] In such simulations, Newton's equations of motion are used with a model of the forces within a molecular system to generate trajectories of the atomic fluctuations. The 1986 work was made possible in part by the growth of computer power, by roughly a factor of 100 over the 10 years from 1977 according to Moore's law. The 1977 molecular dynamics simulation was limited to a very small protein with no explicit solvent, but the 1986 work utilized a molecular dynamics simulation of the large enzyme trypsin in a bath of explicitly represented water molecules. But, as with many advances in computational chemistry, the key factor leading to the 1986 breakthrough was a new theoretical element, in this case the concept of using thermodynamic cycles to relate the desired free energy change to those for two nonphysical processes: computational "alchemical" transformations of one inhibitor into another one, in solution and in the binding site.[11]

While free energy calculations for the interaction of prospective drugs and targets have proven helpful in drug discovery,[12] computer power has generally not allowed the extensive sampling of the rough, many-dimensional potential energy surfaces required for accurate treatment of such systems. This is gradually changing as a result of steady increases in computer power and, again, as a result of theoretical and algorithmic advances. Some of this progress is outlined briefly below.

A newly emerging frontier in computer-aided drug discovery is taking shape with the advent of physically-based simulations of supramolecular and subcellular systems. There are often advantages in treating patients with multiple drugs in coordinated fashion, including simultaneous administration.[13] The availability of physical models of more complete physiological systems is expected to be helpful in planning such treatment regimens.

2. Free Energy Calculations

For calculations of relative free energies of binding, the theoretical framework outlined by Tembe and McCammon[11] has often been used essentially without change. This framework recognizes that brute force calculations of standard free energies of binding will encounter convergence problems related to the dramatic changes in solvation of the binding partners, conformational changes that require physical times longer that those that can be explored by simulation, etc. Tembe and McCammon[11] introduced the use of thermodynamic cycle analyses that allow the desired relative free energies to be computed in terms of "alchemical" transformations, as described above. The advantage is that only relatively localized changes occur in the simulated system, at least in favorable cases.

Calculation of the standard free energy of binding itself can be viewed as a special case of the above, in which one of the pair of ligands contains no atoms.[14] Some care is required to be sure that such calculations yield answers that actually correspond to the desired standard state.[15,16]

It has been mentioned that perhaps the greatest limitation to the precision of free energy calculations to date has been the often-inadequate sampling of a representative set of configurations of the system. Increases in computer power of course increase the "radius of convergence" of such calculations. Such increases come not only from the Moore's Law improvements in hardware, but also from algorithmic advances for parallelization and for increasing time steps in molecular dynamics.[17] New theoretical methods have also been developed to speed convergence. One such method is the use of soft-core solute models, so that one simulation can generate an adequate reference ensemble for a family of alchemical changes.[18,19] Hamelberg *et al.* have recently introduced an "accelerated molecular dynamics" method that substantially improves sampling, while preserving the ability to recover thermodynamic data.[20] It has been shown that such simulations can be used to speed the calculations of entropies, in particular.[21]

More rapid convergence of free energy calculations can also be obtained by replacing part of the system with a simpler model, such as a continuum model for the solvent. This has the advantage of obviating the need for sampling the configurations of this part of the system, reducing both noise and the computation time required. In view of the important role that specific hydrogen bonds may play, the combination of fully atomistic simulations with subsequent continuum analyses is probably a more reliable procedure than using a continuum solvent model exclusively. The Kollman group demonstrated impressive success with this approach to calculations of free energies of binding.[22] The rigor and accuracy of such calculations can be increased by explicit treatment of solute translational and rotational motions to provide proper reference to standard states.[15,23]

3. Large Changes of Conformation

It is sometimes true that the binding of a ligand to a protein is associated with large changes in conformation of one or both molecules. Our group has recently developed a "relaxed complex" approach to the particularly challenging case in which the protein undergoes conformational changes.[24,25] The basic idea is simple. One selects a large number of "snapshot" conformations from a molecular dynamics simulation of the unliganded protein. Methods such as accelerated molecular dynamics can alternatively be used to generate snapshots of a more diverse set of protein conformations.[20] One or a set of ligand molecules can then be docked to these snapshots, using any convenient rapid docking algorithm. The most tightly bound complexes can then be rescored using higher accuracy methods, ideally based on rigorous statistical mechanical foundations.[23]

Early applications of the relaxed complex approach have already yielded a dramatic success in drug discovery. Most of the antiviral drugs currently available for treatment of HIV/AIDS are inhibitors of two of the three enzymes associated with HIV, namely, the reverse transcriptase and the protease. The third enzyme, the integrase that catalyzes integration of the virally encoded, reverse-transcribed DNA into the genome of host cells, has been much more difficult to target. Molecular dynamics simulations as part of a relaxed complex study revealed surprisingly large fluctuations in the enzyme, intermittently exposing a cryptic binding site not seen in the starting crystal structure; see Fig. 1.[26,27] These unexpected results suggested an approach to the development of integrase inhibitors with unique resistance profiles, an approach that led to the first in a new class of drugs for HIV/AIDS, the compound raltegravir, which was approved by the US Food and Drug Administration in 2007.[28]

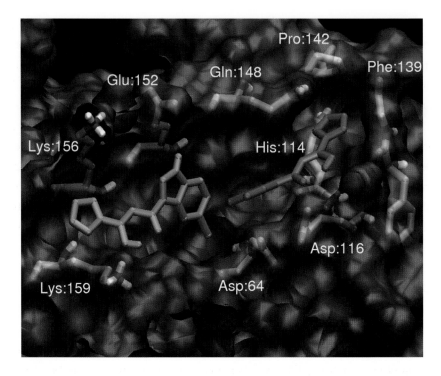

Fig. 1. Two computed binding orientations of an HIV-1 integrase inhibitor to a molecular dynamics snapshot of the protein.[26] The green conformation is similar to that in a crystal structure; the magenta is in an unexpected binding site discovered by our computational Relaxed Complex Scheme. The discovery of "this unexpected result … suggests an approach to the development of integrase inhibitors with unique resistance profiles,"[28] leading to the invention of raltegravir by Merck & Co. This is the first in a new class of antiviral drugs for HIV/AIDS; it was approved by the US Food and Drug Administration in October, 2007.

4. The Future of Computer-Aided Drug Discovery

While much remains to be done to improve the computational tools for drug discovery that are focused on single molecular targets, advances in computer power, in our understanding of cellular anatomy and physiology, and in theoretical and mathematical areas are opening the way for a more integrated approach to drug discovery at the supramolecular level. Some existing drugs are known to act on more than one macromolecular target. An example is galantamine, which appears to restore some cognitive function in Alzheimer's patients by simultaneously inhibiting acetylcholinesterase and activating acetylcholine receptors in the hippocampal synapses.[29] Optimizing the effects of single drugs with multiple sites of action, or of multiple drugs administered in combination, is likely to benefit from models that describe the various targets in integrated models. Our group has recently employed modern methods for solving reaction-diffusion equations to simulate synaptic activity using such integrated models.[30] In the future, it should be possible to upload the results of simulations of the molecular components of such systems into these integrated models to allow analyses of drug effects at higher scales.

Acknowledgments

This work has been supported in part by grants from NIH, NSF, the NSF Center for Theoretical Biological Physics, the National Biomedical Computing Resource, the NSF Supercomputer Centers, the W.M. Keck Foundation, and Accelrys, Inc.

References

1. Hansch C. (1969) A quantitative approach to biochemical structure-activity relationships. *Acc Chem Res* **2:** 232–239.
2. Stroud R (ed). (2007) *Computational and Structural Approaches to Drug Discovery: Ligand-Protein Interactions.* Royal Society of Chemistry, London.
3. Marrone TJ, Briggs JM, McCammon JA. (1997) Structure-based drug design: Computational advances. *Annu Rev Pharmacol Toxicol* **37:** 71–90.
4. Ulm EH, Greenlee WJ. (1989) Inhibitors of the renin-angiotensin system enzymes. In: Sandler M & Smith HJ (eds), *Design of Enzyme Inhibitors as Drugs*, pp. 146–177. Oxford University Press, New York.

5. Platzer KEB, Momany FA, Scheraga HA. (1972) Conformational energy calculations of enzyme-substrate interactions 2. Computation of binding-energy for substrates in active-site of alpha-chymotrypsin. *Int J Peptide Protein Res* **4:** 201–219.

6. Dovstrovsky I, Hughes ED, Ingold CK. (1946) Mechanism of substitution at a saturated carbon atom. Part XXXII. The role of steric hindrance. (Section G) magnitude of steric effects, range of occurrence of steric and polar effects, and place of the Wagner rearrangement in nucleophilic substitution and elimination. *J Chem Soc* 173–194.

7. Hill TL. (1946) On steric effects. *J Chem Phys* **14:** 465.

8. Westheimer FH, Mayer JE. (1946) The theory of the racemization of optically active derivatives of diphenyl. *J Chem Phys* **14:** 733–738.

9. Wong CF, McCammon JA. (1986) Dynamics and design of enzymes and inhibitors. *J Amer Chem Soc* **108:** 3830–3833.

10. McCammon JA, Gelin BR, Karplus M. (1977) Dynamics of folded proteins. *Nature* **267:** 585–590.

11. Tembe BL, McCammon JA. (1984) Ligand-receptor interactions. *Comput Chem* **8:** 281–283.

12. Reddy MR, Varney JMD, Kalish V, Viswanadhan VN, Appelt K. (1994) Calculation of relative differences in the binding free energies of HIV-l protease inhibitors: A thermodynamic cycle perturbation approach. *J Med Chem* **37:** 1145–1152.

13. Stommel JM, Kimmelman AC, Ying H, Nabioullin R, Ponugoti AH, Wiedemeyer R, Stegh AH, Bradner JE, Ligon KL, Brennan C, Chin L, DePinho RA. (2007) Coactivation of receptor tyrosine kinases affects the response of tumor cells to targeted therapies. *Science* **318:** 287–290.

14. Jorgensen WL, Buckner JK, Boudon S, Tirado-Rives J. (1988) Efficient computation of absolute free-energies of binding by computer-simulations — application to the methane dimer in water. *J Chem Phys* **89:** 3742–3746.

15. Gilson MK, Given JA, Bush BL, McCammon JA. (1997) The statistical-thermodynamic basis for computation of binding affinities: A critical review. *Biophys J* **72:** 1047–1069.

16. Hermans J, Wang L. (1997) Inclusion of loss of translational and rotational freedom in theoretical estimates of free energies of binding.

Application to a complex of benzene and mutant T4 lysozyme. *J Amer Chem Soc* **119**: 2707–2714.

17. Schlick T, Skeel RD, Brunger AT, Kale LV, Board JA, Hermans J, Schulten K. (1999) Algorithmic challenges in computational molecular biophysics. *J Comp Phys* **151**: 9–48.

18. Liu H, Mark AE, van Gunsteren WF. (1996) Estimating the relative free energy of different molecular states with respect to a single reference state. *J Phys Chem* **100**: 9485–9494.

19. Mordasini TZ, McCammon JA. (2000) Calculations of relative hydration free energies: A comparative study using thermodynamic integration and an extrapolation method based on a single reference state. *J Phys Chem* **B104**: 360–367.

20. Hamelberg D, Mongan J, McCammon JA. (2004) Accelerated molecular dynamics: A promising and efficient simulation method for biomolecules. *J Chem Phys* **120**: 11919–11929.

21. Minh DLD, Hamelberg D, McCammon JA. (2007) Accelerated entropy estimates with accelerated dynamics. *J Chem Phys* **127**: 154105-1–154105-5.

22. Massova I, Kollman PA. (2000) Combined molecular mechanical and continuum solvent approach (MM-PBSA/GBSA) to predict ligand binding. *Perspect Drug Discov* **18**: 113–135.

23. Swanson JMJ, Henchman RH, McCammon JA. (2004) Revisiting free energy calculations: A theoretical connection to MM/PBSA and direct calculation of the association free energy. *Biophys J* **86**: 67–74.

24. Lin JH, Perryman AL, Schames JR, McCammon JA. (2002) Computational drug design accommodating receptor flexibility: The relaxed complex scheme. *J Amer Chem Soc* **124**: 5632–5633.

25. Lin JH, Perryman AL, Schames JR, McCammon JA. (2003) The relaxed complex method: Accommodating receptor flexibility for drug design with an improved scoring scheme. *Biopolymers* **68**: 47–62.

26. Schames J, Henchman RH, Siegel JS, Sotriffer CA, Ni H, McCammon JA. (2004) Discovery of a novel binding trench in HIV integrase. *J Med Chem* **47**: 1879–1881.

27. Goldgur Y, Craigie R, Cohen GH, Fujiwara T, Yoshinaga T, Fujishita T, Sugimoto H, Endo T, Murai H, Davies DR. (1999) Structure of the HIV-1 integrase catalytic domain complexed with an inhibitor: A

platform for antiviral drug design. *Proc Natl Acad Sci USA* **96:** 13040–13043.

28. Hazuda DJ, Anthony NJ, Gomez RP, Jolly SM, Wai JS, Zhuang L, Fisher TE, Embrey M, Guare, Jr. JP, Egbertson MS, Vacca JP, Huff JR, Felock PJ, Witmer MV, Stillmock KA, Danovich R, Grobler J, Miller MD, Espeseth AS, Jin L, Chen IW, Lin JH, Kassahun K, Ellis JD, Wong BK, Xu W, Pearson PG, Schleif WA, Cortese R, Emini E, Summa V, Holloway MK, Young SD. (2004) A naphthyridine carboxamide provides evidence for discordant resistance between mechanistically identical inhibitors of HIV-1 integrase. *Proc Natl Acad Sci USA* **101:** 11233–11238.

29. Woodruff-Pak DS, Vogel RW, Wenk GL. (2001) Galantamine: Effect on nicotinic receptor binding, acetylcholinesterase inhibition, and learning. *Proc Natl Acad Sci USA* **98:** 2089–2094.

30. Cheng Y, Suen JK, Radic Z, Bond SD, Holst MJ, McCammon JA. (2007) Finite element simulations of acetylcholine diffusion and reaction in neuromuscular junction models. *Biophys Chem* **127:** 129–139.

chapter 17

Precision Measurements in Biology

Stephen R. Quake[*]

Is biology a quantitative science like physics? In this chapter I will discuss the role of precision measurement in both physics and biology, and argue that in fact both fields can be tied together by the use and consequences of precision measurement.

1. Introduction

"Measure what is measurable, and make measurable what is not so."
Galileo Galilei

There is a great tradition of discovery through precision measurement in physics. This dates to the 17th century, when Galileo Galilei built the most powerful telescope of his day. He pointed it to the night sky, discovered a large number of heavenly bodies that were not visible to the naked eye, and showed that the Milky Way was in fact made up of many stars. He discovered the phases of Venus, whose pattern provided strong evidence that the Copernican model of a heliocentric universe was correct. Another important discovery was that the moon had structure — mountains and valleys — and was not the "perfect sphere" that Aristotle had claimed. These and other astronomical discoveries were a direct consequence of the

[*]Departments of Bioengineering and Applied Physics, Stanford University and Howard Hughes Medical Institute, Stanford, CA 94305, USA, e-mail: quake@stanford.edu.

precision of his instrument. The concept of precision measurement informed virtually all aspects of Galileo's scientific research — he also used careful, quantitative measurements of balls rolling down inclined planes to discover that acceleration is independent of mass and that the distance traveled by an accelerating object is proportional to the square of time. He is widely recognized as *the* pioneer who demonstrated how quantitative experiments, mathematics, and theoretical physics combine to advance the cause of science.[1] Albert Einstein regarded him as "the father of modern science".[2]

This defining tradition continued unabated as the history of physics unfolded and the 20th century provided many wonderful examples of discovery through precision measurement, such as the story of hydrogen. Continuing improvements in the ability to measure the properties of this simple atom have led to major discoveries which forced theorists time and time again to revise fundamental theories of matter. Otto Stern and Immanuel Estermann pushed the limits of existing technology to make the first measurement of the magnetic moment of the proton, and discovered that it was substantially larger than Dirac's theory of relativistic quantum mechanics predicted. I.I. Rabi and his colleagues invented the magnetic resonance method to improve the precision of such measurements, and they discovered that the deuteron has a quadrupole moment, necessitating major revisions of our understanding of what holds together the atomic nucleus. Willis Lamb used a variation of this approach to perform microwave spectroscopy of the difference between two nominally equal states of hydrogen and discovered a small energy splitting, which in turn inspired the development of the theory of quantum electrodynamics. These are all justifiably regarded as major achievements in the history of physics, and they are tied together by ever increasing precision of measurement.[3]

In addition to direct scientific discovery, there is another thread that runs through the history of precision measurement and physics: such measurements are usually enabled by clever development of new instruments and new technologies. The technologies themselves often have far-reaching applications well beyond their original purpose. Galileo also made telescopes for Venetian merchants, who used them for shipping and trade. He used his knowledge of optics to develop one of the first compound microscopes, which was used to study insects. He proposed that knowledge of the position of the orbits of the moons of Jupiter could be used as a universal clock that would allow one to determine longitude — a method that was later used by Lewis and Clark as they surveyed the American West. Galileo also discovered that the period of a pendulum is independent of its amplitude, and used this principle to design a clock. Rabi realized that his magnetic resonance method could be used to create extraordinarily accurate clocks, and indeed it became an essential component for the atomic clocks which are the basis for today's global positioning system (GPS). The magnetic resonance method also led to magnetic resonance imaging (MRI), the widely used medical imaging technology. Time and time again, the quest for higher precision measurement has paid unexpected dividends in the invention of new technologies with practical application.

What is the role of precision measurement in biology? As the old joke goes, of course biology is a quantitative science: gene expression either goes up or it goes down! There are now methods to measure the expression of nearly all the genes in an organism, but this is simply parallelism and should not be confused with precision. Similarly, genome sequencing is a triumph of biological automation but not necessarily of precision. My approach to physical biology will be to ask how the concepts of precision measurements in physics can be brought to bear on biological systems. What I will discuss here are precision measurement techniques that allow one to analyze the behavior of single molecules and single cells with unprecedented sensitivity, and the science that has flowed from them. I will describe how this approach enabled new types of measurements and the invention of new technologies for biology, just as it has with physics. This discussion will be largely populated with examples from my own research, so it should not be considered a dispassionate review of the field.

2. Single Molecule Biophysics

"I was fortunate at Columbia, there was no publish or perish thing."
I.I. Rabi

The forces created by individual biological macromolecules are exquisitely small, and can derive from a number of sources.[4] There is the intrinsic stiffness of the molecule, which gives it a mechanical bending force that is surprisingly well described by the same kind of continuum mechanics developed to understand beams and girders. If the molecule is flexible enough, there is an entropic force generated by interaction with a thermal bath — this is the same kind of statistical physics used to describe how rubber bands and other polymers work. In some cases there are also chemical forces — either from chemical bonds that can associate and dissociate (such as by pulling apart the two strands making up the double helix of DNA) or by catalysis of some fuel such as ATP which converts chemical energy into work. Over the last two decades, an enormous amount of effort has gone into making precision measurements of these forces, and the results have been profound.[5]

I was lucky enough to get in on the ground floor of single molecule biophysics. As an undergraduate at Stanford, I was inspired by a course from Steve Chu to join his lab for a senior thesis. I started working with him the spring of my junior year, in 1990. He and Art Ashkin had invented optical tweezers about five years earlier, and Steve's dream was to use them to manipulate individual molecules such as DNA, and to watch various enzymes such as DNA and RNA polymerase interacting with DNA. This was really early days; single molecule biophysics was in its infancy. Analytical chemists had just worked out that it was possible to image the fluorescence from single organic fluorophores, and a few biochemists had demonstrated that they could study individual motor proteins such as myosin and kinesin by staining large macromolecular filaments of actin or microtubules and observing their motion across surfaces as they were driven by individual molecular motors. Steve Chu and Steve Kron had demonstrated that they could glue DNA molecules onto micron sized latex beads, and manipulate them with optical tweezers — this was probably the first direct manipulation of single molecules. When I joined the lab Steve Chu asked me to try to measure the force required to extend a single DNA molecule.

The work was great fun, and it is hard to describe the intense fascination I felt while pulling on a bead with optical tweezers, only to see it pop out under the force of an invisible DNA spring that kept it tethered to the surface. After months of work with Libby Sunderman, we finally were able to make precise measurements of the bead as the DNA pulled it. By plotting the bead position as a function of time, I could estimate the spring constant of the DNA molecule and test whether or not it obeyed Hooke's law, i.e., whether it had a linear force versus extension curve (Fig. 1). In fact it was linear except at the highest extensions, at which there were deviations. I was

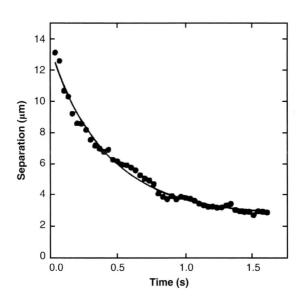

Fig. 1. The relaxation of a single molecule of lambda phage DNA after it has been stretched out with optical tweezers. The solid line shows the expected behavior if the restoring force is linear, a valid approximation for small and moderate displacements. From Ref. 6.

incredibly excited by my first "real" science, even though the results were only published in my thesis in June 1991 and in a review article that Steve Chu wrote later that year.[6]

A year later I was back at Stanford and spent a considerable amount of time refining the experiment to make static measurements of the forces, by holding the DNA at a fixed extension and measuring the tiny displacements of the bead in the optical tweezers. From these measurements it became clear that the deviations we had seen before were in fact significant effects and that DNA at high extension was not a Hooke's law spring. I found predictions in the literature that this would happen, but our data did not agree exactly with the theory. As we were trying to sort through the reasons why, one morning in November 1992 I picked up the latest issue of *Science* and found that Carlos Bustamante and colleagues had just published a paper in which they reported force extension curves from single molecules of DNA, which they had measured using a combination of magnetic and fluid forces applied to a bead.[7] They also found deviations from the theoretical predictions, but did not wait for theory to agree with their experiment in order to publish! A few years later Eric Siggia and John Marko worked out the correct theory for highly extended polymers and were able to explain the Bustamante (and my) data sets.[8]

This sequence of events made a strong impression on me, and illustrates the continuing interplay between theory and experiment. In this case, the development of new precision techniques for measuring forces on individual molecules pushed theorists to rethink what most people accepted as a "solved" problem. More generally, these techniques of measuring forces on single molecules have now been successfully applied by many researchers to numerous areas of biology, including various motor proteins, DNA replication, RNA polymerase, and chromatin structure.[5]

A practical application of precision measurement in single molecule biophysics was the development of the first single molecule DNA sequencer. Sequencing single molecules of DNA had been a significant goal since the early days of single molecule biophysics, dating back to the mid-1980s.

People imagined that single molecule techniques would permit radically improved sequencing throughput and orders of magnitude lower cost than conventional approaches. However, achieving both the sensitivity and the resolution to detect single bases proved elusive. Many different ideas were attempted — various forms of force spectroscopy, enzymatic digestion of DNA followed by spectroscopy of the bases, impedance spectroscopy of DNA through small pores — and none succeeded.[9] Rather than measure forces, I decided to take an imaging approach in which we measured the fluorescence of single labeled nucleotide molecules as DNA polymerase inserted them in a growing DNA strand, replicating the sequence of the original template. This avoids the need to directly image the DNA at sub-nanometer resolution because it substitutes temporal resolution for spatial

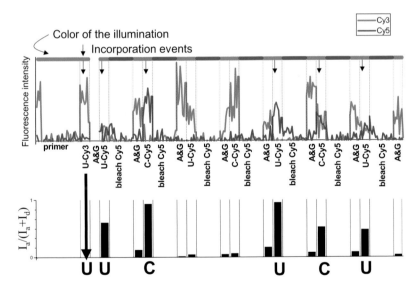

Fig. 2. *Top*: Fluorescence signals from the first single molecule DNA sequencing experiment. The green trace shows signal from the donor fluorophore, which is attached to the DNA template. The red trace shows signal from the acceptor fluorophore, which when present indicates that a nucleotide was incorporated. *Bottom*: The fluorescence resonance transfer efficiency between the green and red traces provides a clear signal indicating the relative positions of U's and C's in a progressively sequenced DNA molecule. From Ref. 10.

resolution (Fig. 2). My postdoc Ido Braslavsky worked for three years on this project along with graduate students Emil Kartalov and Ben Hebert, and in 2003 we ultimately succeeded in making the first demonstration of single molecule DNA sequencing.[10]

While our first sequencer did not have impressive throughput, the results were promising enough to convince some visionary investors to form a company called Helicos BioSciences to commercialize the technology (full disclosure: I am a co-founder and shareholder). The successive generations of instruments developed by this company have made single molecule sequencing a practical reality, and it is now performed on a daily basis. It has been used to resequence everything from viral genomes[11] to cancer-related exons in the human genome, to discover micro RNAs, and to perform digital gene expression assays. The most recent instrument follows hundreds of millions of individual DNA templates with single molecule and single base resolution as they are sequenced, and its phenomenal sequencing throughput might enable the first US$1,000 human genome. While there are many interesting aspects to the Helicos story, perhaps the most relevant to the theme of this chapter is that it illustrates the close connection between precision measurement and new technology — the commercial development of single molecule DNA sequencing was a direct consequence of years of work doing esoteric biophysics, connected by a thread of ever improving precision measurement techniques.

3. Single Cell Analysis

"It is because simplicity and vastness are both beautiful that we seek by preference simple facts and vast facts; that we take delight, now in scrutinizing with a microscope that prodigious smallness which is also a vastness..." Henri Poincaré

Another example of the connections between precision measurement, technology development and biological discovery is given by the story of the biological equivalent of the integrated circuit. The silicon integrated circuit changed our world because it automated mathematical calculation to an extent never before possible, and it is natural to ask if one could achieve similar results in biology by automating wet bench experiments. Throughout the 1990s, many scientists and engineers around the world were struggling to figure out how to apply the ideas used to fabricate computer chips to biological analysis. Most people took the electrical analogy literally, and tried to manipulate biological molecules with electric fields in passive channel networks — this is the equivalent of analog electronics. Their success was limited for the same reasons that integrated circuits do not use analog electronics — it is difficult to scale such circuits to any kind of complexity. The essence of modern electrical engineering is digital electronics, which permits modular design and therefore allows one to create arbitrarily complex circuits. My group and I eventually invented what is now called microfluidic large scale integration (LSI): the ability to fabricate thousands of micromechanical valves on a single chip.[12] This let us create any "plumber's nightmare" of tiny pipes, chambers, valves and pumps and was the first practical equivalent of digital electronics.[13]

Microfluidic LSI is useful for more than automation: it also enables new types of precision measurements. This is due in part to the fact that there exist a number of interesting physical and chemical effects which only occur in the nanoliter scale volumes that these devices manipulate.[14] Exploiting these effects has led to discoveries in emerging fields such as systems biology[15] and synthetic biology.[16] Such effects have also been used to invent new technologies[17–20] for what is perhaps the original precision measurement in biology: protein crystallography. These technologies have in turn been applied to solve a growing number of novel protein structures.[21–29] Here I will focus only on the application of microfluidic LSI to a particular topic: single cell analysis.

Single cells that are nominally genetically identical are in fact chemically quite heterogeneous. Measurements on bulk distributions of cells mask many important aspects about their behavior. One of the more interesting has to do with expression of genes. There has been quite a bit of work to show that protein expression levels are highly heterogeneous,[30] but it is challenging to make absolute measurements and therefore difficult to rigorously understand the fundamental statistical and biological processes governing this phenomenon. We have spent some effort making absolute measurements of gene expression at the mRNA level.

With conventional PCR it is quite challenging to make absolute measurements: the results are influenced by the quality of the reagents, which can fluctuate on a daily basis, and it is not easy to calibrate them. There is always considerable uncertainty in the final result since one is essentially measuring the log of the expression level. Because of this, it is often difficult to compare the results measured by different groups, and it can be difficult to interpret or replicate results. To circumvent these difficulties, people often resort to relative measurements and compare the expression of the gene of interest to some internal standard, such as a housekeeping gene. However, this has a number of limitations, not least of which is that the quality of the reagents can still be gene-specific.

There is a powerful way to make absolute, linear quantitation of PCR, which is called digital PCR.[31] This method involves partitioning the sample into many small subsamples such that there is no more than one template molecule per subsample. By performing PCR on all the subsamples and counting the number of positive results, one can actually count the number of starting template molecules in a manner that is essentially independent of the amplification efficiency, and hence reagent quality. This method lends itself particularly well towards microfluidic automation, and in fact benefits from improved PCR sensitivity in small volumes.[32] Fluidigm used microfluidic LSI to create a practical and sensitive digital PCR technology (full disclosure: I am a co-founder and shareholder). My student Luigi Warren showed that Fluidigm's device could be used to make absolute measurements of gene expression levels in individual hematopoietic progenitor cells — the stem cells of the blood and immune system.[33] One of the genes we studied is a transcription factor, present at only a handful of

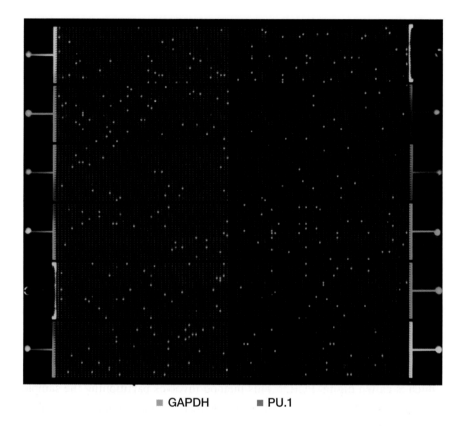

■ GAPDH   ■ PU.1

Fig. 3. Digital PCR allows one to count individual molecules with high sensitivity and precision. This fluorescent image shows the gene expression of the transcription factor PU.1 and the housekeeping gene GAPDH from single murine hematopoietic stem cells. Each of the twelve panels are the results of a single cell, and each bright spot shows the amplification from a single starting cDNA molecule (GAPDH are green, PU.1 are red) from that cell. From Ref. 33.

copies per cell. The other gene is a housekeeping gene which is present at much higher copy number (Fig. 3).

The transcription factor was the interesting stem cell biology story, but from the point of view of physical biology I would like to focus on the humble housekeeping gene GAPDH. We found that the distribution of

GAPDH expression levels for a given cell type is surprisingly broad — tending to be more lognormal than Gaussian. We found that the distributions depend on the cell type: both the means and the widths of the distributions vary significantly. At the very least, this means that one has to be careful what one is normalizing by. In this particular case, the distributions changed in such a way that they suppressed the effect we were interested in. That is, in the cell type for which the transcription factor was more highly expressed, so was GAPDH. Normalizing to GAPDH tends to artificially suppress the true expression level of the transcription factor. Think of how much more powerful each biological experiment would be if gene expression results were measured and stated as an absolute quantity — number of messages per cell — and the results from lab to lab and publication to publication could be compared and synthesized with ease. The connection between theory and experiment would become much closer and collectively we would create an accurate atlas not only of the behavior of the average cell, but also of the range of behaviors from cell to cell.

There is another situation where the precision of performing single cell analysis is important, and it is somewhat the opposite of what we were just discussing. We have been talking about how nominally identical cells in a pure population in fact display surprising heterogeneity. Now let us consider a population which is highly heterogeneous — containing not only different species, but cells from divergent phyla. This is in fact the case for nearly every environmental ecosystem, whether it is a spoonful of soil, a thimble of pond water, or even the human mouth. Each of these ecosystems has literally thousands of different microbes, and 99% of them cannot be grown in laboratory culture. It would therefore be incredibly useful to be able to perform single cell genetic analyses to characterize the microbes that make up these environmental ecosystems.

Jared Leadbetter is interested in the bacteria that live inside the guts of termites. Termites eat wood and make acetate, and they cannot do that alone. Crucial biochemical steps are performed by the commensal bacteria that live in their guts. This is known because people have studied the biochemistry of this process in great detail, and key enzymes are missing from the termite genome. Rather, those genes are found in what is now called the termite gut "metagenome." That is simply a fancy way of saying that DNA from all the microbes living in the termite gut was prepared and analyzed — in this case as a big mess so that all the genomes of all the different microbe species are mixed together. Unfortunately, this process loses track of the fundamental unit of biology — the cell.

Jared has spent many years trying to culture the mysterious microbes which carry the FTHFS gene, to no avail. He was working on a specific example of a very general problem in microbiology: global studies show that the gene for an enzyme of interest is harbored somewhere in the environmental ecosystem, but there is no clue as to which microbes may have it. We used digital PCR as a general solution to this problem: to map gene function to organism identity. In this case Jared's student Liz Otteson loaded the actual microbes from the termite gut into the digital PCR chip, and used it as a ten thousand well parallel PCR experiment. She used two primer sets: one for the gene of interest (FTHFS) and another for the 16S ribosomal RNA gene, which is used as the universal definition of identity for microorganisms. By analyzing the contents of the chambers with amplification from both primer sets, we discovered thirteen new species which live in the gut of the termite and carry the gene FTHFS.[34]

I decided to push a bit harder and see if it was possible to get more than just two genes at a time. My postdoc Yann Marcy designed a new type of microfluidic device that functions as a "genome amplifier" (Fig. 4). This device combines on-chip cell sorting in order to select a single bacterium, which is then sorted to one of eight processing chambers. After the eight chambers are filled, a succession of biochemical steps are performed in parallel on each of the isolated cells which lyse the bacteria, denature their genomes, and then amplify their genomes more than a million-fold. Thus, from an individual bacterium's genome, one creates enough DNA that it can be removed from the device, amplified again in microliter volumes, and then sequenced. In this manner one can in principle sequence the genome from an individual bacterium without the need for culture.

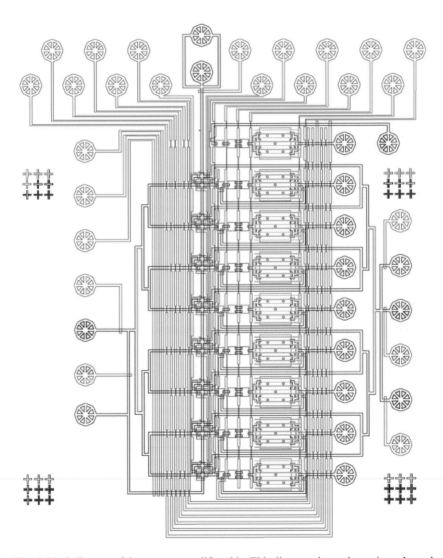

Fig. 4. Mask diagram of the genome amplifier chip. This diagram shows the various channel (blue) and valve (red) structures used to isolate individual microbes into chambers, and subsequently to amplify their genomes. From Ref. 36.

We practiced quite a bit using *E. coli*, and discovered that in fact the biochemical amplifier we used (called multiple displacement amplification) works better in small volumes, probably for the same reasons PCR does (fewer parasitic side reactions due to primer self-interaction and contaminant DNA). We did a side by side test, in which *E. coli* genomes amplified from individual bacteria in 60 nanoliter volumes by Yann were compared to individual bacterial genomes amplified in 50 microliter volumes by Roger Lasken's postdoc Thomas Ishoey.[35] The nanoliter amplifications had much better representation and lower bias than the microliter amplifications — providing an example of greatly improved precision as one scales down volume by orders of magnitude.

Once we were convinced that the genome amplifier was giving good signal from single *E. coli*, we decided to try it out on a real environmental sample. David Relman has been quite interested in bacteria that live in the human oral cavity, in particular those that grow in tooth plaque. There is a species called TM7 that exists in virtually all human mouths, and yet no member of its phylum has ever been cultured or sequenced. We decided to go after that particular bug and managed to sequence the better part of its genome.[36] We discovered more than 1000 new genes, whereas before the only genetic information about TM7 was simply one gene (16S ribosomal RNA). Continued use of such precision measurements will allow us not only to dissect environmental ecosystems, but will also allow us to gain a deeper understanding of the evolutionary relationships between microbes, and will ultimately lead to a deeper understanding of the tree of life.

4. Conclusion

"Where the telescope ends, the microscope begins. Which of the two has the grander view?" Victor Hugo

We have seen that in all the ways that the philosophy of precision measurement has advanced the field of physics, so too can it advance biology. Making absolute measurements of gene expression that are referenced to a universal standard is not unlike using an atomic clock as a universal time standard. Precise measurements of entropic forces on single molecules required revisions to the theory of polymer forces, just as Galileo's experiments with rolling balls led to the mathematics of acceleration, and just as the spectroscopy of hydrogen led to new theories of matter. Using single cell gene and genome analysis to discover and characterize new microbial species is not unlike using the telescope to discover that the Milky Way is composed of individual stars.

It is important to appreciate that precision measurement is not simply about adding a few more decimal places to a physical constant. Rather, it often concerns how to measure a quantity of interest in the presence of competing systematic or physical effects. The magnetic moment of the proton disagreed with theoretical predictions by nearly a factor of 3, and it was still extremely challenging to measure! In other cases, a precision measurement can enable new qualitative discoveries — such as the astronomical surveys of Galileo or the modern surveys of microbial ecology.

The elementary quanta of biology are twofold: the macromolecule and the cell. Cells are the fundamental unit of life, and macromolecules are the fundamental elements of the cell. This chapter has described how precision measurements have been used to explore the basic properties of these quanta, and more generally how the quest for higher precision almost inevitably leads to the development of new technologies, which in turn catalyze further scientific discovery. In the 21st century, there are no remaining experimental barriers to biology becoming a truly quantitative and mathematical science.

Acknowledgments

I thank Rob Phillips and Scott Delp for helpful criticism of this manuscript.

References

1. Stillman D. (1978) *Galileo at Work*. University of Chicago Press, Chicago.
2. Einstein A. (1954) *Ideas and Opinions*, translated by Sonja Bargmann, Crown Publishers, London.
3. Rigden JS. (2002) *Hydrogen: The Essential Element*. Harvard University Press, Cambridge, MA.
4. Phillips R, Quake SR. (2006) The biological frontier of physics. *Phys Today* **59(5):** 38–43.
5. Mehta AD, Rief M, Spudich JA, Smith DA, Simmons RM. (1999) Single-molecule biomechanics with optical methods. *Science* **283:** 1689–1695.
6. Chu S. (1991) Laser manipulation of atoms and particles. *Science* **253(5022):** 861–866.
7. Smith SB, Finzi L, Bustamante C. (1992) Direct mechanical measurements of the elasticity of single DNA molecules by using magnetic beads. *Science* **258(5085):** 1122–1126.
8. Bustamante C, Marko JF, Siggia ED, Smith S. (1994) Entropic elasticity of lambda-phage DNA. *Science* **265(5178):** 1599–1600.
9. Marziali A, Akeson M. (2001) New DNA sequencing methods. *Ann Rev Biomed Eng* **3:** 195–223.
10. Braslavsky I, Hebert B, Quake SR. (2003) Sequence information can be obtained from single DNA molecules. *Proc Natl Acad Sci* **100:** 3960–3964.
11. Harris TD, et al. Single molecule DNA sequencing of a viral genome. *Science* in press.
12. Thorsen T, Maerkl SJ, Quake SR. (2002) Microfluidic large scale integration. *Science* **298:** 580–584.
13. Melin J, Quake SR. (2007) Microfluidic large-scale integration: The evolution of design rules for biological automation. *Annu Rev Biophys Biomol Struct* **36:** 213–231.
14. Squires TM, Quake SR. (2005) Microfluidics: Fluid physics at the nanoliter scale. *Rev Mod Phys* **7(3):** 977–1026.

15. Maerkl SJ, Quake SR. (2007) A systems approach to measuring the binding energy landscapes of transcription factors. *Science* **315**: 233–237.

16. Balagadde FK, You LC, Hansen CL, Arnold FH, Quake SR. (2005) Long-term monitoring of bacteria undergoing programmed population control in a microchemostat. *Science* **309(5731)**: 137–140.

17. Hansen CL, Skordalakes E, Berger JM, Quake SR. (2002) A robust and scalable microfluidic metering method that allows protein crystal growth by free interface diffusion. *Proc Natl Acad Sci* **99**: 16531–16536.

18. Hansen CL, Sommer MOA, Quake SR. (2004) Systematic investigation of protein phase behavior with a microfluidic formulator. *Proc Natl Acad Sci* **101(40)**: 14431–14436.

19. Hansen CL, Classen S, Berger JM, Quake SR. (2006) A microfluidic device for kinetic optimization of protein crystallization and *in situ* structure determination. *J Am Chem Soc* **128(10)**: 3142–3143.

20. Anderson MJ, Hansen CL, Quake SR. (2006) Phase knowledge enables rational screens for protein crystallization. *Proc Natl Acad Sci* **103**: 16746–16751.

21. Xiao T, Takagi J, Coller BS, Wang JH, Springer TA. (2004) Structural basis for allostery in integrins and binding to fibrinogen-mimetic therapeutics. *Nature* **432(7013)**: 59–67.

22. Rowland P, Blaney FE, Smyth MG, *et al.* (2006) Crystal structure of human cytochrom P450 2D6. *J Biol Chem* **281(11)**: 7614–7622.

23. Stevens J, Blixt O, Tumpey TM, Taubenberger JK, Paulson JC, Wilson, IA. (2006) Structure and receptor specificity of the hemagglutinin from an H5N1 influenza virus. *Science* **312(5772)**: 404–410.

24. English CM, Adkins MW, Carson JJ, Churchill ME, Tyler JK. (2006) Structural basis for the histone chaperone activity of Asf1. *Cell*.

25. Kothe M, Kohls D, Low S, *et al.* (2007) Structure of the catalytic domain of human polio-like kinase 1. *Biochemistry* **46(20)**: 5960–5971.

26. Hu SH, Latham CF, Gee CL, Martin JL. (2007) Structure of the Munc18c/Syntaxin4 N-peptide complex defines universal features of the N-peptide binding mode of Sec1/Munc18 proteins. *Proc Natl Acad Sci* **104(21)**: 8773–8778.

27. Latham CF, Hu SH, Gee CL *et al.* (2007) Crystallization and preliminary X-ray diffraction of the Munc18c-syntaxin4 1-29 complex. *Acta Crys Sec F Struct Bio Cryst Comm* **63(pt 6):** 524–528.

28. Yasui N, Nogi T, Nakano Y, Hattori M, Takagi J. (2007) Structure of a receptor-binding fragment of reelin and mutational analysis reveal a recognition mechanism similar to endocytic receptors. *Proc Natl Acad Sci* **104(24):** 9988–9993.

29. Anderson MJ, DeLaBarre B, Raghunathan A, Palsson BO, Brunger AT, Quake SR. (2007) Crystal structure of a hyperactive *Escherichia coli* gylcerol kinase mutant Gly230Asp obtained using microfluidic crystallization devices. *Biochemistry* **46:** 5277–5731.

30. Raser JM, O'Shea EK. (2005) Noise in gene expression: Origins, consequences, and control. *Science* **309(5743):** 2010–2013.

31. Blow N. (2007) PCR's next frontier. *Nature Methods* **4:** 869–875.

32. McBride L, Facer G, Unger M. (2005) USPTO 20050252773.

33. Warren L, Bryder D, Weissman IL, Quake SR. (2006) Transcription factor profiling in individual hematopoietic progenitors by digital RT-PCR. *Proc Natl Acad Sci* **103:** 17807–17812.

34. Ottesen EA, Hong JW, Quake SR, Leadbetter JR. (2006) Microfluidic digital PCR enables multigene analysis of individual environmental bacteria. *Science* **314:** 1464–1467.

35. Marcy Y, Oshoey T, Lasken RS, Stockwell TB, Walenz BP, Halpern AL, Beeson KY, Goldberg SMD, Quake SR. Nanoliter reactors improve multiple displacement amplification of genomes from single cells. *PLoS Genet* **3(9):** e155 [doi:10.1371/journal.pgen.0030155].

36. Marcy Y, Ouverney C, Blik EM, Losekann T, Ivanova N, Martin HG, Szeto E, Platt D, Hugenholtz P, Relman DA, Quake SR. (2007) Dissecting biological "dark matter" with single cell genetic analysis of rare and uncultivated TM7 microbes from the human mouth. *Proc Natl Acad Sci* **104(29):** 11889–11894.

chapter 18

Potassium Channels and the Atomic Basis of Selective Ion Conduction[†]

Roderick MacKinnon[*]

1. Introduction

All living cells are surrounded by a thin, approximately 40 Å thick lipid bilayer called the cell membrane. The cell membrane holds the contents of a cell in one place so that the chemistry of life can occur, but it is a barrier to the movement of certain essential ingredients including the ions Na^+, K^+, Ca^{2+} and Cl^-. The barrier to ion flow across the membrane — known as the dielectric barrier — can be understood at an intuitive level: the cell membrane interior is an oily substance and ions are more stable in water than in oil. The energetic preference of an ion for water arises from the electric field around the ion and its interaction with neighboring molecules. Water is an electrically polarizable substance, which means that its molecules rearrange in an ion's electric field, pointing negative oxygen atoms in the direction of cations and positive hydrogen atoms toward anions. These electrically stabilizing interactions are much weaker in a less polarizable substance such as oil. Thus, an ion will tend to stay in the water on either side of a cell membrane rather than enter and cross the membrane. And yet numerous cellular processes, ranging from electrolyte transport across

[*]Howard Hughes Medical Institute, Laboratory of Molecular Neurobiology and Biophysics, Rockefeller University, 1230 York Avenue, New York, NY 10021, USA.
[†]Due to unforeseen circumstances, Professor MacKinnon could not contribute to this volume. Included instead is his Nobel paper reprinted here with permission of the Nobel Foundation. © The Nobel Foundation 2003.

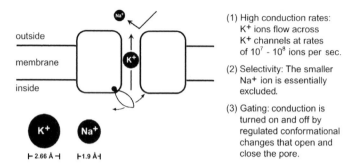

(1) High conduction rates: K^+ ions flow across K^+ channels at rates of 10^7 - 10^8 ions per sec.

(2) Selectivity: The smaller Na^+ ion is essentially excluded.

(3) Gating: conduction is turned on and off by regulated conformational changes that open and close the pore.

Fig. 1. Ion channels exhibit three basic properties depicted in the cartoon. They conduct specific ions (for example K^+) at high rates, they are selective (a K^+ channel essentially excludes Na^+), and conduction is turned on and off by opening and closing a gate, which can be regulated by an external stimulus such as ligand-binding or membrane voltage. The relative size of K^+ and Na^+ ions is shown.

epithelia to electrical signal production in neurons, depend on the flow of ions across the membrane. To mediate the flow, specific protein catalysts known as ion channels exist in the cell membrane. Ion channels exhibit the following three essential properties: (1) they conduct ions rapidly, (2) many ion channels are highly selective, meaning only certain ion species flow while others are excluded, and (3) their function is regulated by processes known as gating, that is, ion conduction is turned on and off in response to specific environmental stimuli. Figure 1 summarizes these properties.

The modern history of ion channels began in 1952 when Hodgkin and Huxley published their seminal papers on the theory of the action potential in the squid giant axon.[1-4] A fundamental element of their theory was that the axon membrane undergoes changes in its permeability to Na^+ and K^+ ions. The Hodgkin–Huxley theory did not address the mechanism by which the membrane permeability changes occur: ions could potentially cross the membrane through channels or by a carrier-mediated mechanism. In their words: "Details of the mechanism will probably not be settled for some time."[1] It is fair to say that the pursuit of this statement has accounted for much ion channel research over the past fifty years.

As early as 1955 experimental evidence for channel mediated ion flow was obtained when Hodgkin and Keynes measured the directional flow of K^+ ions across axon membranes using the isotope $^{42}K^+$.[5] They observed that K^+ flow in one direction across the membrane depends on flow in the opposite direction, and suggested that "the ions should be constrained to move in single file and that there should, on average, be several ions in a channel at any moment." Over the following two decades Armstrong and Hille used electrophysiological methods to demonstrate that Na^+ and K^+ ions cross cell membranes through unique protein pores — Na^+ channels and K^+ channels — and developed the concepts of selectivity filter for ion discrimination and gate for regulating ion flow.[6–12] The patch recording technique invented by Neher and Sakmann then revealed the electrical signals from individual ion channels, as well as the extraordinary diversity of ion channels in living cells throughout nature.[13]

The past twenty years have been the era of molecular biology for ion channels. The ability to manipulate amino acid sequences and express ion channels at high levels have opened up entirely new possibilities for analysis. The advancement of techniques for protein structure determination and the development of synchrotron facilities also created new possibilities. For me, a scientist who became fascinated with understanding the atomic basis of life's electrical system, there could not have been a more opportune time to enter the field.

2. My Early Studies: The K^+ Channel Signature Sequence

The cloning of the *Shaker* K^+ channel gene from *Drosophila melanogaster* by Jan, Tanouye, and Pongs revealed for the first time a K^+ channel amino acid sequence and stimulated efforts by many laboratories to discover which of these amino acids form the pore, selectivity filter, and gate.[14–16] At Brandeis University in Chris Miller's laboratory I had an approach to find the pore amino acids. Chris and I had just completed a study showing that charybdotoxin, a small protein from scorpion venom, inhibits a K$^+$ channel isolated from skeletal muscle cells by plugging the pore and obstructing the flow of ions.[17] In one of those late night "let's see what happens if" experiments while taking a molecular biology course at Cold Spring Harbor I found that the toxin — or what turned out to be a variant of it present in the charybdotoxin preparation — inhibited the Shaker K$^+$ channel.[18,19] This observation meant I could use the toxin to find the pore, and it did not take very long to identify the first site-directed mutants of the Shaker K$^+$ channel with altered binding of toxin.[20] I continued these experiments at Harvard Medical School where I began as assistant professor in 1989. Working with my small group at Harvard, including Tatiana Abramson, Lise Heginbotham, and Zhe Lu, and sometimes with Gary Yellen at Johns Hopkins University, we reached several interesting conclusions concerning the architecture of K$^+$ channels. They had to be tetramers in which four subunits encircle a central ion pathway.[21] This conclusion was not terribly surprising but the experiments and analysis to reach it gave me great pleasure since they required only simple measurements and clear reasoning with binomial statistics. We also deduced that each subunit presents a "pore loop" to the central ion pathway (Fig. 2).[22] This "loop" formed the binding sites for scorpion toxins[20,23,24] as well as the small-molecule inhibitor tetraethylammonium ion,[25,26] which had been used by Armstrong and Hille decades earlier in their pioneering analysis of K$^+$ channels.[9,27] Most important to my thinking, mutations of certain amino acids within the "loop" affected the channel's ability to discriminate between K$^+$ and Na$^+$, the selectivity hallmark of K$^+$ channels.[28,29] Meanwhile, new K$^+$ channel genes were discovered and they all had one obvious feature in common: the very amino acids that we had found to be important for K$^+$ selectivity were conserved (Fig. 3). We called these amino acids the K$^+$ channel signature sequence, and imagined four pore loops somehow forming a selectivity filter with the signature sequence amino acids inside the pore.[22,29]

When you consider the single channel conductance of many K$^+$ channels found in cells you realize just how incredible these molecular devices are.

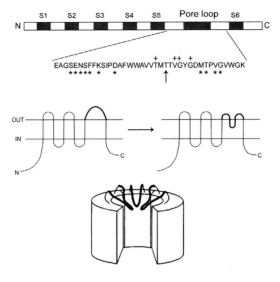

Fig. 2. Early picture of a tetramer K$^+$ channel with a selectivity filter made of pore loops. A linear representation of a Shaker K$^+$ channel subunit on top shows shaded hydrophobic segments S1 to S6 and a region designated the pore loop. A partial amino acid sequence from the Shaker K$^+$ channel pore loop highlights amino acids shown to interact with extracellular scorpion toxins (*), intracellular tetraethylammonium (↑) and K$^+$ ions (+). The pore loop was proposed to reach into the membrane (middle) and form a selectivity filter at the center of four subunits (bottom).

```
Bacteria 2TM : TATTVGYG
Archea   6TM : TATTVGYG
Plant    6TM : TLTTVGYG
Fruitfly 6TM : TMTTVGYG
Worm     6TM : TMTTVGYG
Mouse    6TM : SMTTVGYG
Human    2TM : TQTTIGYG
Human    6TM : TMTTVGYG
```

Fig. 3. The K$^+$ channel signature sequence shown as single letter amino acid code (blue) is highly conserved in organisms throughout the tree of life. Some K$^+$ channels contain six membrane-spanning segments per subunit (6TM) while others contain only two (2TM). 2TM K$^+$ channels correspond to 6TM K$^+$ channels without the first four membrane-spanning segments (S1–S4 in Fig. 2).

With typical cellular electrochemical gradients, K^+ ions conduct at a rate of 10^7 to 10^8 ions per second. That rate approaches the expected collision frequency of K^+ ions from solution with the entryway to the pore. This means that K^+ ions flow through the pore almost as fast as they diffuse up to it. For this to occur the energetic barriers in the channel have to be very low, something like those encountered by K^+ ions diffusing through water. All the more remarkable, the high rates are achieved in the setting of exquisite selectivity: the K^+ channel conducts K^+, a monovalent cation of Pauling radius $1.33\,\text{Å}$, while essentially excluding Na^+, a monovalent cation of Pauling radius $0.95\,\text{Å}$. And this ion selectivity is critical to the survival of a cell. How does nature accomplish high conduction rates and high selectivity at the same time? The answer to this question would require knowing the atomic structure formed by the signature sequence amino acids, that much was clear. The conservation of the signature sequence amino acids in K^+ channels throughout the tree of life, from bacteria[30] to higher eukaryotic cells, implied that nature had settled upon a very special solution to achieve rapid, selective K^+ conduction across the cell membrane. For me, this realization provided inspiration to want to directly visualize a K^+ channel and its selectivity filter.

3. The KcsA Structure and Selective K^+ Conduction

I began to study crystallography, and although I had no idea how I would obtain funding for this endeavor, I have always believed that if you really want to do something then you will find a way. By happenstance I explained my plan to Torsten Wiesel, then president of Rockefeller University. He suggested that I come to Rockefeller where I would be able to concentrate on the problem. I accepted his offer and moved there in 1996. In the beginning I was joined by Declan Doyle and my wife Alice Lee MacKinnon and within a year others joined including João Morais Cabral, John Imredy, Sabine Mann and Richard Pfuetzner. We had to learn as we went along, and what we may have lacked in size and skill we more than compensated for with enthusiasm. It was a very special time. At first I did not know how we would ever reach the point of obtaining enough K^+ channel protein to attempt crystallization, but the K^+ channel signature sequence continued to appear in a growing number of prokaryotic genes, making expression in *Escherichia coli* possible. We focused our effort on a bacterial K^+ channel called KcsA from *Streptomyces lividans*, discovered by Schrempf.[31] The KcsA channel has a simple topology with only two membrane spanning segments per subunit corresponding to the Shaker K^+ channel without S1 through S4 (Fig. 2). Despite its prokaryotic origin KcsA closely resembled the Shaker K^+ channel's pore amino acid sequence, and even exhibited many of its pharmacological properties, including inhibition by scorpion toxins.[32] This surprised us from an evolutionary standpoint, because why should a scorpion want to inhibit a bacterial K^+ channel! But from the utilitarian point of view of protein biophysicists we knew exactly what the scorpion toxin sensitivity meant: that KcsA had to be very similar in structure to the Shaker K^+ channel.

The KcsA channel produced crystals but they were poorly ordered and not very useful in the X-ray beam. After we struggled for quite a while I began to wonder whether some part of the channel was intrinsically disordered and interfering with crystallization. Fortunately my neighbor Brian Chait and his postdoctoral colleague Steve Cohen were experts in the analysis of soluble proteins by limited proteolysis and mass spectrometry, and their techniques applied beautifully to a membrane protein. We found that KcsA was as solid as a rock, except for its C-terminus. After removing disordered amino acids from the c-terminus with chymotrypsin, the crystals improved dramatically, and we were able to solve an initial structure at a resolution of 3.2 Å.[33] We could not clearly see K^+ in the pore at this resolution, but my years of work on K^+ channel function told me that Rb^+ and Cs^+ should be valuable electron dense substitutes for K^+, and they were. Rubidium and Cs^+ difference Fourier maps showed these ions lined up in the pore — as Hodgkin and Keynes might have imagined in 1955.[5]

The KcsA structure was altogether illuminating, but before I describe it, I will depart from chronology to explain the next important technical step. A very accurate description of the ion coordination chemistry inside the selectivity filter would require a higher resolution structure. With 3.2 Å data we could infer the positions of the main-chain carbonyl oxygen atoms by applying our knowledge of small molecule structures, that is, our chemical intuition, but we needed to see the selectivity filter atoms in detail. A high-resolution structure was actually quite difficult to obtain. After more than three additional years of work by João and then Yufeng (Fenny) Zhou, we finally managed to produce high-quality crystals by attaching monoclonal Fab fragments to KcsA. These crystals provided the information we needed, a structure at a resolution of 2.0 Å in which K^+ ions could be visualized in the grasp of selectivity filter protein atoms (Fig. 4).[34] What did the K^+

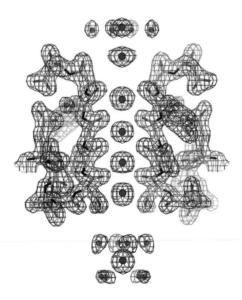

Fig. 4. Electron density ($2F_o$-F_c contoured at 2σ) from a high-resolution structure of the KcsA K^+ channel is shown as blue mesh. This region of the channel features the selectivity filter with K^+ ions and water molecules along the ion pathway. The refined atomic model is shown in the electron density. Adapted from Ref. 34.

channel structure tell us and why did nature conserve the K^+ channel signature sequence amino acids?

Not all protein structures speak to you in an understandable language, but the KcsA K^+ channel does. Four subunits surround a central ion pathway that crosses the membrane [Fig. 5(a)]. Two of the four subunits are shown in Fig. 5(b) with electron density from K^+ ions and water along the pore. Near the center of the membrane the ion pathway is very wide, forming a cavity

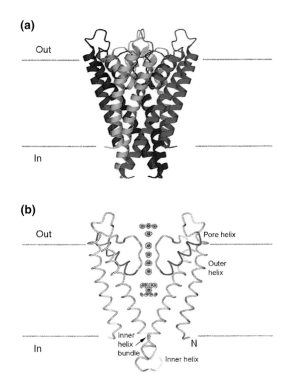

Fig. 5. (a) A ribbon representation of the KcsA K^+ channel with its four subunits colored uniquely. The channel is oriented with the extracellular solution on top. **(b)** The KcsA K^+ channel with front and back subunits removed, colored to highlight the pore-helices (red) and selectivity filter (yellow). Electron density in blue mesh is shown along the ion pathway. Labels identify the pore, outer, and inner helices and the inner helix bundle. The outer and inner helices correspond to S5 and S6 in Fig. 2.

about 10 Å in diameter with a hydrated K^+ ion at its center. Each subunit directs the C-terminal end of a "pore helix," shown in red, toward the ion. The C-terminal end of an α-helix is associated with a negative "end charge" due to carbonyl oxygen atoms that do not participate in secondary structure hydrogen bonding, so the pore helices are directed as if to stabilize the K^+ ion in the cavity. At the beginning of this lecture I raised the fundamental issue of the cell membrane being an energetic barrier to ion flow because of its oily interior. KcsA allows us to intuit a simple logic encoded in its structure, and electrostatic calculations support the intuition[35]: the K^+ channel lowers the membrane dielectric barrier by hydrating a K^+ ion deep inside the membrane, and by stabilizing it with α-helix end charges.

How does the K^+ channel distinguish K^+ from Na^+? Our earlier mutagenesis studies had indicated that the signature sequence amino acids would be responsible for this most basic function of a K^+ channel. Figure 6 shows the structure formed by the signature sequence — the selectivity filter — located in the extracellular third of the ion pathway. The glycine

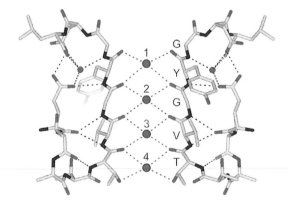

Fig. 6. Detailed structure of the K^+ selectivity filter (two subunits). Oxygen atoms coordinate K^+ ions (green spheres) at positions 1 to 4 from the extracellular side. Single letter amino acid code identifies select signature sequence amino acids. Yellow, blue and red correspond to carbon, nitrogen and oxygen atoms, respectively. Green and gray dashed lines show oxygen-K^+ and hydrogen bonding interactions.

amino acids in the sequence TVGYG have dihedral angles in or near the left-handed helical region of the Ramachandran plot, as does the threonine, allowing the main-chain carbonyl oxygen atoms to point in one direction, toward the ions along the pore. It is easy to understand why this sequence is so conserved among K^+ channels: the alternating glycine amino acids permit the required dihedral angles, the threonine hydroxyl oxygen atom coordinates a K^+ ion, and the side-chains of valine and tyrosine are directed into the protein core surrounding the filter to impose geometric constraint. The end result when the subunits come together is a narrow tube consisting of four equal spaced K^+ binding sites, labeled 1 to 4 from the extracellular side. Each binding site is a cage formed by eight oxygen atoms on the vertices of a cube, or a twisted cube called a square antiprism (Fig. 7). The binding sites are very similar to the single alkali metal site in nonactin, a K^+ selective antibiotic with nearly identical K^+-oxygen distances.[36,37] The principle of K^+ selectivity is implied in a subtle feature of the KcsA crystal structure. The oxygen atoms surrounding K^+ ions in the selectivity filter are arranged quite like the water molecules surrounding the hydrated K^+ ion in the cavity. This comparison conveys a visual impression of binding sites in the filter paying for the energetic cost of K^+ dehydration. The Na^+ ion is apparently too small for these K^+-sized binding sites, so its dehydration energy is not compensated.

The question that compelled us most after seeing the structure was, Exactly how many ions are in the selectivity filter at a given time? To begin to understand how ions move through the filter, we needed to know the stoichiometry of the ion conduction reaction, and that meant knowing how many ions can occupy the filter. Four binding sites were apparent, but are they all occupied at once? Four K^+ ions in a row separated by an average center-to-center distance of 3.3 Å seemed unlikely for electrostatic reasons. From an early stage we suspected that the correct number would be closer to two, because two ions more easily explained the electron density we observed for the larger alkali metal cations Rb^+ and Cs^+.[33,38] Quantitative evidence for the precise number of ions came with the high-resolution structure and

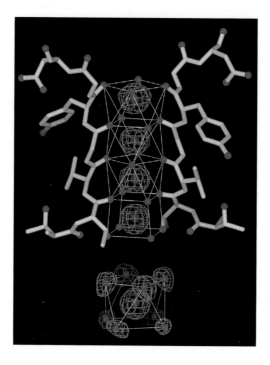

Fig. 7. A K^+ channel mimics the hydration shell surrounding a K^+ ion. Electron density (blue mesh) for K^+ ions in the filter and for a K^+ ion and water molecules in the central cavity are shown. White lines highlight the coordination geometry of K^+ in the filter and in water. Adapted from Ref. 34.

with the analysis of Tl^+.[39] Thallium is the most ideally suited "K^+ analog" because it flows through K^+ channels, has a radius and dehydration energy very close to K^+, and has the favorable crystallographic attributes of high electron density and an anomalous signal. The one serious difficulty in working with Tl^+ is its insolubility with Cl^-. Fenny meticulously worked out the experimental conditions and determined that on average there are between two and two and a half conducting ions in the filter at once, with an occupancy at each position around one half.

We also observed that if the concentration of K^+ (or Tl^+) bathing the crystals is lowered sufficiently (below normal intracellular levels) then a collapsed   conductive

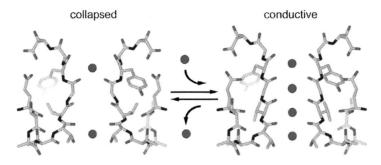

Fig. 8. The selectivity filter can adopt two conformations. At low concentrations of K^+ on average one K^+ ion resides at either of two sites near the ends of the filter, which is collapsed in the middle. At high concentrations of K^+ a second ion enters the filter as it changes to a conductive conformation. On average, two K^+ ions in the conductive filter reside at four sites, each with about half occupancy.

reduction in the number of ions from two to one occurs and is associated with a structural change to a "collapsed" filter conformation, which is pinched closed in the middle.[34,39] At concentrations above 20 mM the entry of a second K^+ ion drives the filter to a "conductive" conformation, as shown in Fig. 8. Sodium on the other hand does not drive the filter to a "conductive" conformation even at concentrations up to 500 mM.

The K^+-induced conformational change has thermodynamic consequences for the affinity of two K^+ ions in the "conductive" filter. It implies that a fraction of the second ion's binding energy must be expended as work to bring about the filter's conformational change, and as a result the two ions will bind with reduced affinity. To understand this statement at an intuitive level, recognize that for two ions to reside in the filter they must oppose its tendency to collapse and force one of them out, i.e., the two-ion "conductive" conformation is under some tension, which will tend to lower K^+ affinity. This is a desirable property for an ion channel because weak binding favors high conduction rates. The same principle, referred to as the "induced fit" hypothesis, had been proposed decades earlier by enzymologists to explain high specificity with low substrate affinity in enzyme catalysis.[40]

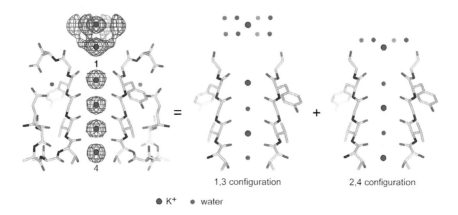

1,3 configuration   2,4 configuration

● K$^+$ ● water

Fig. 9. Two K$^+$ ions in the selectivity filter are hypothesized to exist predominantly in two specific configurations 1,3 and 2,4 as shown. K$^+$ ions and water molecules are shown as green and red spheres, respectively. Adapted from Ref. 34.

In the "conductive" filter if two K$^+$ ions were randomly distributed then they would occupy four sites in six possible ways. But several lines of evidence hinted to us that the ion positions are not random. For example Rb$^+$ and Cs$^+$ exhibit preferred positions with obviously low occupancy at position 2.[38,39] In K$^+$ we observed an unusual doublet peak of electron density at the extracellular entryway to the selectivity filter, shown in Fig. 9.[34] We could explain this density if K$^+$ is attracted from solution by the negative protein surface charge near the entryway and at the same time repelled by K$^+$ ions inside the filter. Two discrete peaks implied two distributions of ions in the filter. If K$^+$ ions tend to be separated by a water molecule for electrostatic reasons then the two dominant configurations would be 1,3 (K$^+$ ions in positions 1 and 3 with a water molecule in between) and 2,4 (K$^+$ ions in positions 2 and 4 with a water molecule in between). A mutation at position 4 (threonine to cysteine) was recently shown to influence K$^+$ occupancy at positions 2 and 4 but not at 1 and 3, providing strong evidence for specific 1,3 and 2,4 configurations of K$^+$ ions inside the selectivity filter.[41]

(a)   **(b)**

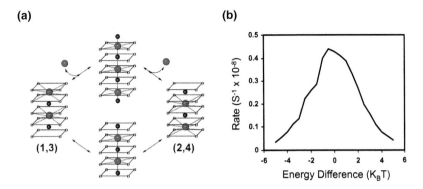

Fig. 10. (a) Throughput cycle for K^+ conduction invoking 1,3 and 2,4 configurations. The selectivity filter is represented as five square planes of oxygen atoms. K^+ and water are shown as green and red spheres, respectively. **(b)** Simulated K^+ flux around the cycle is graphed as a function of the energy difference between the 1,3 and 2,4 configurations. Adapted from Ref. 38.

Discrete configurations of an ion pair suggested a mechanism for ion conduction [Fig. 10(a)].[38] The K^+ ion pair could diffuse back and forth between 1,3 and 2,4 configurations (bottom pathway), or alternatively an ion could enter the filter from one side of the membrane as the ion-water queue moves and a K^+ exits at the opposite side (the top pathway). Movements would have to be concerted because the filter is no wider than a K^+ ion or water molecule. The two paths complete a cycle: in one complete cycle each ion moves only a fraction of the total distance through the filter, but the overall electrical effect is to move one charge all the way. Because two K^+ ions are present in the filter throughout the cycle, we expect there should be electrostatic repulsion between them. Together with the filter conformational change that is required to achieve a "conductive" filter with two K^+ ions in it, electrostatic repulsion should favor high conduction rates by lowering K^+ affinity.

Absolute rates from 10^7 to 10^8 ions per second are truly impressive for a highly selective ion channel. One aspect of the crystallographic data suggests that very high conductance K^+ channels such as KcsA might operate near the maximum rate that the conduction mechanism will allow. All four positions in the filter have a K^+ occupancy close to one half, which implies that the 1,3 and 2,4 configurations are equally probable, or energetically equivalent, but there is no *a priori* reason why this should be. A simulation of ions diffusing around the cycle offers a possible explanation: maximum flux is achieved when the energy difference between the 1,3 and 2,4 configurations is zero because that is the condition under which the "energy landscape" for the conduction cycle is smoothest [Fig. 10(b)]. The energetic balance between the configurations therefore might reflect the optimization of conduction rate by natural selection.[38] It is not so easy to demonstrate this point experimentally but it is certainly fascinating to ponder.

4. Common Structural Principles Underlie K^+ and Cl^- Selectivity

The focus of this chapter is K^+ channels, but for a brief interlude I would like to show you a Cl^- selective transport protein. By comparing a K^+ channel and a Cl^- "channel" we can begin to appreciate familiar themes in nature's solutions to different problems: getting cations and anions across the cell membrane. ClC Cl^- channels are found in many different cell types and are associated with a number of physiological processes that require Cl^- ion flow across lipid membranes.[42,43] As is the case for K^+ channels, ClC family genes are abundant in prokaryotes, a fortunate circumstance for protein expression and structural analysis. When Raimund Dutzler joined my laboratory he, Ernest Campbell and I set out to address the structural basis of Cl^- ion selectivity. We determined crystal structures of two bacterial members of the ClC Cl^- channel family, one from *Escherichia coli* (EcClC) and another from *Salmonella typhimurium* (StClC).[44] Recent studies by Miller on the function of EcClC have shown that it is actually a Cl^--proton exchanger.[45] We do not yet know why certain members of this family of Cl^- transport proteins function as channels and others as exchangers, but the crystal structures are fascinating and give us a view of Cl^- selectivity.

Potassium

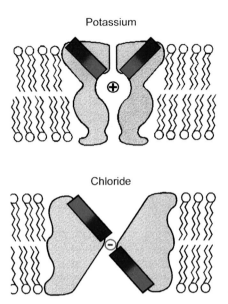

Chloride

Fig. 11. The overall architecture of K$^+$ channels and ClC Cl$^-$ transport proteins is very different but certain general features are similar. One similarity shown here is the use of α-helix end charges directed toward the ion pathway. The negative C-terminal end charge (red) points to K$^+$. The positive N-terminal end charge (blue) points to Cl$^-$.

Architecturally the ClC proteins are unrelated to K$^+$ channels, but if we focus on the ion pathway certain features are similar (Fig. 11). As we saw in K$^+$ channels, the ClC proteins have α-helices pointed at the ion pathway, but the direction is reversed with the positive charge of the N-terminus close to Cl$^-$. This makes perfect sense for lowering the dielectric barrier for a Cl$^-$ ion. In ClC we see that ions in its selectivity filter tend to be coordinated by main chain protein atoms, with amide nitrogen atoms surrounding Cl$^-$ instead of carbonyl oxygen atoms surrounding K$^+$ (Fig. 12). We also see that both the K$^+$ and Cl$^-$ selectivity filters contain multiple close-spaced binding sites and appear to contain more than one ion, perhaps to exploit electrostatic repulsion between ions in the pore. I find these similarities fascinating. They tell us that certain basic physical principles are important,

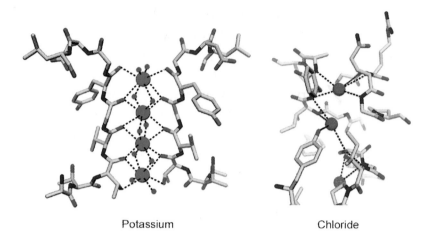

Potassium   Chloride

Fig. 12. K^+ and Cl^- selectivity filters make use of main chain atoms to coordinate ions: carbonyl oxygen atoms for K^+ ions (green spheres) and amide nitrogen atoms for Cl^- ions (red spheres). Both filters contain multiple close-spaced ion binding sites. The Cl^- selectivity filter is that of a mutant ClC in which a glutamate amino acid was changed to glutamine.[47]

such as the use of α-helix end charges to lower the dielectric barrier when ions cross the lipid membrane.

5. Trying to See a K⁺ Channel Open and Close

Most ion channels conduct when called upon by a specific stimulus such as the binding of a ligand or a change in membrane voltage.[46] The processes by which ion conduction is turned on are called gating. The conduction of ions occurs on a time scale that is far too rapid to involve very large protein conformational changes. That is undoubtedly one of the reasons why a single KcsA structure could tell us so much about ion selectivity and conduction. Gating on the other hand occurs on a much slower time scale and can involve large protein conformational changes. The challenge for a structural description of gating is to capture a channel in both opened (on) and closed (off) conformations so that they can be compared.

In the KcsA K$^+$ channel gating is controlled by intracellular pH and lipid membrane composition, but unfortunately the KcsA channel's open probability reaches a maximum value of only a few percent in functional assays.[48,49] At first we had no definitive way to know whether a gate was open or closed in the crystal structures. In the 1970s Armstrong had proposed the existence of a gate near the intracellular side of the membrane in voltage dependent K$^+$ channels because he could "trap" large organic cations inside the pore between a selectivity filter near the extracellular side and a gate near the intracellular side.[9,50] Following these ideas we crystallized KcsA with a heavy atom version of one of his organic cations, tetrabutyl antimony (TBA), and found that it binds inside the central cavity of KcsA.[51] This was very interesting because the ~10 Å diameter of TBA far exceeds the pore diameter leading up to the cavity: in KcsA the intracellular pore entryway is constricted to about 3.5 Å by the inner helix bundle [Fig. 5(b)]. Seeing TBA "trapped" in the cavity behind the inner helix bundle evoked Armstrong's classical view of K$^+$ channel gating, and implied that the inner helix bundle serves as a gate and is closed in KcsA. Mutational and spectroscopic studies in other laboratories also pointed to the inner helix bundle as a possible gate-forming structural element.[52,53]

Youxing Jiang and I hoped we could learn more about K$^+$ channel gating by determining the structures of new K$^+$ channels. From gene sequence analysis we noticed that many prokaryotic K$^+$ channels contain a large C-terminus that encodes what we called RCK domains, and we suspected that these domains control pore opening, perhaps through binding of an ion or a small-molecule. Initially we determined the structure of isolated RCK domains from an *Escherichia coli* K$^+$ channel, but by themselves they were not very informative beyond hinting that a similar structure exists on the C-terminus of eukaryotic Ca^{2+}-dependent "BK" channels.[54] We subsequently determined the crystal structure of MthK, complete K$^+$ channel containing RCK domains, from *Methanobacterium thermoautotrophicus* (Fig. 13).[55] This structure was extremely informative. The RCK domains form a "gating ring" on the intracellular side of the pore. In clefts between domains we

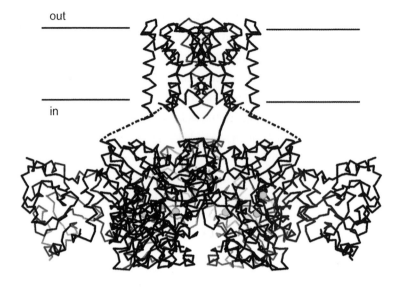

out

in

Fig. 13. The MthK K⁺ channel contains an intracellular gating ring (*bottom*) attached to its ion conduction pore (*top*). Ca^{2+} ions (yellow spheres) are bound to the gating ring in clefts in between domains. The connections between the gating ring and the pore, which were poorly ordered in the crystal, are shown as dashed lines.

could see what appeared to be divalent cation binding sites, and the crystals had been grown in the presence of Ca^{2+}. In functional assays we discovered that the open probability of the MthK channel increased as Ca^{2+} or Mg^{2+} concentration was raised, giving us good reason to believe that the crystal structure should represent the open conformation of a K⁺ channel.

In our MthK structure the inner helix bundle is opened like the aperture of a camera (Fig. 14).[56] As a result, the pathway leading up to the selectivity filter from the intracellular side is about 10 Å wide, explaining how Armstrong's large organic cations can enter the cavity to block a K⁺ channel, and how K⁺ ions gain free access to the selectivity filter through aqueous diffusion. By comparing the KcsA and MthK channel structures it seemed that we were looking at examples of closed and opened K⁺ channels, and could easily imagine the pore undergoing a conformational change from closed to open. To open, the inner helices would have to bend at a point halfway across the membrane as their C-terminus is displaced laterally away from the pore axis by conformational changes in the gating ring. A glycine amino acid facilitates the bending in MthK by introducing a hinge point in the middle of the inner helix. Like MthK, KcsA and many other K^+ channels contain a glycine at the very same location; its conservation suggests that the inner helices move in a somewhat similar manner in many different K^+ channels (Fig. 14).

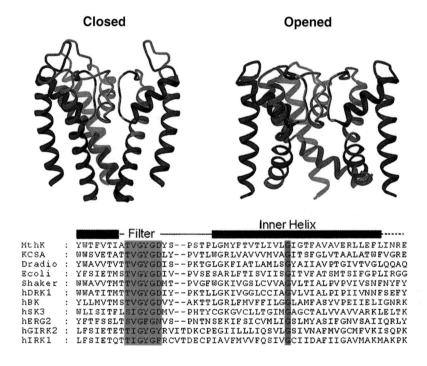

Fig. 14. KcsA and MthK represent closed and opened K^+ channels. Three subunits of the closed KcsA K^+ channel (*left*) and opened MthK K^+ channel (*right*) are shown. The inner helices of MthK are bent at a glycine gating hinge (red), allowing the inner helix bundle to open. Partial amino acid sequences from a variety of K^+ channels with different gating domains are compared. Colors highlighting the selectivity filter sequence (gold) and inner helix glycine hinge (red) match colors used in the structures. Adapted from Ref. 56.

Gating domains convert a stimulus into pore opening. Further studies are needed to understand how the free energy of Ca^{2+} binding is converted into pore opening in the MthK channel. And the mechanistic details of ligand gating will vary from one channel type to the next because nature is very modular with ion channels, just like with other proteins. Gene sequences show us that a multitude of different domains can be found attached to the inner helices of different K^+ channels, allowing ions such as Ca^{2+} or Na^+, small organic molecules, and even regulatory proteins to control the conformational state of the pore and so gate the ion channel (http://www.ncbi.nlm.nih.gov/BLAST/).[57–60]

A fundamentally different kind of gating domain allows certain K^+, Na^+, Ca^{2+} and nonselective cation channels to open in response to membrane voltage changes. Referred to as voltage sensors, these domains are connected to the outer helices of the pore and form a structural unit within the membrane. The basic principle of operation for a voltage sensor is the movement of protein charges through the membrane electric field coupled to pore opening.[62–64] Like transistors in an electronic device, voltage-dependent channels are electrical switches. They are a serious challenge for crystallographic analysis because of their conformational flexibility. Youxing Jiang and I working with Alice Lee and Jiayun Chen solved the structure of a voltage-dependent K^+ channel, KvAP, from the thermophilic Archea *Aeropyrum pernix*[61,65] (Fig. 15). In the crystal of KvAP the voltage sensors, held by monoclonal Fab fragments, adopted a non-native conformation. This observation in itself is meaningful as it underscores the intrinsic flexibility of voltage sensors: in contrast Fab fragments had little effect on the more rigid KcsA K^+ channel and ClC Cl^- channel homolog, both of which we determined in the presence and absence of Fab fragments.[33,34,44,47] KvAP's voltage sensors contain a hydrophobic helix-turn-helix element with arginine residues beside the pore,[61] and functional experiments using tethered biotin and avidin show that this element moves relative to the plane of the membrane.[66] Additional structures revealing different channel conformations will be needed to better understand the mechanistic details of

Fig. 15. Crystal structure of the KvAP K$^+$ channel in complex with monoclonal antibody Fab fragments. The channel is viewed along the pore axis from the intracellular side of the membrane, with α-helical subunits colored in blue, yellow, cyan, and red. One Fab fragment (green) is bound to the helix-turn-helix element of the voltage sensor on each sub-unit. From Ref. 61.

voltage-dependent gating. But the KvAP structure and associated functional studies have provided a conceptual model for voltage-dependent gating — one in which the voltage sensors move at the protein-lipid interface in response to a balance between hydrophobic and electrostatic forces. Rees and colleagues at the California Institute of Technology determined the structure of a voltage regulated mechanosensitive channel called MscS, and although it is unrelated to traditional voltage-dependent channels, it too contains hydrophobic helix-turn-helix elements with arginine residues apparently against the lipid membrane.[67] MscS and KvAP are fascinating membrane protein structures. They do not fit into the standard category of membrane proteins with rigid hydrophobic walls against the lipid membrane core. I find such proteins intriguing.

We are only just beginning to understand the structural principles of ion channel gating and regulation. Electrophysiological studies have uncovered a multitude of connections between cellular biochemical pathways and ion channel function.[46] New protein structures are now beginning to do the same. Beta subunits of certain eukaryotic voltage-dependent channels are structurally related to oxido-reductase enzymes.[68,69] PAS domains on other K^+ channels belong to a family of sensory molecules,[70] and a specialized structure on G protein-gated channels forms a binding site for regulatory G protein subunits.[71] The interconnectedness of ion channel function with many aspects of cell function is beginning to reveal itself as complex and fascinating.

6. Concluding Remarks

I think the most exciting time in ion channel studies is just beginning. So many of the important questions are waiting to be answered and we have the tools in hand to answer them. I am very optimistic about the future, and for the great possibilities awaiting young scientists who are now setting out to study ion channels and other membrane proteins. I consider myself very fortunate to have contributed to some small part of the knowledge we have today.

Acknowledgments

My contributions would never have been possible without the efforts and enthusiasm of the young scientists who have come from around the world to study ion channels with me (Fig. 16). I also owe thanks to the Rockefeller University, the Howard Hughes Medical Institute, and the National Institutes of Health for supporting my scientific research.

Thanks also to Rockefeller University, HHMI, NIH, to the synchrotrons CHESS, NSLS, ALS, APS and ESRF and to my assistant Wendell Chin.

MacKinnon Laboratory: 1989–2003.

Postdoctoral	Students	Staff Scientists	Collaborators
Laura Escobar	Lise Heginbotham	Tatiana Abramson	Gary Yellen
Zhe Lu	Michael Root	John Lewis	Maria Garcia
Adrian Gross	Patricia Hidalgo	Alice Lee MacKinnon	Gerhard Wagner
Kenton Swartz	Sanjay Aggarwal	Sabine Mann	Andrzej Krezel
Chul-Seung Park	James Morrell	Richard Pfuetzner	Brian Chait
Rama Ranganathan	Alexander Pico	Anling Kuo	Steve Cohen
Chinfei Chen	Vanessa Ruta	Minhui Long	Martine Cadene
Declan Doyle	Ian Berke	Amelia Kaufman	Benoit Roux
John Imredy		Ernest Campbell	Tom Muir
João Morais Cabral		Jiayun Chen	
Youxing Jiang			
Jacqueline Gulbis			
Raimund Dutzler			
Francis Valiyaveetil			
Xiao-Dan Pfenninger-Li			
Ming Zhou			
Ofer Yifrach			
Yufeng Zhou			
Sebastien Poget			
Motohiko Nishida			
Uta-Maria Ohndorf			
Steve Lockless			
Qiu-Xing Jiang			
Seok-Yong Lee			
Stephen Long			

Fig. 16. MacKinnon laboratory from 1989 to 2003.

References

1. Hodgkin AL, Huxley AF. (1952a) A quantitative description of membrane current and its application to conduction and excitation in nerve. *J Physiol* **117**: 500–544.
2. Hodgkin AL, Huxley AF. (1952b) Currents carried by sodium and potassium ions through the membrane of the giant axon of *Loligo*. *J Physiol* **116**: 449–472.
3. Hodgkin AL, Huxley AF. (1952c) The components of membrane conductance in the giant axon of *Loligo*. *J Physiol* **116**: 473–496.

4. Hodgkin AL, Huxley AF. (1952d) The dual effect of membrane potential on sodium conductance in the giant axon of *Loligo*. *J Physiol* **116:** 497–506.

5. Hodgkin AL, Keynes RD. (1955) The potassium permeability of a giant nerve fibre. *J Physiol (Lond)* **128:** 61–88.

6. Hille B. (1970) Ionic channels in nerve membranes. *Prog Biophys Mol Biol* **21:** 1–32.

7. Hille B. (1971) The permeability of the sodium channel to organic cations in myelinated nerve. *J Gen Physiol* **58:** 599–619.

8. Hille B. (1973) Potassium channels in myelinated nerve. Selective permeability to small cations. *J Gen Physiol* **61:** 669–686.

9. Armstrong CM. (1971) Interaction of tetraethylammonium ion derivatives with the potassium channels of giant axons. *J Gen Physiol* **58:** 413–437.

10. Armstrong CM, Bezanilla F, Rojas E. (1973) Destruction of sodium conductance inactivation in squid axons perfused with pronase. *J Gen Physiol* **62:** 375–391.

11. Armstrong CM, Bezanilla F. (1977) Inactivation of the sodium channel. II. Gating current experiments. *J Gen Physiol* **70:** 567–590.

12. Armstrong CM. (1981) Sodium channels and gating currents. *Physiol Rev* **61:** 645–683.

13. Neher E, Sakmann B. (1976) Single-channel currents recorded from membrane of denervated frog muscle fibres. *Nature* **260:** 799–802.

14. Tempel BL, Papazian DM, Schwarz TL, Jan YN, Jan LY. (1987) Sequence of a probable potassium channel component encoded at Shaker locus of *Drosophila*. *Science* **237:** 770–775.

15. Kamb A, Iverson LE, Tanouye MA. (1987) Molecular characterization of Shaker, a *Drosophila* gene that encodes a potassium channel. *Cell* **50:** 405–413.

16. Pongs O, Kecskemethy N, Muller R, Krah-Jentgens I, Baumann A, Kiltz HH, Canal I, Llamazares S, Ferrus A. (1988) Shaker encodes a family of putative potassium channel proteins in the nervous system of *Drosophila*. *EMBO J* **7:** 1087–1096.

17. MacKinnon R, Miller C. (1988) Mechanism of charybdotoxin block of the high-conductance, Ca^{2+}-activated K^+ channel. *J Gen Physiol* **91:** 335–349.

18. MacKinnon R, Reinhart PH, White MM. (1988) Charybdotoxin block of Shaker K^+ channels suggests that different types of K^+ channels share common structural features. *Neuron* **1**: 997–1001.

19. Garcia ML, Garcia-Calvo M, Hidalgo P, Lee A, MacKinnon R. (1994) Purification and characterization of three inhibitors of voltage-dependent K^+ channels from *Leiurus quinquestriatus var. hebraeus* venom. *Biochemistry* **33**: 6834–6839.

20. MacKinnon R, Miller C. (1989) Mutant potassium channels with altered binding of charybdotoxin, a pore-blocking peptide inhibitor. *Science* **245**: 1382–1385.

21. MacKinnon R. (1991) Determination of the subunit stoichiometry of a voltage-activated potassium channel. *Nature* **350**: 232–235.

22. MacKinnon R. (1995) Pore loops: An emerging theme in ion channel structure. *Neuron* **14**: 889–892.

23. Hidalgo P, MacKinnon R. (1995) Revealing the architecture of a K^+ channel pore through mutant cycles with a peptide inhibitor. *Science* **268**: 307–310.

24. Ranganathan R, Lewis JH, MacKinnon R. (1996) Spatial localization of the K^+ channel selectivity filter by mutant cycle-based structure analysis. *Neuron* **16**: 131–139.

25. MacKinnon R, Yellen G. (1990) Mutations affecting TEA blockade and ion permeation in voltage-activated K^+ channels. *Science* **250**: 276–279.

26. Yellen G, Jurman ME, Abramson T, MacKinnon R. (1991) Mutations affecting internal TEA blockade identify the probable pore-forming region of a K^+ channel. *Science* **251**: 939–942.

27. Armstrong CM, Hille B. (1972) The inner quaternary ammonium ion receptor in potassium channels of the node of Ranvier. *J Gen Physiol* **59**: 388–400.

28. Heginbotham L, Abramson T, MacKinnon R. (1992) A functional connection between the pores of distantly related ion channels as revealed by mutant K^+ channels. *Science* **258**: 1152–1155.

29. Heginbotham L, Lu Z, Abramson T, MacKinnon R. (1994) Mutations in the K^+ channel signature sequence. *Biophys J* **66**: 1061–1067.

30. Milkman, R. (1994). An *Escherichia coli* homologue of eukaryotic potassium channel proteins. *Proc Natl Acad Sci* **91(9)**: 3510–3514.

31. Schrempf H, Schmidt O, Kummerlen R, Hinnah S, Muller D, Betzler M, Steinkamp T, Wagner R. (1995) A prokaryotic potassium ion channel with two predicted transmembrane segments from *Streptomyces lividans*. *EMBO J* **14:** 5170–5178.

32. MacKinnon R, Cohen SL, Kuo A, Lee A, Chait BT. (1998) Structural conservation in prokaryotic and eukaryotic potassium channels. *Science* **280:** 106–109.

33. Doyle DA, Morais Cabral JH, Pfuetzner RA, Kuo A, Gulbis JM, Cohen SL, Chait BT, MacKinnon R. (1998) The structure of the potassium channel: Molecular basis of K^+ conduction and selectivity. *Science* **280:** 69–77.

34. Zhou Y, Morais-Cabral JH, Kaufman A, MacKinnon R. (2001b) Chemistry of ion coordination and hydration revealed by a K^+ channel-Fab complex at 2.0 Å resolution. *Nature* **414:** 43–48.

35. Roux B, MacKinnon R. (1999) The cavity and pore helices in the KcsA K^+ channel: Electrostatic stabilization of monovalent cations. *Science* **285:** 100–102.

36. Dobler v M, Dunitz JD, Kilbourn BT. (1969) Die struktur des KNCS-Komplexes von nonactin. *Helvetica Chimica Acta* **52:** 2573–2583.

37. Dunitz JD, Dobler M. (1977) Structural studies of ionophores and their ion-complexes. In: Addison AW, Cullen WR, Dolphin D, James BR (eds), *Biological Aspects of Inorganic Chemistry*, pp. 113–140. John Wiley & Sons, Inc.

38. Morais-Cabral JH, Zhou Y, MacKinnon R. (2001) Energetic optimization of ion conduction rate by the K^+ selectivity filter. *Nature* **414:** 37–42.

39. Zhou Y, MacKinnon R. (2003) The occupancy of ions in the K^+ selectivity filter: Charge balance and coupling of ion binding to a protein conformational change underlie high conduction rates. *J Mol Biol* **333:** 965–975.

40. Jencks WP. (1987) *Catalysis in Chemistry and Enzymology*. Dover Publications, Inc.

41. Zhou M, MacKinnon R. (2004) A mutant KcsA K^+ channel with altered conduction properties and selectivity filter ion distribution. *J Mol Biol* **338:** 839–846.

42. Jentsch TJ, Friedrich T, Schriever A, Yamada H. (1999) The ClC chloride channel family. *Pflugers Arch* **437**: 783–795.

43. Maduke M, Miller C, Mindell JA. (2000) A decade of ClC chloride channels: Structure, mechanism, and many unsettled questions. *Annu Rev Biophys Biomol Struct* **29**: 411–438.

44. Dutzler R, Campbell EB, Cadene M, Chait BT, MacKinnon R. (2002) X-ray structure of a ClC chloride channel at 3.0 Å reveals the molecular basis of anion selectivity. *Nature* **415**: 287–294.

45. Accardi A, Miller C. (2004) Proton-coupled chloride transport mediated by ClC-ec1, a bacterial homologue of the ClC chloride channels. *Biophys J* **86(1)**: 286a.

46. Hille B. (2001) *Ion Channels of Excitable Membranes*. Sinauer Associates, Inc., Sunderland, MA.

47. Dutzler R, Campbell EB, MacKinnon R. (2003) Gating the selectivity filter in ClC chloride channels. *Science* **300**: 108–112.

48. Cuello LG, Romero JG, Cortes DM, Perozo E. (1998) pH-dependent gating in the *Streptomyces lividans* K^+ channel. *Biochemistry* **37**: 3229–3236.

49. Heginbotham L, Kolmakova-Partensky L, Miller C. (1998) Functional reconstitution of a prokaryotic K^+ channel. *J Gen Physiol* **111**: 741–749.

50. Armstrong CM. (1974) Ionic pores, gates, and gating currents. *Q Rev Biophys* **7**: 179–210.

51. Zhou M, Morais-Cabral JH, Mann S, MacKinnon R. (2001a) Potassium channel receptor site for the inactivation gate and quaternary amine inhibitors. *Nature* **411**: 657–661.

52. Perozo E, Cortes DM, Cuello LG. (1999) Structural rearrangements underlying K^+-channel activation gating. *Science* **285**: 73–78.

53. del Camino D, Holmgren M, Liu Y, Yellen G. (2000) Blocker protection in the pore of a voltage-gated K^+ channel and its structural implications. *Nature* **403**: 321–325.

54. Jiang Y, Pico A, Cadene M, Chait BT, MacKinnon R. (2001) Structure of the RCK domain from the *E. coli* K^+ channel and demonstration of its presence in the human BK channel. *Neuron* **29**: 593–601.

55. Jiang Y, Lee A, Chen J, Cadene M, Chait BT, MacKinnon R. (2002a) Crystal structure and mechanism of a calcium-gated potassium channel. *Nature* **417**: 515–522.

56. Jiang Y, Lee A, Chen J, Cadene M, Chait BT, MacKinnon R. (2002b) The open pore conformation of potassium channels. *Nature* **417**: 523–526.

57. Atkinson NS, Robertson GA, Ganetzky B. (1991) A component of calcium-activated potassium channels encoded by the *Drosophila slo* locus. *Science* **253**: 551–555.

58. Schumacher MA, Rivard AF, Bachinger HP, Adelman JP. (2001) Structure of the gating domain of a Ca^{2+}-activated K^+ channel complex with Ca^{2+}/calmodulin. *Nature* **410**: 1120–1124.

59. Yuan A, Santi CM, Wei A, Wang ZW, Pollak K, Nonet M, Kaczmarek L, Crowder CM, Salkoff L. (2003) The sodium-activated potassium channel is encoded by a member of the *Slo* gene family. *Neuron* **37**: 765–773.

60. Kubo Y, Reuveny E, Slesinger PA, Jan YN, Jan LY. (1993) Primary structure and functional expression of a rat G-protein-coupled muscarinic potassium channel. *Nature* **364**: 802–806.

61. Jiang Y, Lee A, Chen J, Ruta V, Cadene M, Chait B, MacKinnon R. (2003a) X-ray structure of a voltage-dependent K^+ channel. *Nature* **423**: 33–41.

62. Armstrong CM, Bezanilla F. (1974) Charge movement associated with the opening and closing of the activation gates of the Na^+ channels. *J Gen Physiol* **63**: 533–552.

63. Sigworth FJ. (1994) Voltage gating of ion channels. *Q Rev Biophys* **27**: 1–40.

64. Bezanilla F. (2000) The voltage sensor in voltage-dependent ion channels. *Physiol Rev* **80**: 555–592.

65. Ruta V, Jiang Y, Lee A, Chen J, MacKinnon R. (2003) Functional analysis of an archeabacterial voltage-dependent K^+ channel. *Nature* **422**: 180–185.

66. Jiang Y, Ruta V, Chen J, Lee A, MacKinnon R. (2003b) The principle of gating charge movement in a voltage-dependent K^+ channel. *Nature* **423**: 42–48.

67. Bass RB, Strop P, Barclay M, Rees DC. (2002) Crystal structure of *Escherichia coli* MscS, a voltage-modulated and mechanosensitive channel. *Science* **298**: 1582–1587.

68. Gulbis JM, Mann S, MacKinnon R. (1999) Structure of a voltage-dependent K^+ channel beta subunit. *Cell* **97:** 943–952.

69. Gulbis JM, Zhou M, Mann S, MacKinnon R. (2000) Structure of the cytoplasmic β subunit-T1 assembly of voltage-dependent K^+ channels. *Science* **289:** 123–127.

70. Morais-Cabral JH, Lee A, Cohen SL, Chait BT, Li M, MacKinnon R. (1998) Crystal structure and functional analysis of the HERG potassium channel N-terminus: A eukaryotic PAS domain. *Cell* **95:** 649–655.

71. Nishida M, MacKinnon R. (2002) Structural basis of inward rectification: Cytoplasmic pore of the G protein-gated inward rectifier GIRKI at 1.8 Å resolution. *Cell* **III:** 957–965.

The 2007
Welch Award Papers

Symmetry Breaking, Delocalization and Dynamics in Electron Transfer Systems

Noel S. Hush[*]

Chemistry abounds with examples of molecules that have broken symmetry, such as ammonia, which is not flat but oscillates between two identical "umbrella" conformations. Benzene in its ground state is a flat symmetrical ring, but has lowered symmetry in its first excited state. What is in evidence here is competition between energy of electronic coupling between nuclei, tending to delocalize, and energy of electronic-vibrational coupling, tending to localize. In the fully delocalized symmetrical state we have a static charge distribution, whereas in the strongly localized case there is dynamic equilibrium between equivalent structures, with a measurable thermal exchange rate constant. Electron transfer processes, including critical biological ones, are governed by these restraints. The history is intriguing.

1. Introduction

In this paper, I am concerned with the roles played by symmetry breaking and charge delocalization in electron transfer processes. I shall confine this to consideration of elementary electron or hole transfers in solvents (either liquid or protein shell). The treatment is general, and my aim is to provide understanding with a minimum of technical detail. The ideas outlined were in fact first developed for interpretation of electron exchange kinetics at metal electrodes. The interpretation of the transfer coefficient in elementary electron exchange reactions as an experimental measure of the electronic

[*]Chemistry School and School of Molecular and Microbial Biosciences University of Sydney, NSW 2006, Australia, e-mail: hush_n@chem.usyd.edu.au.

asymmetry (delocalization) in the heterogeneous transition state, first put forward in 1956,[1] is now well established, and the elaboration of this[2,3] is the basis of the current theory of elementary electrode processes.[4] There are close parallels with related homogeneous processes.

The experimental kinetics of homogeneous electron transfer processes were quite vigorously studied in England during and after the second world war, particularly in the Physical Chemistry Department at Manchester University, which I joined as junior lecturer in 1950, together with theoretical studies of their overall thermodynamics (see, e.g., Evans, Hush and Uri[5]). There was, however, no attempt at basic interpretation of the nature of these processes. What was required above all for this were experimental data on symmetrical elementary electron transfer kinetics in order to be able to test theoretical predictions. The basic chemical investigations accompanying the development of nuclear fission for construction of atomic weapons provided just such information.

2. The Manhattan Project 1942–1946: New Elements, Oxidation States and Electron Transfer Kinetics

The project to discover how to initiate and control nuclear fission, during the World War II, the Manhattan Project, was successful, and it was an important task to investigate the fission products, one mainly carried out at the Lawrence Berkeley Laboratory in California, under the direction of Glenn T. Seaborg. The aims were to isolate the new elements produced in the fission reaction and to ascertain their chemical properties, which were not easily predictable. The most well known of the new elements is plutonium (Pu), and investigation of this element followed a typical course. Firstly, the principal oxidation states were determined for plutonium and plutonyl ions: these were found to be, repectively, Pu^{3+} and Pu^{4+}, and PuO_2 and PuO_2^+.

Clearly, a major concern was the nuclear lifetimes of the various isotopes. However, in order to establish these, it was necessary also to establish the lifetimes of particular oxidation states of an isotope, and the isotopic exchange or self-exchange rates, such as

$$*Pu^{3+} + Pu^{4+} \xrightarrow{k_{11}} *Pu^{4+} + Pu^{3+}, \tag{1}$$

$$*PuO_2 + PuO_2^+ \xrightarrow{k_{22}} *PuO_2^+ + PuO_2. \tag{2}$$

Thus for the first time, elementary self-exchange rates in water for actinide and also first transition series ions were measured.

In addition, and importantly since interest is ultimately mainly in kinetics of reactions with overall energy change, the rates of a number of cross-reactions, such as

$$*Pu^{3+} + *PuO_2^+ \xrightarrow{k_{11}} Pu^{4+} + PuO_2, \tag{3}$$

were also determined (for a review, see Ref. 6). Thus, for the first time, a considerable body of data on the kinetics of elementary electron transfer reactions was available. However, nearly a decade went by before any light was thrown on their basic mechanisms.

After the shutdown of the Manhattan Project, experimental work on isotopic exchange processes was continued in the newly-formed Chemistry Department of the Brookhaven National Laboratory under the energetic direction of Richard Dodson, who transferred from the Berkeley group. That department has continued to this day to be a source of important advances in electron transfer research.

3. The Paradigmatic Reaction: Fe^{2+}/Fe^{3+} Self-Exchange

What was perceived as the main difficulty in accounting for electron self-exchange at quite reasonable rates was the large distance between the reactants. Thus, for the Fe^{2+}/Fe^3 exchange, the center-to-center distance of the reacting ions, allowing for two intervening water molecules, is *ca.* 6.4 Å. Attempts to account for observed rates in terms of simple tunneling were clearly unsatisfactory. Starting around 1956, however (to repeat: *a decade after the end of the Manhattan Project!*), there were several almost simultaneous attacks on the problems both of heterogeneous ion/metal and homogeneous transfers, each of which contributed to the understanding of transfer kinetics, starting from quite different approaches. In the USA, R.A. Marcus[7–9] used semiclassical statistical mechanics and solvent polarization theory as a basis for an exceedingly important interpretation. In the USSR, R.R. Dogonadze (partly in collaboration with V. Levich)[10,11] approached the problem from the point of view of quantum vibronic transitions, and in the UK, I began with a molecular quantum-absolute reaction rate theoretical approach.[12–14] I have compared the insights of these differing approaches, with particular reference to heterogeneous exchanges, elsewhere.[2] In the present instance, my concern is more generally with charge delocalization as a reaction parameter, and with concepts associated with this, which are unique to the third approach, and I will confine my remarks solely to this.

During the earlier period in which I worked in this area, from 1955–1965, I was living through four chemical revolutions of the utmost importance, each of which contributed to my view of the basic nature of electron or hole transfer processes. These have subsequently been followed by a further four major revolutions, bringing us further towards an understanding of these processes.

4. Revolution 1: Molecular Orbital Formalism Replaces Pauling Valence-Bond "Resonance" Description of Molecular Electronic Structure

During the mid-1940s, the molecular orbital method of calculating electronics structures was rapidly replacing the valence-bond method, owing *inter alia* to its more straightforward physical approach and much greater ease of computation. R.A. Mulliken (co-founder of the theory) in the USA, C.A. Coulson, H.C. Longuet-Higgins, and John Pople were among those with whom I interacted, and I worked on a number of problems interpreting energetics of organic electron transfer (including semiquinone stability) and related processes in terms of molecular orbital concepts.[15–19]

The application to the rate problem was obvious. It was known from crystal hydrate structures that both ferrous and ferric ions formed regular octahedral coordination complexes with water. Thus the interacting species in the exchange were not bare ions but coordination complexes of known geometry. The orbitals occupied by the transferring electron would be linear combinations of central metal ion $3d$ and water oxygen $2p\sigma$ orbitals, a typical example being as shown in Fig. 1 for a section through the plane containing a $3d_{x2-y2}$ orbital.

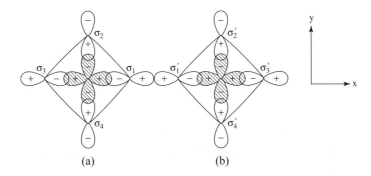

(a)   (b)

Fig. 1 (a) Section in xy plane of molecular orbital formed by overlap of $3d_{x2-y2}$ metal orbital with $2p\sigma$ orbitals on oxygen centers of adjacent water molecules in octahedral complex. **(b)** Identical molecular orbital for second octahedral complex in apical contact with this, showing overlap of $2p\sigma$ functions on nearest neighbor oxygen centers.[20]

The composition of the molecular orbital γ formed by linking the central metal $3d_{x2-y2}$ orbital to the appropriate combination of $2p$-σ type oxygen orbitals is

$$\gamma = \alpha d_{x2-y2} - \frac{1}{2}(1 - \alpha^2)^{1.2}(\sigma_1 + \sigma_2 + \sigma_3 + \sigma_4),$$

where α is the amplitude of the metal function in this orbital.

The corresponding molecular orbital involving a $3d_{xy}$ orbital is

$$\varepsilon = \beta d_{xy} - \frac{1}{2}(1 - \beta^2)^{1/2}(\pi_1 + \pi_2 + \pi_3 + \pi_4).$$

The complexes are in close contact (shown here as apical, but face-to-face orientations will show similar effects) and it is clear the critical distance for electronic interaction is not the large center-to-center distance, but the very much smaller oxygen-oxygen separation of adjacent water molecules, provided that the sum of amplitudes $(1 - \alpha^2)^{1/2}$ of the ligands are appreciable. The distinction between center-to-center and close-contact distance has since been a central one in electron transfer theory.

5. Revolution 2: Electron Paramagnetic Resonance Spectroscopy Shows Spin and Charge Delocalization to Ligands

Electron paramagnetic resonance (EPR) techniques were just beginning to be used in analyzing the spin and charge distribution in magnetic coordination complexes. It was clear from this that in general there was very appreciable delocalization of the central metal orbitals onto the ligands, a surprising discovery at the time. In Fig. 2 an example is shown of the EPR spectrum of the $MnCl_6^{4-}$ ion.

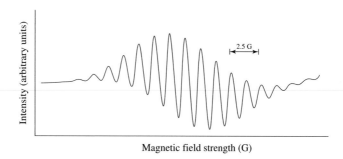

Fig. 2. Electron spin resonance spectrum of $Mn(Cl)_6^{4-}$ ion. The multi-line hyperfine structure results from extensive spin and charge delocalization from central metal $3d$ orbital onto adjacent chloride ligand centers.

Thus it seemed very reasonable to assume that the typical self-exchange would (unless possibly impeded by spin symmetry restrictions) proceed with a transmission coefficient of nearly unity. Somewhat later, data began to appear for metal hydrate complexes. For example, ESR measurements for the $Cr(H_2O)_6^{3+}$ ion yielded a value of 0.3 for the overall ligand occupancy of the vacant $3d$ orbital: together with a quantum calculation of the oxygen orbital overlap energy, this gave an estimate for the electronic coupling J for a chromous-chromic collision pair in apical contact as a function of inter-water separation.[19] This is shown in Fig. 3.

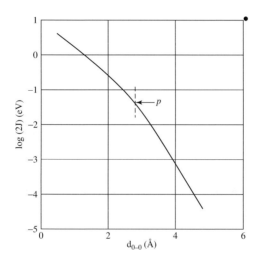

Fig. 3. Calculated energy separation $2\varepsilon^1$ ($= 2J$) between upper and lower potential surfaces of the $Cr(H_2O)_6^{2+}/Cr(H_2O)_6^{3+}$ pair in apical contact as a function of oxygen-oyxgen interatomic distance. This is given by $J = 0.25\,(1 - \alpha^2)\varepsilon$, where ε is the electronic coupling energy of two neighboring oxygen $2p\sigma$ orbitals and α^2 is the square of amplitude of the $3d_{x2-y2}$ orbital in the relevant molecular $Cr(H_2O)_6^{2+}$ ion molecular orbital, equal to 0.7 from the ESR spectrum. The point p indicates the close contact separation at which the oxygen-oxygen distance is twice the radius of a water molecule: at this point J is 0.05 eV, approximately twice the thermal quantum k_BT at 298 K.[20]

This was 0.05 eV, or twice the thermal quantum at 298 K at close-contact separation, amply sufficient to provide a transmission coefficient of unity for the exchange. This was actually the first calculation for the electronic coupling in self-exchange reactions, and the essentially exponential decay with distance has later been seen to be typical.

6. Revolution 3: Transition State (Absolute Reaction Rate) Theory

While absolute reaction rate theory had been formulated considerably earlier, its impact on calculations of reaction rates had been somewhat minimal. The Department of Physical Chemistry at Manchester was an exception, owing to the former presence of Michael Polanyi and the actual presence of his highly talented successor M.G. Evans. I was thoroughly conversant with the theory. The above considerations led me to propose that electron exchange in close-contact collision homogeneous electron transfer will proceed on an energy surface with a well-defined transition state with transmission coefficients not too far removed from unity. This has subsequently been found to be correct — and also for analogous exchange between a metal electrode and an ion.

7. The Activation Energy: Ion-Solvent Contribution

Since in these simplest of reactions, no chemical bands are broken or formed, the activation process involves vibrational, torsional, librational, etc. degrees of freedom both of the medium and internally of the complexes. The activation free energy $\Delta G_s^\dagger$ is that required to reorient the solvent dipoles (in the simplest interpretation) from the initial and final vector orientations, shown schematically in Fig. 4, to the symmetrical position of the transition state.

In current work, we consider this solvent orientation at the molecular level, as in Fig. 5, which shows a snapshot of 192 water molecules around a symmetrical ion pair (actually, the Creutz–Taube ion, considered later[21]).

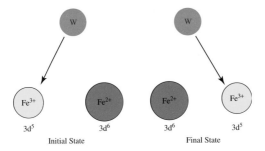

Fig. 4. Schematic of change of solvent (W) orientation vector (arrow) between initial and final states in Fe^{2+}/Fe^{3+} electron transfer process. Large spheres represent aquo complexes: difference in size indicates qualitatively the change in coordination sphere radius accompanying electron transfer (not to scale), which gives rise to "internal" contribution to free energy of activation in the exchange reaction.

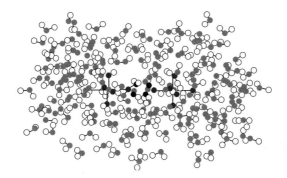

Fig. 5. Molecular dynamics snapshot of 192-water-molecule system configuration surrounding a symmetrical 2.5+ dimeric charge distribution, corresponding qualitatively to the situation in the transition state of the Fe^{2+}/Fe^{3+} transfer in water solution. This actually applies to the ground state of the Creutz–Taube ion (see text and Ref. 21). The solvent contribution to the activation energy for electron transfer is due to the energy of solvent reorientation required to reach such a symmetrical state from the asymmetrical distribution in the initial state of the system.

But quite a good approximation can be obtained using dielectric continuum theory, for which the calculation of $\Delta G_s^{\dagger}$ by a modified Born method is quite trivial.[13]

8. The Activation Energy: Contribution of Internal Vibrational Modes

At this time essentially nothing was known about the frequencies of the vibrational modes in aquo complexes, so any estimate of internal energetics had to be a theoretical one. This was made possible by ...

9. Revolution 4: Crystal Field Theory and the Renaissance of Inorganic Chemistry

The 1929 theory of H. Bethe[22] of the splitting of degenerate levels of transition ions in the presence of surrounding dipole or anions coordinating species was rediscovered in the mid-1950s, and revolutionized inorganic chemistry departments, which sprang up all over the world. It was applied to detailed elucidation of spectroscopic, magnetic and energetic properties of coordination complexes.

It was an area in which I was involved particularly in calculations of vibrational force constants, geometries and bond lengths.[23–25] Figure 6 illustrates how transition ion radii vary systematically in a calculable way as the $3d$ shell is progressively filled.

This provided an immediate way to estimate the free energy of forming a symmetrical transition state from the solvent symmetry-broken initial states. The activation energy for self-exchange transfer was envisaged as involving energy of reorientation of solvent $\Delta G_s^\dagger$ plus internal reorganization $\Delta G_i^\dagger$ plus energy of formation of bimolecular collision or precursor complex $\Delta G_{bimol}^\dagger$.

The results of the first such calculations, which are essentially first principle in nature, are shown in Table 1.[13]

The enthalpy and entropy of activation of the reaction, likewise partitioned, are also shown. These were the first such calculations for an elementary electron self-exchange. The agreement with experiment is clearly reasonable, indicating the essential correctness of the approach.

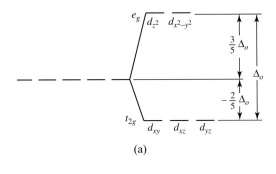

(a)

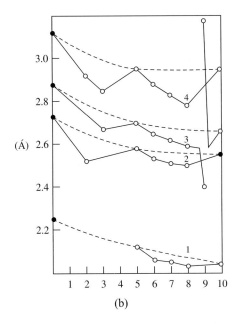

(b)

Fig. 6. (a) Crystal field splitting of fivefold degeneracy of d orbital manifold into lower t_{2g} triplet and upper e_g doublet in octahedral (O_h) symmetry.[22] **(b)** Variation of radius of divalent transition ions in octahedral environments as the $3d$ shell is filled resulting from the crystal field effect.[25] The topmost data are for oxides, and the remaining correlations are for halide lattices. It is this variation which gives rise to variation of "internal" activation energy of electron transfer between +2 and +3 ion $3d$ ions across the first transition series.

Table 1. Theoretical Activation Thermodynamics for Fe^{3+}/Fe^{2+} Self-exchange in Water at 298 K

	$\Delta G^\dagger$	ΔH	$\Delta S^\dagger$
Bimolecular terms	8.0	−1.5	−32
	(6.4)	(−1.8)	(−27.6)
Intramolecular modes	5.0	5.0	0
	(7.7)	(6.4)	(4.1)
Medium modes	7.3	7.3	0.7
	(5.6)	(5.8)	(0.6)
Total	20.3	10.8	−32
	(19.7)	(10.4)	(-31.1)
Experimental	19	> 10	< −30

Energy in kcal/mole, entropy in cal·mole/degree.
First entry: Hush (1960[12], 1961[13])
Second entry: Newton (1980[26])

Values in parentheses in Table 1 are those calculated by M.D. Newton[26] 20 years later. Although the later calculations are more detailed (particularly of precursor formation) the agreement with the first ones are remarkably good. In the later calculations, as had become customary, the inner contribution to the activation energy was calculated from experimental bond lengths and vibration frequencies, which were unobtainable earlier.

These considerations led naturally to the calculation of the rates of *cross-reactions* from those of the constituent self-exchange reactions plus a knowledge of the overall thermodynamics.

In its simplest form, the free energy of activation $\Delta G_{12}^\dagger$ of the cross-reaction is given by[13]

$$\Delta G_{12}^\dagger = \frac{1}{2}(\Delta G_{11}^\dagger + \Delta G_{22}^\dagger + \Delta G_o),\qquad(4)$$

where ΔG_o is the overall reaction free energy. The first cross-reaction for which the energetics of formation of the transition state were calculated was the Pu^{3+}/PuO_2 exchange of Eq. (3) as the few data for actinides were the only ones available at the time. In applying Eq. (4) the experimental value for the overall energy change was used:

$$\Delta G^{\dagger} \text{ (kcalmole}^{-1}\text{: water solution 298 K)}$$

	$\Delta G^{\dagger}$
$*Pu^{3+} + Pu^{4+} \xrightarrow{k_{11}} *Pu^{4+} + Pu^{3+}$	21.2
$*PuO_2 + PuO_2^+ \xrightarrow{k_{22}} *PuO_2^+ + PuO_2$	15.4
$*Pu^{3+} + *PuO_2^+ \xrightarrow{k_{12}} Pu^{4+} + PuO_2$	18.2
	(exp. 16.9)

Henry Taube, commenting on these calculations in his Nobel lecture[27] noted the irony of this first theoretical discussion of cross electron transfer reactions with finite energy of reaction — which are overwhelmingly those of greatest general interest in chemistry, the first to be so discussed — was for a process involving a man-made element.

10. Potential Energy Surfaces

Assuming a two-site system locating orthonormal molecular electronic orbitals ϕ_1 and ϕ_2, the system wavefunction in state s is expressed as

$$\Psi_s = c_s(1)\phi_1 + c_s(2)\phi_2. \tag{5}$$

The fermion occupation number or charge density $\langle n_2 \rangle$, for ϕ_2 (acceptor), the eigenvalue of the electronic number operator n_2, is

$$\langle n_2 \rangle = |c_s(2)c_s^*(2)| \tag{6}$$

and will vary from (approximately) zero to (approximately) 1 during the course of a reaction. For present purposes, we assume that there is linear vibronic coupling to a set of k harmonic oscillators of frequencies ω_k, which represents the environment (inner coordination plus solvent). The electronic coupling energy between ϕ_1 and ϕ_2 is J. With the Hamiltonian for the system of Ref. 28 , we obtain the Born–Oppenheimer potential surfaces illustrated in Fig. 7.

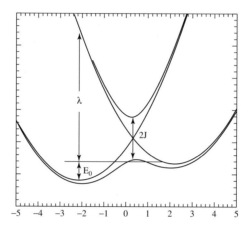

Fig. 7. Potential energy surfaces for two-site electron transfer system linearly coupled to manifold of harmonic vibrations: abscissa is representative distance (arbitrary units). The overall free energy change is E_0, the separation of upper and lower surfaces at the conical intersection is $2J$, and the reorganization energy is λ.

The difference between electron-vibrational coupling energies in initial and vertical final states is λ; the reorganization energy; and the energy difference between initial and final states is E_0.

For a symmetrical system ($E_0 = 0$), the three possible potential energy profiles for a generalized coordinate with respect to delocalization are shown in Fig. 8.

At the extreme left of Fig. 8(a), we have solvent-broken symmetry, with a double-well potential and thus the possibility of kinetics, and at right, there is complete delocalization, with single minima ground and excited state energies. The conditions for these are, respectively, $2J > \lambda$ and $2J > \lambda$.[29] The intermediate case, with an almost quartic ground state potential curve, represents the situation where $2J$ and λ are approximately equal: this case has until very recently been presumed to be extremely uncommon.

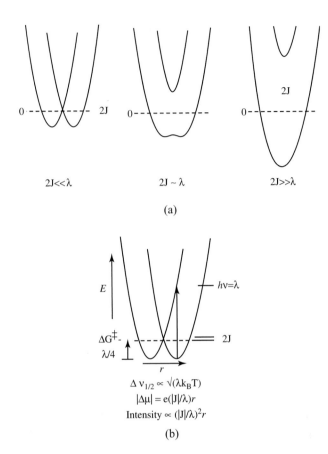

Fig. 8. (a) The three limiting ground and excited state potential energy surface patterns for two-site symmetrical electron transfer. For a ratio of $2J$ to λ much less than 1 (*left panel*) the ground state is strongly localized. Where this is much greater than 1, the system is strongly delocalized (*right panel*). Where the ratio is close to unity (*center*), the ground state is "trembling on the brink" of localization.[29] **(b)** The strongly localized symmetrical case [left panel of (a)]. Optical transition from the ground state to the upper state of the two-site electron transfer system will occur with an absorption maximum equal to the reorganization energy λ . This is termed an *intervalence transition*.[30] The Gaussian absorption envelope has half-width proportional to $\lambda^{1/2}$, and the band intensity is proportional to $J^{1/2}$. Thus the two parameters determining the thermal rate (J and λ) can be obtained from the properties of the optical intervalence transition. In the corresponding case of an unsymmetrical transfer, the free energy asymmetry E_0 can also be obtained, thus permitting calculation of the thermal rate in this more general case also.

11. Intervalence Transfer: Connection between Thermal Kinetics and Optical Absorption Probability

Consider now the case where the electronic coupling energy J is sufficiently small to be ignored in the estimation of activation energy, as has been assumed in the above of thermal rates. The ground and first excited state potential curves are as shown in Fig. 8(b) [same as Fig. 8(a), extreme left panel]. In addition to the thermal transition (passage over the potential barrier of the lower surface) one sees that there is the possibility of a vertical optical transition to the upper potential surface, with energy very close to the reorganization energy λ. In the linear vibronic coupling model assumed here, the thermal activation energy will thus be $1/4\lambda$, and the Gaussian band half-width $\Delta v_{1/2}$ is also a function of λ. In addition, the electronic coupling J can be obtained from the intensity (oscillator strength) of the transition. These predicted absorption bands, which I termed *intervalence* bands, Hush[30] would thus yield experimental data for the two parameters of the self-exchange rate. One can regard the thermal transition as the optical transition in the limit of zero frequency. Intervalence absorption for the symmetrical case is unusual, because although it involves an electronic transition, the energies of initial and final states are equal. In the general case for an unsymmetrical transfer, the corresponding intervalence band will yield the three parameters λ, J and E_0 necessary to characterize the thermal kinetics. This is a rather remarkable linkage between optical and thermal transfer probabilities. The expressions quoted are for the "high-temperature" limit in which it is assumed that vibration frequencies are less than or of order of twice the thermal quantum. This usually suffices for discussion of inorganic rates, but more detailed analysis is required when precise data are available for vibronic coupling (see below in the account of photosynthetic mechanisms).

Intervalence transfer theory was first applied to analysis of the electronic spectra of naturally-occurring or synthetic mixed-valence systems, all of the unsymmetric type,[31] where there is a finite E_0. The intervalence absorption frequency in the weak-coupling case is $E_0 + \lambda$, and this is generally large enough to result in absorption in the visible region, and so give rise to color: this explained the most striking feature of mixed-valence systems. Among the former, the most important are rock-forming minerals such as hornblendes, or asbestos minerals such as crocidolite (the toxic "blue asbestos") whose color is largely due to Fe^{2+}/Fe^{3+} optical transfer. Further application to naturally-occurring minerals has been explored extensively by Burns.[32] Among the many synthetic cases, probably the best-known is the pigment Prussiuan Blue, the deep color of which was explained as resulting from the transition:

$$Fe(III)t_2^3 e_g^2(^6A_{1g}) + Fe(II)t_{2g}^3(^1A_{1g}) + h\nu$$
$$\longrightarrow Fe(II)t_{2g}^3(^1A_{1g}) + Fe(III)t_{2g}^3 e_g^2(^6A_{1g})$$

between low-spin and high-spin iron ions. The extent of localization of the transferring t_{2g} electron in the initial state is calculated (99.03%) as can also λ and E_0. The thermal transfer rate can thus also be calculated. This is an example of information that is yielded by intervalence spectra, the study of which, as will be shown, has subsequently yielded crucial information about dynamics and delocalization in wider chemical and biological contexts.

The current state of the interpretation of optical, transport and magnetic properties of mixed-valence systems is the topic of a recent Royal Society Discussion edited by Clark, Day and Hush.[33]

12. Revolution 5: Ligand-Bridged Electron and Hole Transfer

Henry Taube at Stanford University ushered in a new era in the study of electron transfer processes by synthesizing complexes in which the reacting ions were both bound covalently to a bifunctional ligand. In such a species, the geometry of the transfer assembly could be explicitly measured, thus providing a firmer foundation for comparison with theory. The first one synthesized in 1969 together with Carol Creutz[34] is shown in Fig. 9.

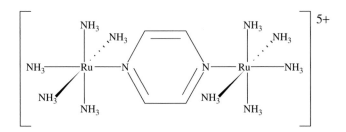

Fig. 9. The Creutz–Taube ion.[34]

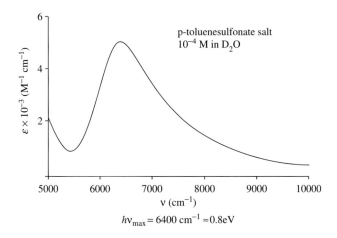

$$h\nu_{max} = 6400 \text{ cm}^{-1} \approx 0.8 \text{eV}$$

Fig. 10. Intervalence absorption spectrum of the Creutz–Taube ion.[34]

Its synthesis was inspired in part by the intention to search for the intervalence band which should be displayed by this mixed-valence ion. Now known as the Creutz–Taube ion, it did in fact[34] show an absorption in the predicted infrared region.[32] Taube's work revolutionized the experimental study of electron transfer kinetics, and initiated an explosion of work in the area, in which the connection between optical and thermal rates has played a key role.

Typical intervalence spectra (typically for symmetrical systems in the near infra-red region) involving (as in the Creutz–Taube ion) low spin Ru^{2+}/Ru^{3+} transfer[35] are shown in Fig. 11.

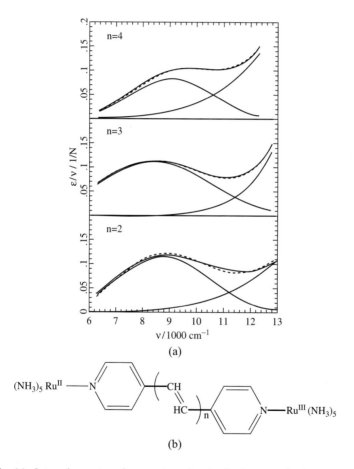

(a)

(b)

Fig. 11. (a) Intervalence transfer spectra of mixed-valence ruthenium pentammine complexes bridged by olefin linkages of varying number n of double bonds.[35] These have been deconvoluted to isolate the intervalence band from the tail of the charge transfer to metal band at higher frequency.[36] **(b)** The formulae of the complexes.

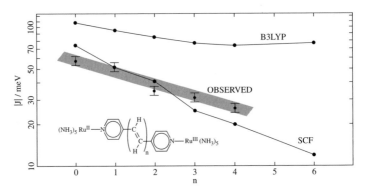

Fig. 12. The energy J of electronic coupling between ruthenium hexammine complexes calculated from the intensity of the intervalence spectra of Fig. 10 as a function of n, the number of double bonds in the olefin linkage.[36] The transmission coefficient for thermal electron exchange can be obtained from this and the reorganization energy. The observed decay of J with distance is clearly exponential. Calculation of J by a self-consistent field (SCF) method reproduces the observed behavior of J quite well. Calculations using the B3LYP density functional (DFT) method overestimate the extent of the coupling through the bridge.[36,37]

Here, the metal complexes are separated by conjugated olefin bridges of varying length. The electronic coupling J through these π-bridges can be estimated from the intervalence intensities[36] and are shown as a function of length in Fig. 12.

Also shown are calculations of J. A self-consistent field (SCF) molecular orbital calculation yields quite a satisfactory agreement. However, the density-functional (DFT) calculation which was used badly overestimated the strength of the interaction. Care has to be taken in the choice of DFT codes for electron transfer systems (see, e.g., Ref. 37).

The dependence of J on bridge length in these systems is clearly exponential. This is in accord with the theory of McConnell for superexchange coupling[38] illustrated in Fig. 13.

According to this approach, the electronic interaction J is related to the coupling a (not to be confused with orbital coupling amplitude) between elements of the bridge and β for coupling of metal orbitals to the

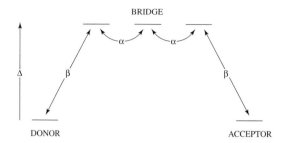

Fig. 13. The mechanism of superexchange in a symmetrical two-site system coupled through a bridge of n identical units (as in Figs. 11 and 12).[38] The off-diagonal elements in the bridge Hamiltonian are indicated by α (not to be confused with amplitude in wavefunction) and that for interaction of donor or acceptor to the bridge end unit is β. The energy separation of donor/acceptor and lowest bridge level is Δ.

ends of the bridge:

$$J = \alpha^2 \beta^{-1} \left(\frac{-\beta}{\Delta} \right)^n ,\qquad (7)$$

where the donor/acceptor-to-bridge energy gap is Δ, and the number of bridge elements is n. Exponential dependence of electronic coupling through a bridge is expected when the energy gap between transferring orbitals and the bridge levels is sufficiently large. This is the most commonly observed situation. However, as this gap narrows, more complex behavior (non-monotonic, oscillatory etc.) is predicted as resonance is approached. The general theory is discussed in Ref. 37.

13. Revolution 6: Introduction of Ultrafast Spectroscopic Techniques

The study of excited state electron dynamics has been transformed by the advent of ultrafast spectroscopic techniques, ranging from nanosecond to picosecond and to femtosecond. The latter was introduced and extensively developed in the fundamental, groundbreaking work of Zewail.[40] Atto-seconds are also now just accessible.

An area in which picosecond spectroscopy has yielded fundamental information about bridged electron transfer is photochemical donor–acceptor optical and thermal transfer through non-conjugated (σ-type) bridges. An example[41] is shown in Fig. 14, where the bridge is a rigid norbornyl molecule of length up to 14 Å.

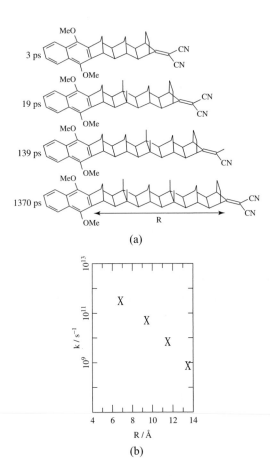

(a)

(b)

Fig. 14. (a) Picosecond excited state electron transfer rates between 1:4 dimethoxynaphthalene and 1:1 dicyanoethylene via a rigid non-conjugated σ-bonded norbornyl bridge of varying length.[41] (b) Exponential dependence of logarithm of electron transfer rate on donor–acceptor separation.

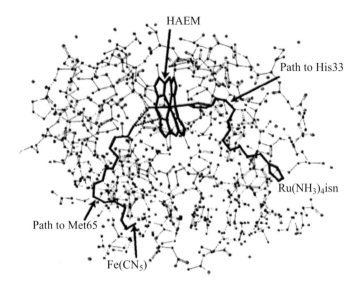

HAEM

Path to His33

Ru(NH$_3$)$_4$isn

Path to Met65

Fe(CN$_5$)

Fig. 15. Protein electron pathways for electron transfer between Fe(II) in modified Horse Heart Cytochrome C and two different bound acceptors at 15.3–16.0 Å distances.[42] Switching acceptor groups between the two acceptor sites results in marked change in rate, showing that it is mediated by protein paths rather than being a simple "through space" process.

The announcement of picosecond rates for photochemical electron transfer was at first greeted with incredulity, since σ-type molecules were after all like paraffins and were thought to obviously function as insulators. Indeed, when I reported on the pioneering experimental work of Paddon-Row, Verhoeven, Warman and collaborators and my theoretical interpretations (Ref. 41 and references therein) at the *First Bioinorganic Conference* in Florence in 1983, the response was deeply sceptical and the address was reported in the *New Scientist* as "provocative". However, this type of transfer is now well-established and is of fundamental importance for biological electron or hole transfer. An example is shown in Fig. 15 for transfer across Horse Heart Cytochrome C.[42] The detailed interpretation of transfer via protein molecules, such as this, has been studied extensively by David Beratan.

14. Revolution 7: Computers

The earliest work discussed here was carried out in the period when computers were just starting to be developed. In particular, no computer (not even of the innovative Harvard type above) was yet available at Manchester. An irony of this is that one of my senior colleagues at this time was Alan Turing, who had invented them. By this time, Turing had many other interests, including a theory of biological morphogenesis: since it was based on coupled diffusion-reaction mechanisms, which I was also studying in the context of heterogeneous electron transfer, we had several discussions about the basic theory. The paper that Turing subsequently wrote remained essentially unread for about twenty years, after which it was hailed as a cornerstone of this important aspect of biological growth.

Fig. 16. Harvard College's ten computers in 1917.

With the advent of advanced computing and supercomputing complexes, it has been possible to extend theoretical examination of the fundamentals of electron transfer to quite large systems, and to interpret the often very complex structure of the reorganization and electron coupling energies. This applies particularly to biological systems.

15. Delocalization in Biological Systems: Interplay of *J*, λ and E_0 in the Bacterial Photosynthetic Reaction Center

An illustrative example of the way in which the basic approach I have outlined can be usefully applied to the analysis of systems involved in more complex biological situations is provided by consideration of the oxidized special-pair in bacterial photosynthetic reaction centers. The "solvent" in which the reaction centers are located is a protein sheath, and for a number of species X-ray measurements have yielded reasonably good information about all atomic positions. (Parenthetically, I might remark that an improved *ab initio* molecular orbital procedure has been developed for optimizing X-ray crystal data: this has been applied to optimization of the structure of the 150,000-atom trimeric structure of Photosytem I.[43])

The structure of the reaction center for the bacterial photosynthetic reaction center of *Rhodobacter Sphaeroides* is shown schematically in Fig. 17(a).[44]

The "special-pair" of bacteriochorophylls whose left-hand and right-hand components are named P_M and P_L, respectively, is indicated at the top of the complex.

This is the "engine-room" of the photosynthetic array. It absorbs light from the antenna system into excited electronic states of the dimer; then by photochemical transfer spits out an electron in a few picoseconds into the biosynthetic pathway.

The perspective in Fig. 17(a) might suggest that the two component bacteriochlorophylls lie in a symmetrical arrangement. However, this is, importantly, not the case, and in fact solvent symmetry-breaking has reduced the overlap from four pyrrole rings, as it would be in the symmetrical eclipsed arrangement, to only one of each ring, as the "slipped" configuration of Fig. 17(b) indicates.[44]

This greatly reduces the inter-ring coupling *J* from the value of *ca.* 0.5 eV found in eclipsed porphyrin-type dimers to a much smaller value

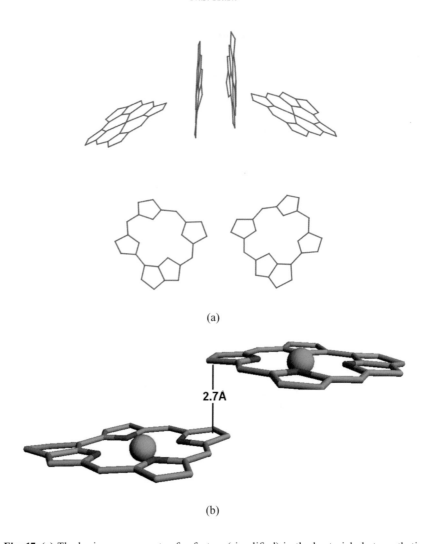

(a)

(b)

Fig. 17. (a) The basic arrangements of cofactors (simplified) in the bacterial photosynthetic reaction center (in this case that of *Rhodobacter Sphaeroides*) showing approximate bilateral symmetry.[44] The two bacteriochlorophylls at the top, approximately normal to the plane of the page, constitute the special-pair P. The right-hand component is P_L and the left-hand one is P_M. **(b)** The P_L and P_M components of the special-pair overlap only through one pyrrole ring on each molecule, with an approximately parallel separation of 1.7 Å.

(estimated around 0.15 eV as a simplistic guess). The competition with reorganization energy, which tends to localize the unpaired electron on one or the other component of the special-pair together with variations in energy asymmetry E_0, could result in a variety of structures for the special-pairs in the oxidized bacterial photosynthetic special-pair mixed-valence cations.

The schematic potential-energy surface for the oxidized special-pair[45] is shown in Fig. 18. Calcuations of properties of the system can start either from a localized or a delocalized adiabatic formalism: either, however, leads to the same Born–Oppenheimer states.

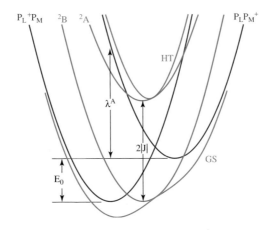

Fig. 18. Potential energy surface for the oxidized special-pair $P_L^+P_M/P_LP_M^+$. Both the localized diabatic states (blue) and the delocalized diabatic states (green) lead to the same Born–Oppenheimer adiabatic states (red). The ground state is GS and the hole transfer state is HT. Other symbols have their previous significance.[45]

By comparison with the observed properties of a model for an oxidized special-pair, Binstead and Hush[46] predicted an intervalence band for the system at quite low energy, *ca.* 0.2 eV — actually slightly later than the very important work of Breton and colleagues, in which such a band was observed for oxidized *Rhodobacter Sphaeroides*.[47] This is shown in the central panel of Fig. 19.

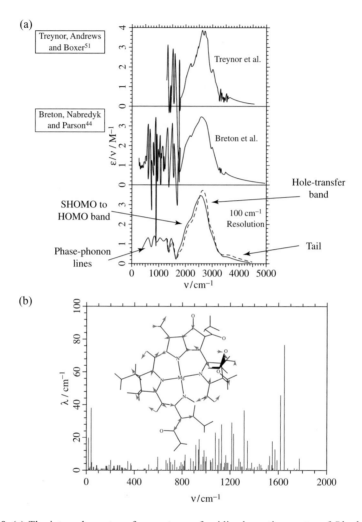

Fig. 19. (a) The intervalence transfer spectrum of oxidized reaction center of *Rhodobacter Sphaeroides*. The top and second spectra are experimental (Refs. 48 and 47, respectively) and the lowest is simulated (Ref. 45). In the latter spectra, the components of the band are indicated by arrows as tail (short wavelength edge), main hole-transfer band, an additional shoulder due to a separate transition from Second HOMO (S) to HOMO level, and phase-phonon vibrational structure. The resolution has been smoothed to $100\,cm^{-1}$. **(b)** The calculated phase-phonon transitions. The contribution of each mode to the reorganization energy is indicated on the left-hand ordinate. The vibrational mode depicted is that most strongly coupled (highest intensity) to the electronic system.

The upper panel shows a later measurement by Treynor *et al.*[48] At frequencies below the main "electronic" portion of the intervalence spectrum, there is a very interesting series of vibrational lines. This is the phase-phonon structure. Each line represents a vibrational mode of the special pair complex which contributes to the reorganization energy: its intensity is a measure of the extent of its individual contribution.

I mentioned that one might expect quite a variation in the properties of oxidized special-pairs in various bacterial photosynthetic species. This is in fact so,[49] as is evidenced by the intervalence hole-transfer spectra of ten such species shown in Fig. 20.

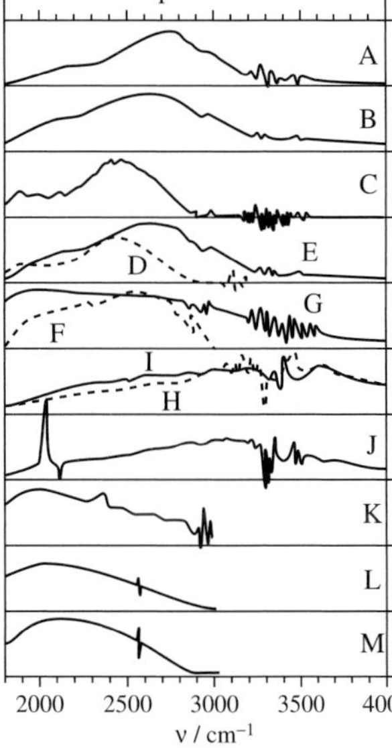

Spectrum		Species	RC	Qualitative Appearance
	A	Rh. viridis	BChl-b	Narrow delocalized HT absorption with SH shoulder and long tail
	B	Rb. sphaeroides	BChl-a	Narrow delocalized HT absorption with SH shoulder and long tail
	C	Cl. linicola	BChl-a	Narrow delocalized HT absorption with SH shoulder
	D E	Cl. tepidum	BChl-a	Narrow delocalized HT absorption with SH shoulder and long tail
	F G	C. reinhardttii	Chl-a/ Chl-a'	E: Possible shoulder, unexpected narrow width F: Broad localized HT absorption
	H I	Synochocystis S 6803	Chl-a	Broad localized HT absorption
	J	Spinach PS-I	Chl-a	Broad localized HT absorption (ferricyanide used as oxidant)
	K	Spinach PS-II	Chl-a	intermediate bandwidth, band possibly extends well below 1800 cm^{-1}
	L	Hb. mobilis	BChl-g	intermediate bandwidth, band possibly extends well below 1800 cm^{-1}
	M	H. modesticaldum	BChl-g	intermediate bandwidth, band possibly extends well below 1800 cm^{-1}

2000   2500   3000   3500   4000
v / cm^{-1}

Fig. 20. Intervalence hole-transfer spectra of oxidized special-pairs in a series of naturally-occurring bacterial species.[49] These range from those of narrow bands of delocalized systems (e.g., S) to broad bands of localized systems (e.g., J), through intermediately localized systems (e.g., K).

The striking features are (1) a considerable variation of the energy asymmetry E_0 of the hole transfer, and (2) a very marked variation in spectral bandwidth, ranging from relatively narrow to very broad. Qualitatively, the latter indicates that the structure ranges from near delocalization in the former case to marked asymmetry in the latter. All have evolved at different evolutionary periods, in which nature has carried out the experiment of juggling electronic coupling, reorganization energy and asymmetry to modify the efficiency of the bacterial photosynthetic apparatus. A study of the historical unfolding of this evolutionary process could be revealing.

In addition to naturally-occurring variants, there is also a considerable amount of data on systems produced by site-directed mutagenesis. The data for intervalence hole-transfer bands of all systems known[50] are listed in Table 2.

Table 2. Observed Properties of Mutant Reaction Centers of *Rhodobacter Sphaeroides*.[51] The midpoint potential is E_m (in Volts) and the deduced energy asymmetry is E_0 (eV). The spin densities on P_L and P_M are, respectively, ρ_L and ρ_M. Theoretical analysis is given in Refs. 50 and 51

Mutant	E_m / V	$\rho_L \rho_M$	ρ_L / e	E_0 / eV	$\bar{\nu}$ / cm^{-1}
LH(L131)+LN(M160)	0.591	0.82	0.45	−0.018	
LH(L131)+LS(M160)	0.584	0.83	0.45	−0.017	
LH(L131)	0.578	0.89	0.47	−0.010	2550
LE(L131)	0.555	1.00	0.50	0.000	
LH(L131+LD(M160)	0.606	1.10	0.52	0.009	
LD(L131)	0.544	1.37	0.58	0.028	
LQ(L131)	0.551	1.43	0.59	0.032	
LS(L131)	0.535	1.49	0.60	0.036	
LH(L131)+LQ(M160)	0.619	1.51	0.60	0.037	
LN(L131)	0.535	1.57	0.61	0.041	
LH(L131)+LE(M160)	0.623	1.86	0.65	0.056	

Table 2. (Continued)

Mutant	E_m / V	$\rho_L\rho_M$	ρ_L / e	E_0 / eV	$\bar{v}$ / cm^{-1}
LK(M160)	0.511	1.92	0.66	0.060	
Wild Type	0.503	2.11	0.68	0.069	2560
LS(M160)	0.514	2.11	0.68	0.069	
LH(L131)+LH(M160)	0.621	2.21	0.69	0.073	2610
LN(M160)	0.527	2.31	0.70	0.078	
LD(M160)	0.539	2.77	0.73	0.097	
LN(L131)+LH(M160)	0.583	3.89	0.80	0.136	
LS(L131)+LH(M160)	0.594	4.08	0.80	0.142	
LY(M160)	0.558	4.71	0.82	0.161	
LH(M160)	0.563	4.94	0.83	0.168	2780
RE(M164)	0.471	1.29	0.56	0.023	2450
RL(M164)	0.487	1.55	0.61	0.039	
RL(L135)	0.483	2.90	0.74	0.102	2450

The table also contains the deduced asymmetries and degrees of delocalization of the electron hole in the mixed-valence oxidized special-pair. In agreement with the above qualitative conclusions, we observe variation of hole localization ranging from almost equal to totally delocalized charge [$q_L = 0.47$, $q_M = 0.53$ in LH(L131) and $q_L = 0.74$, $q_M = 0.26$ in the very asymmetric species RL(M164)].

The lower panel of Fig. 19 shows the result of a simulation of the HT spectrum for oxidized *Rhodobacter Sphaeroides*.[45] This involved firstly optimization of the geometry of the reaction center. The calculation of critical parameters involved four electronic states, 50 antisymmetric vibrational modes, 20 symmetric vibrational modes, and 4,000,000 individual vibronic levels, using mostly B3LYP DFT theory. Seven parameters were optimized to within 15% to fit experiment.[45]

Examples of detailed charge distribution for various mutants are shown in Figs. 21(a)–(d).

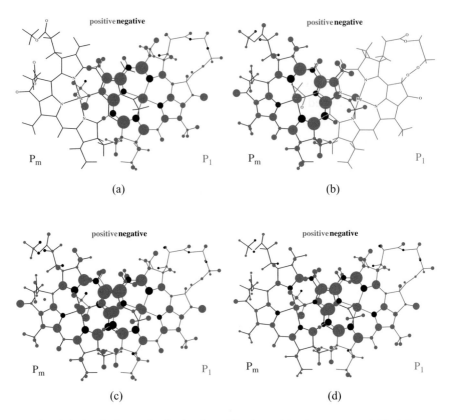

Fig. 21. Calculated charge distribution in oxidized special-pairs mutants of *Rhodobacter Sphaeroides* genetically modified systems: **(a)** charge localized on P_L, **(b)** charge localized on P_M, **(c)** charge delocalized over both moieties, and **(d)** P_L/P_M ratio approximately 2:1, as also deduced for the wild type.

These range from strong localization either on P_L or P_M, through to near symmetrical, to the 2:1 charge distribution found in the wild type and in two mutants.

Finally, it is of interest to note a further experimental method which is valuable in obtaining information about the symmetry of the oxidized special-pair in the excited intervalence state: this is the electronic Stark spectroscopy of the hole-transfer band: one property deducible from this is the dipole moment of the excited state, a very important quantity. An example is shown in Fig. 22, in which a predicted[51] and an observed[52] Stark spectrum for the HT intervalence band of an oxidized special-pair is shown. The agreement is very satisfactory. The general theory of the electronic Stark effect, as applied in particular to such biological systems, has recently been discussed in detail by Treynor, Boxer, Reimers, Hush *et al.*[53]

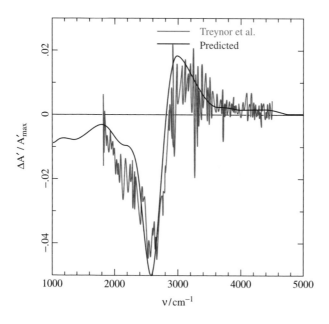

Fig. 22. The observed electroabsorption Stark spectrum[52] of oxidized wild type *Rhodobacter Sphaeroides* compared with the spectrum calculated by the quantum vibronic method outlined in Ref. 45 (full line). The form of the electroabsorption response for intervalence transitions is intermediate between the Liptay classical second derivative (with two nodes) and first derivative (with one node): both theory and experiment are in agreement with prediction of one node in this case, and the amplitudes are in agreement within experimental error. The structure around $2200 \, cm^{-1}$ is largely due to response of the SHOMO-LUMO transition appearing as a shoulder in the main band (*cf.* Fig. 19): theory and experiment are again in good agreement. Calculations for this band are extremely sensitive to the extent of vibronic coupling considered: those here result from the full vibronic analysis.

16. Back to Origins: The Creutz–Taube Ion 39 Years on: Multiple Lifetimes and New Vistas in Chemical Physics

So far, little has been said about the iconic Creutz–Taube ion, the first synthetic bridged mixed-valence species for which the predicted appearance of an infra-red intervalence band was indeed observed. A glance at the spectrum in Fig. 1 shows that this band is very markedly narrower than those shown by the undoubtedly localized ions in Fig. 2. This suggests that the Creutz–Taube ion may have a special structure. In fact its electronic structure was debated for many years until electrochemical/vibrational measurements by R.J.H. Clark et al.[54] comparing of oxidized, mixed-valence and reduced state vibrational frequencies were finally accepted to have settled the question in favor of a totally delocalized ground state, as in the last example of Fig. 8. These agreed in fact with similar measurements for spectator mode frequencies reported by Beattie, Hush and Taylor much earlier.[53] This is a long and interesting history, the earlier part of which is discussed in Ref. 28.

For this interpretation to be the correct one, it is necessary, in light of the relationships of previous theory, for the reorganization energy to be very close to twice the electronic coupling energy. There was no independent measurement of λ, however, to test this.

The long consensus was disturbed by the proposal of Desmadis et al.[56] in 2001 that there is growing evidence (particularly in osmium complexes) for mixed-valence systems which have structures almost on the localization-delocalization limit, with extremely small but non-vanishing energy barrier. They made the tentative proposal that the Creutz–Taube ion was also of this unusual type — exemplified by the central structure in Fig. 8 with an almost quartic ground state potential energy profile.

Jeffrey Reimers and I have returned to this problem. The surprising fact is that detailed calculation so far (not possible until recently) does support this proposal. We know that the internal reorganization energy is very small (almost equal to Ru ion radii) and can be accurately estimated. It is the ion-solvent interaction that controls the possible symmetry-breaking which is the main question. Firstly, a detailed quantum calculation was carried out,[57] which required computer resources unavailable at previous times when the structure was considered. Importantly, the geometry of the ion was optimized *in the water solvent*, a requirement whose critical importance had not been previously realized. The conclusion was that, within the estimated limits of accuracy, the value of the reorganization energy was indeed very close to $2J$.[57]

More recently, in an additional collaboration with B. Wallace,[21] a molecular dynamics calculation involving 192 water molecules was carried out. Figure 5 shows a calculation of a snapshot of a particular simulation, showing high symmetry. From these simulations, it is possible to obtain the thermally averaged solvent density of states $\rho(\omega)$, where ω is the solvent frequency: from this, weighted by the chromophore-solvent coupling, one obtains the integrated solvent reorganization energy λ_s:

$$\lambda_s = \eta \int_0^\infty \omega\rho(\omega)d\omega \tag{8}$$

which is obtainable from the Stokes shift function $S(t)$,

$$\omega\rho(\omega) = \frac{2}{\pi}\int_0^\infty S(t)\cos\omega t \, d\omega \tag{9}$$

a function that depicts the dependence of the non-equilibrium energy difference $\Delta E(t)$ between ground and the excited states:

$$S(t) = \frac{\Delta\bar{E}(t) - \Delta\bar{E}(\infty)}{\Delta\bar{E}(0) - \Delta\bar{E}(\infty)} \tag{10}$$

By differentiation of Eq. (8) one obtains the dependence of solvation energy on solvent frequency. For the Creutz–Taube ion in water solution, this is shown in Fig. 23.

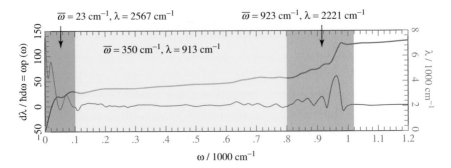

Fig. 23. The calculated reorganization energy distribution as a function of frequency (left scale, full line) and its cumulative integral (dashed line, right scale) for the Creutz–Taube ion in water solution.[21] The shading for the latter is a simplified summary of the solvent response, in which $2567\ \mathrm{cm}^{-1}$ of reorganization energy is partitioned to coupling to averaged solvent modes of frequency $23\ \mathrm{cm}^{-1}$, $913\ \mathrm{cm}^{-1}$ to those of average frequency $350\ \mathrm{cm}^{-1}$, and $2221\ \mathrm{cm}^{-1}$ to coupling with modes of average frequency $923\ \mathrm{cm}^{-1}$.

These reveals the build-up of reorganization energy as the solvent modes (libration, translation etc.) are progressively coupled in over time. Thus a detailed picture of the structure of the reorganization interaction is made explicit.

If the cumulative reorganization modes were coupled just a little more strongly with the electronic system of the Creutz–Taube ion, it would be tipped quickly into a clearly more localized mode, and simulations were also made of such hypothetical scenarios.

It has thus taken 39 years to come closer to an understanding of the symmetry of this fundamental species. But in doing so, we have entered a surprising new area of chemical physics in which an electronic structure "trembling on the brink" of localization-delocalization has astonishing and quite unexpected vibrational properties.[21]

17. Revolution 8: Nanotechnology — Venturing into Feynman's "Room at the Bottom": Molecular Electronics

I have so far dealt with clearly identifiable chemical and biological electron or hole transfer systems. There are of course wider implications of the basic theoretical approach. An obvious one is Molecular Electronics (the ultimate form of nanotechnology). A landmark in this area was the proposal in 1973 by Aviram and Ratner[58] of a design for a molecular diode, which was based on what was known both theoretically and experimentally about the dynamics of charge transfer. The theory of the electric-field operated molecular switch was put forward in Ref. 59, again a straightforward application of basic ideas. In general, proposed designs for molecular devices build in large part on the concepts discussed above. More recently, as a member of a group concerned with development of the theory towards detailed design of practical molecular devices, I have been concerned with properties such as ballistic transport,[60] channels[61] and phonon structure and noise[62] in molecular wires. There are hopes of a practical ultradense molecular memory.

I have discussed the early development of Molecular Electronics in an overview[63]: predictably, the first person mentioned in the half-century history of this exponentially expanding area, bringing us full circle with the origins of understanding of the basics of electron transfer, is Robert Mulliken.

References

1. Hush NS. (1961) *Fourth Moscow Conference of Electrochemistry 1956*, p. 99. Eng. Trans. Consultant's Bureau, New York.
2. Hush NS. (1957) *Z Elektrochem* **61**: 734
3. Hush NS. (1999) *J Electronal Chem* **460**: 5
4. Schmickler W. (1986) *J Electroanal Chem* **204**: 201.

5. Evans MG, Hush NS, Uri N. (1952) *Quart Rev Chem Soc (London)* **6:** 186.
6. Newton TP, Montag T. (1975) *Inorg Chem* **15:** 2856.
7. Marcus RA. (1956) *J Chem Phys* **24:** 96.
8. Marcus RA. (1956) *J Chem Phys* **24:** 979.
9. Marcus RA. (1964) *Ann Rev Phys Chem* **15:** 155.
10. Levich V, Dodonadze RR. (1959) *Dokl Akad Nauk SSSR Ser Fiz Khim* **124:** 123.
11. Dogonadze RR. (1959) *Dokl Akad Nauk SSSR Ser Fiz Khim* **124:** 1108.
12. Hush NS. (1960) *Disc Faraday Soc* **29:** 113, 116, 133, 249.
13. Hush NS. (1961) *Trans Faraday Soc* **57:** 557.
14. Hush NS. (1957) *Z Elektrochem* **61:** 734.
15. Hush NS. (1952) *J Chem Phys* **20:** 1343.
16. Hush NS. (1952) *J Chem Phys* **20:** 1660.
17. Hush NS. (1953) *J Chem Soc* 275.
18. Hush NS. (1955) *J Chem Phys* **23:** 514.
19. Hush NS, Pople JA. (1955) *Trans Faraday Soc* **51:** 75
20. Hush NS. (1968) *Electrochimica Acta* **13:** 1005.
21. Reimers JR, Cai Z-L, Hush NS. (2008) *Phil Trans Roy Soc* **A366:** 5.
22. Bethe H. (1929) *Ann Physik* **3:** 133.
23. Hush NS, Pryce, MHL. (1957) *J Chem Phys* **26:** 143.
24. Hush NS. (1958) *J Chem Phys* **28:** 244.
25. Hush NS. (1958) *Disc Faraday Soc* **26:** 145.
26. Newton MD. (1980) *Int J Quantum Chem Symp* **14:** 363.
27. Taube H. (1984) *Science* **226:** 1028.
28. Hush NS. (1980) In: Brown DB, Reidel D (eds), *Mixed-Valence Compounds*, p. 151.
29. Hush NS. (1975) *Chem Phys* **10:** 361.
30. Hush NS. (1967) *Prog Inorg Chem* **8:** 391.
31. Allen GC, Hush NS. (1967) *Prog Inorg Chem* **8:** 391.
32. Burns RG. (1970) *Mineralogical Applications of Crystal Field Theory.* Cambridge University Press, UK.
33. Clark RJH, Day P, Hush NS (eds). (2008) Mixed Valency. *Phil Trans Roy Soc A.*

34. Creutz C, Taube H. (1969) *J Amer Chem Soc* **91**: 39883.
35. Woitellier S, Launay JP, Spangler CW. (1989) *Inorg Chem* **28**: 758.
36. Reimers JR. Hush NS. (1990) *Inorg Chem* **29**: 3686.
37. Cai Z-L, Sendt K, Reimers JR. (2002) *J Chem Phys* **117**: 2543.
38. McConnell HM. (1961) *J Chem Phys* **35**: 508.
39. Reimers JR, Hush NS. (1989) *Chem Phys* **134**: 323.
40. Zewail AH. (1994) *Femtochemistry.* World Scientific, Singapore.
41. Oevering H, Paddon-Row MN, Verhoeven J, Hush NS. (1987) *J Amer Chem Soc* **109**: 3258.
42. Isied S (ed). (1997) *Electron Transfer Reactions: Inorganic, Organo-metallic, and Biological Applications.* ACS Advances in Chemistry Series, No. 253. Oxford University Press, UK.
43. Canfield P, Dahlbom MG, Hush NS, Reimers JR. (2006) *J Chem Phys* **124**: 024301.
44. Arnoux B, Gaucher J-F, Ducruix A. (1995) *Acta Cryst* **1251**: 368.
45. Reimers JR, Hush NS. (2003) *J Chem Phys* **119**: 3262.
46. Binstead R, Hush NS. (1993) *J Phys Chem* **97**: 1317.
47. Breton J, Nabredyk E, Parson WW. (1992) *Biochemistry* **31**: 7503.
48. Treynor TP, Andrews SS, Boxer SG. (2003) *J Phys Chem* **B107**: 11230.
49. Reimers JR, Shapley W, Hush NS. (2003) *J Chem Phys* **119**: 3240.
50. Hughes JM, Hutter MC, Reimers JR, Hush NS. (2001) *J Amer Chem Soc* **123**: 8550.
51. Reimers JR, Hush NS. (2004) *J Amer Chem Soc* **126**: 4132.
52. Treynor TP, Andrews S, Boxer SG. (2003) *J Phys Chem* **B107**: 11230.
53. Kanchanawong P, Dahlbom MG, Treynor TP, Hush NS, Boxer, SG. (2006) *J Phys Chem* **B110**: 18688.
54. Best SP, Clark RJH, McQueen RCS, Joss S. (1989) *J Amer Chem Soc* **111**: 548.
55. Hush NS, Beattie JK, Taylor PR. (1976) *Inorg Chem* **15**: 992.
56. Desmadis KD, Hartshorn CM, Meyer TJ. (2001) *Chem Rev* **101**: 2055.
57. Reimers JR, Cai Z-L, Hush NS. (2005) *Chem Phys* **319**: 39.
58. Aviram A, Ratner MA. (1974) *Chem Phys Lett* **29**: 277.
59. Hush NS, Wong AT, Bacskay GB, Reimers, JR. (1998) *J Amer Chem Soc* **112**: 4192.

60. Hall LE, Reimers JR, Hush NS, Silverbrook K. (2000) *J Chem Phys* **112:** 1510.
61. Solomon GC, Gagliardi A, Pecchia A, Frauenheim T, di Carlo A, Reimers JR, Hush NS. (2006) *J Chem Phys* **125:** 1847.
62. Solomon GC, Gagliardi A, Pecchia A, Frauenheim T, di Carlo A. Reimers JR, Hush NS. (2006) *Nano Lett* **6:** 243.
63. Hush NS. (2003) *Ann New York Acad Science* **1006:** 1.

Chapter 20

The Initial Value Representation of Semiclassical Theory
A Practical Way for Adding Quantum Effects to Classical Molecular Dynamics Simulations of Complex Molecular Systems

William H. Miller[*]

It has been known since the 1970s how one can *in principle* use classical molecular dynamics (i.e., numerically computed classical trajectories) as input to a semiclassical (SC) theory that provides a good description of all quantum mechanical effects in the dynamics of molecular systems. Over the last decade, various initial value representations of SC theory have been shown to provide a *practical* way for implementing these SC approaches for large molecular systems, those of interest for applications in bio-molecular and molecular-materials areas.

1. Introduction

With so many exciting things happening nowadays in the bio-molecular and molecular-materials areas, there is a clear need for theoretical calculations to model these large, complex molecular systems, to aid in the interpretation of these phenomena and to carry out exploratory calculations to suggest new directions for experimental studies. The only generally available approach heretofore for treating the dynamics of such large systems is classical molecular dynamics (MD), i.e., classical trajectory simulations, but this of

[*]Department of Chemistry and K.S. Pitzer Center for Theoretical Chemistry, University of California, and Chemical Sciences Division, Lawrence Berkeley National Laboratory, Berkeley, CA 94720-1460, USA, e-mail: millerwh@berkeley.edu.

course precludes the possibility of describing any of the quantum mechanical aspects of the molecular dynamics. Though for many purposes quantum effects will be unimportant, there is no doubt that they will be significant in some situations, most obviously when the dynamics of hydrogen atom motion is a significant part of the process of interest. Furthermore, unless one's theoretical treatment is able to incorporate quantum effects, even approximately, one will not know whether they are important or not.

It has been known since the earliest days of quantum mechanics that semiclassical approximations, such as the WKB approximation,[1,2] describe quantum effects in molecules quite well, but the WKB approach relies on an analytic solution of the corresponding classical problem, which is in general available only for systems of one degree of freedom (e.g., a diatomic molecule). In the late 1960s and early 1970s, however, it was shown (see Refs. 3–7; for reviews, see Refs. 8 and 9) how such approaches could be generalized to use *numerically computed* classical trajectories of multidimensional systems as input to a general semiclassical (SC) description of molecular dynamics. Applications (for reviews, see Refs. 8 and 9) of this "classical S-matrix" theory to treat inelastic and reactive scattering of small molecular systems, e.g., $A + BC \rightarrow AB + C$, demonstrated that in fact *all* quantum effects — interference/coherence, tunneling (and all other types of "classically forbidden" processes), symmetry based selection rules, etc. — are correctly described by this type of semiclassical theory, at least qualitatively, and typically quite quantitatively. In the late 1970s it was furthermore shown[10,11] how electronically non-adiabatic processes can be incorporated within this framework.

So it has been known for some time how one can *in principle* use numerically computed classical trajectories as input to a SC theory for general molecular systems, and that such a treatment provides a good description of essentially all quantum effects in molecular dynamics to a very useful level of accuracy. The way that numerically computed trajectories are used in SC theory, however, is more complicated than the way they are used in ordinary classical mechanics. Thus what has been lacking is a practical way for implementing SC theory for large molecular systems, and it is in this regard that various initial value representations (IVRs) of SC theory[4] have emerged[12-20] as providing a starting point for these purposes.

I have reviewed SC-IVR approaches[21-23] for adding quantum effects to classical MD simulations several times in recent years, so my survey below will be brief. There are several features of SC-IVR calculations that make them more difficult than the corresponding classical MD calculation, but these difficulties have now been largely overcome, as will be discussed below. The main point I would like to make is that SC-IVR calculations are now not extraordinarily more difficult than the corresponding classical MD calculation. My "goal" is to convince the classical MD simulation community of this, so that they can bring to bear on SC-IVR calculations all of the computational expertise that has been accumulated in MD and Monte Carlo methodology over several decades.

2. SC-IVR Calculation of Time Correlation Functions

Since most quantities of interest in the dynamics of complex systems can be expressed in terms of time correlation functions[24] of the form

$$C_{AB}(t) = tr[e^{-\beta \hat{H}/2} \hat{A} e^{-\beta \hat{H}/2} e^{i\hat{H}t/\hbar} \hat{B} e^{-i\hat{H}t/\hbar}], \tag{1}$$

I will focus the discussion on such quantities. Here $\hat{H}$ is the (time-independent) Hamiltonian of the complete molecular system, and $\hat{A}$ and $\hat{B}$ are operators relevant to the specific property of interest. [The Boltzmann operator need not be factored as it is in Eq. (1), but it is convenient in many cases to do so.] For example, if $\hat{A} = \hat{B}$ is the dipole moment operator, then the Fourier transform of the correlation function is the absorption spectrum; if it is the velocity operator of a tagged particle, or the flux operator related to a chemical reaction, then its time integral gives the diffusion coefficient and the chemical reaction rate, respectively.

The SC-IVR approximates the time evolution operator, $\exp(-i\hat{H}t/\hbar)$ — which determines all quantum dynamics — as a phase space average over the initial conditions of classical trajectories,

$$e^{-i\breve{H}t/\hbar} = \int d\mathbf{p}_0 \int d\mathbf{q}_0 \sqrt{M_{qp}/(2\pi i\hbar)^F} \; e^{iS_t(\mathbf{p}_0,\mathbf{q}_0)/\hbar} |\mathbf{q}_t\rangle\langle\mathbf{q}_0| , \tag{2a}$$

where F is the number of degrees of freedom, $(\mathbf{p}_0,\mathbf{q}_0)$ are the initial coordinates and momenta for a classical trajectory, $\mathbf{q}_t = \mathbf{q}_t(\mathbf{p}_0,\mathbf{q}_0)$ is the coordinate (in the F-dimensional space) at time t which evolves from this trajectory, $S_t(\mathbf{p}_0,\mathbf{q}_0)$ is the classical action (the time integral of the Lagrangian) along the trajectory, and M_{qp} is the determinant of the Jacobian (or monodromy) matrix relating the final position and initial momentum,

$$M_{qp} = \det\left[\frac{\partial \mathbf{q}_t(\mathbf{p}_0,\mathbf{q}_0)}{\partial \mathbf{p}_0}\right] . \tag{2b}$$

[Equation (2) is the original coordinate space,[12] or Van Vleck IVR; a popular alternative is the coherent state, or Herman–Kluk IVR,[13] whereby the initial and final states are coherent states, and the pre-exponential Jacobian factor is also modified.] For the correlation function one needs to insert two such representations of the propagator into Eq. (1), yielding the following double phase space average for the correlation function:

$$C_{AB}(t) = (2\pi\hbar)^{-F} \int d\mathbf{p}_0 \int d\mathbf{q}_0 \int d\mathbf{p}_0' \int d\mathbf{q}_0' (M_{qp} M_{qp}')^{1/2} \langle\mathbf{q}_0|\breve{A}_\beta|\mathbf{q}_0'\rangle$$
$$\times e^{iS_t(\mathbf{p}_0,\mathbf{q}_0)/\hbar} e^{-iS_t(\mathbf{p}_0',\mathbf{q}_0')/\hbar} \langle\mathbf{q}_t'|\breve{B}|\mathbf{q}_t\rangle , \tag{3a}$$

where

$$\hat{A}_\beta = e^{-\beta\hat{H}/2} \hat{A} e^{-\beta\hat{H}/2} . \tag{3b}$$

For comparison, the correlation function is given in classical mechanics by the following single phase space average over initial conditions:

$$C_{AB}(t) = (2\pi\hbar)^{-F} \int d\mathbf{p}_0 \int d\mathbf{q}_0 A_\beta(\mathbf{p}_0,\mathbf{q}_0) B(\mathbf{p}_t,\mathbf{q}_t) , \tag{4}$$

where $A_\beta(\mathbf{p},\mathbf{q})$ and $B(\mathbf{p},\mathbf{q})$ are the classical functions corresponding to operators $\hat{A}_\beta$ and $\hat{B}$.

Here one sees the two essential "extra difficulties" which must be dealt with in carrying out SC-IVR calculations compared to a standard classical approach:

(i) Equation (3) requires a double phase space average rather than the single one in Eq. (4), but more serious than that is the phase factor of the integrand in Eq. (3) which results from the difference in the action integrals from the two trajectory beginning at $(\mathbf{p}_0, \mathbf{q}_0)$ and at $(\mathbf{p}_0', \mathbf{q}_0')$. This introduces an oscillatory character to the integrand that makes Monte Carlo evaluation of these phase space averages very inefficient.

(ii) Equation (3) also requires the monodromy matrix of Eq. (2b), the calculation of which requires the Hessian of the potential surface (the matrix of second derivatives),

$$\frac{\partial^2 V(\mathbf{q}_t)}{\partial \mathbf{q}_t \partial \mathbf{q}_t} \equiv \mathbf{K}(\mathbf{q}_t) \tag{5}$$

along the trajectory $\mathbf{q}_t$. The classical expression, Eq. (4), requires only the gradient of the potential surface (the vector of first derivatives), $\partial V(\mathbf{q}_t)/\partial \mathbf{q}_t$, in order to compute the trajectory, so requiring the Hessian is a major escalation of effort necessary to implement the SC-IVR approach. Below I will sketch the ways we have developed for overcoming both of these bottlenecks.

2.1. The Linearization Approximation

At the beginning of our efforts to make SC-IVR approaches practical for large molecular systems, we introduced a very primitive approximation in order to get started[25,26]: namely, we assumed that the dominant contribution to the double phase space average in Eq. (3) comes from phase points $(\mathbf{p}_0, \mathbf{q}_0)$ and $(\mathbf{p}_0', \mathbf{q}_0')$ — and thus the two trajectories emanating from them — that are close to one another. To effect this approximation one changes to the sum and difference variables

$$\bar{\mathbf{p}}_0 = \frac{1}{2}(\mathbf{p}_0 + \mathbf{p}_0'), \quad \bar{\mathbf{q}}_0 = \frac{1}{2}(\mathbf{q}_0 + \mathbf{q}_0'), \quad \Delta\mathbf{p}_0 = \mathbf{p}_0 - \mathbf{p}_0', \quad \Delta\mathbf{q}_0 = \mathbf{q}_0 - \mathbf{q}_0' \quad (6)$$

and then all quantities in the integrand of Eq. (3) are expanded to first order in $\Delta\mathbf{p}_0$ and $\Delta\mathbf{q}_0$; the integrals over $\Delta\mathbf{p}_0$ and $\Delta\mathbf{q}_0$ thus become Fourier integrals (since the phase of the integrand is linear in them), giving the *linearized* SC-IVR (LSC-IVR), or *classical Wigner model* for the correlation function,

$$C_{AB}(t) = (2\pi\hbar)^{-F} \int d\mathbf{p}_0 \int d\mathbf{q}_0 \, A_w^\beta(\mathbf{p}_0, \mathbf{q}_0) \, B_w(\mathbf{p}_t, \mathbf{q}_t) \, . \quad (7)$$

Here $(\mathbf{p}_0, \mathbf{q}_0)$ are the average values (i.e., the "bars" have been removed), and A_w and B_w are the Wigner functions corresponding to these operators,

$$O_w(\mathbf{p}, \mathbf{q}) \equiv \int d\Delta\mathbf{q} \, e^{i\mathbf{p}\cdot\Delta\mathbf{q}/\hbar} \langle \mathbf{q} - \Delta\mathbf{q}/2 | \hat{O} | \mathbf{q} + \Delta\mathbf{q}/2 \rangle \, , \quad (8)$$

for any operator $\hat{O}$.

The double phase space average of Eq. (3) has thus now become the single phase space average of Eq. (7), and in the process the monodromy matrices in Eq. (3) have completely disappeared (they have not been neglected, but rather explicitly cancel out in the course of carrying through the linearized approximation). Thus both of the two "extra difficulties" noted above [after Eq. (4)] are eliminated by the linearized approximation. Equation (7) is in fact seen to have *precisely the same form as the classical correlation function*, Eq. (4), the only difference being that the Wigner functions for operators $\hat{A}_\beta$ and $\hat{B}$ replace the corresponding classical functions.

The classical Wigner model has been obtained many times before, by a variety of formulations. One such early paper is Ref. 27, but it surely goes back further than this. Heller[28,29] discussed the approximation many years ago (including an illuminating discussion of its limitations), and it was used by Lee and Scully[30] to describe quantum effects in a collinear model of inelastic scattering. More recently it has been obtained from a different approach by Pollak,[31] and also by Rossky *et al.*[32] directly from a path integral representation of the two time evolution operators in Eq. (1) (again by linearizing in the difference between the two paths).

The importance of the above derivation is thus not the result itself, for as noted, the classical Wigner approximation has been around a long time, having been obtained from a variety of approaches. The important point is realizing that the classical Wigner model is *contained within* the SC-IVR description, resulting from a very well defined approximation to it. This also makes it clear that if the SC-IVR can be implemented with less drastic approximations, it will be even more accurate than the classical Wigner model.

However, as drastic as the linearization approximation (LSC-IVR) seems, it is surprising that it can in fact describe some quantum effects quite well, even when they are large. The thermal rate constant for a chemical reaction, for example, is given by the long time limit of the flux-side correlation function,[33,34] i.e., Eq. (1) with operator $\hat{A}$ being the flux operator (with respect to some dividing surface) and operator $\hat{B}$ being a Heaviside function that is 1 (0) on the reactant (product) of the dividing surface. Figure 1[35] shows how it describes tunneling for a standard model of the fundamental hydrogen atom transfer reaction, $H + H_2 \rightarrow H_2 + H$. The Arrhenius plot of the rate shows the expected good agreement with the exact quantum rate at higher temperature, where tunneling corrections are small, but even at lower temperature where tunneling corrections become significant it does reasonably well: at 300 K, where the tunneling correction factor is ~20, the rate given by the LSC-IVR is only 10% too small, and at the lowest temperature shown (200 K), where the tunneling correction is a factor of ~2000, it is only 35% too small. (The full SC-IVR calculation, on the other hand, is accurate to a few percent even down to 200 K.)

The only non-trivial task required to implement the LSC-IVR for calculating thermal correlation functions of complex molecular systems, i.e., beyond what is required for an ordinary classical MD calculation, is evaluation of the Wigner function for operator $\hat{A}_\beta$. We have recently found the thermal Gaussian approximation (TGA), which Mandelshtam and

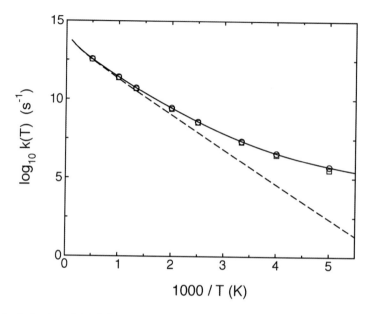

Fig. 1. Arrhenius plot of the rate constant for a 1d Eckart barrier (with parameters correspondingly approximately to the $H + H_2$ reaction). The solid line is the exact quantum value and the dashed line that given by classical mechanics. The circles are values given by the full SC-IVR and the squares its linearized approximation (LSC-IVR).

Frantsuzov[36,37] developed for approximating the Boltzmann operator, to be a very effective for this purpose. Use of the TGA allows the Fourier transform in Eq. (12) to be evaluated analytically, so that calculation of thermal time correlation functions becomes almost as simple as a standard classical MD calculation. Figure 2 shows recent results[38] for the force-force correlation function of the Ne_{13} cluster at three low temperatures. As the temperature is lowered, one sees that the classical correlation function (solid lines) shows the onset of freezing (i.e., structured behavior), while the SC result (dashed lines) shows only the faintest hint of structure. This is clearly a zero point energy effect, i.e., the classical cluster is beginning to freeze at the lower temperatures, while the quantum zero point energy prevents this. We note that in molecular liquids, such as water, one may very well

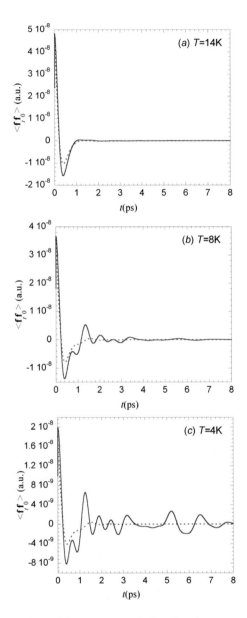

Fig. 2. The Kubo-transformed force autocorrelation function per article for the Ne_{13} LJ cluster system. Solid line: the classical result. Dotted line: the TGA-LSC-IVR result. Temperatures in three panels are, respectively, (a) $T = 14$ K; (b) $T = 8$ K; and (c) $T = 4$ K.

see zero point energy effects even at room temperature because of the high frequency of H atom motion. Similar calculations have been carried out for the velocity-velocity correlation function of liquid *para*-hydrogen to obtain its diffusion coefficient, and also the correlation functions related to neutron scattering. Thus the linearized approximation to the SC-IVR approach is an "operational" methodology that can be applied to essentially any problem for which classical MD simulations are feasible, and it should describe tunneling and zero point energy effects in them to a good approximation.

It should be noted that there are several other approaches that have been developed for calculating the Wigner function for operator $\hat{A}_\beta$, and thus applying the LSC-IVR. Geva *et al.*[39] have developed a local harmonic approximation that also makes it possible to evaluate the multi-dimensional Fourier transform analytically, and have carried out some impressive applications for vibrational relaxation in liquids (where the relevant quantity is a force-force correlation function). Rossky *et al.*[32] (using a variational harmonic approximation to obtain the Wigner function involving the Boltzmann operator) have treated a 1-dimensional chain of helium atoms, and also liquid oxygen (32 O_2 molecules in a box) at low temperature (70 K), and Coker *et al.*[40] have extended the linearized approximation to be able to describe electronically non-adiabatic dynamics.

It should also be noted that there are several other approaches that are very similar in character to the LSC-IVR/classical Wigner model, though not identical to it. For example, the "forward-backward" approximation to the SC-IVR correlation function used very effectively by Makri *et al.*[41] is closely related to it. The centroid molecular dynamics approach developed by Voth *et al.*[42] and the ring polymer molecular dynamics (RPMD) method of Manolopoulos *et al.*[43] also have very similar behavior to the LSC-IVR in that all of these approaches share the ability to describe *some* of the quantum mechanical aspects of molecular dynamics. They are not, however, able to describe quantum coherence effects. Coherence effects arise (in a semiclassical picture) from the interference between *different* trajectories; and since the LSC-IVR only considers trajectories in the double phase space average [Eq. (3)] that are infinitesimally close to one another, such coherence effects are explicitly excluded within this approximation.

2.2. A Forward-Backward SC-IVR

Going beyond the linearized approximation and being able to describe true quantum coherence effects in the dynamics via the SC-IVR approach requires that one deals more accurately with the "two difficulties" noted above [following Eq. (4)], namely that of the oscillatory integrand and the need for the Hessian of the potential along the trajectory. The "oscillatory integrand" problem is very effectively dealt with by the "forward-backward" (FB) approach,[44–46] an idea that was suggested by some earlier work of Makri *et al.*[47] for different purposes.

The motivation of the FB idea is as follows: suppose for the moment that operator $\hat{B}$ in Eq. (1) involved only one degree of freedom, e.g., were a function of coordinate q_1, and furthermore that this degree of freedom was separable from the remaining (many) $F - 1$ degrees of freedom. The Hamiltonian H would thus be separable, $\hat{H} = \hat{H}_1 + \hat{H}_{F-1}$, and the propagators factorable:

$$e^{\pm i\hat{H}t/\hbar} = e^{\pm i\hat{H}_1 t/\hbar} e^{\pm i\hat{H}_{F-1} t/\hbar} \ . \tag{9}$$

The propagator from 0 to t of the $F - 1$ degrees of freedom, $\exp(-i\hat{H}_{F-1}t/\hbar)$, and that from t back to 0, $\exp(+i\hat{H}_{F-1}t/\hbar)$, would then exactly cancel each other [since they commute with $B(q_1)$], and the only time dependence would be that from degree of freedom 1. Similarly, in the IVR expression Eq. (3), the action integrals from the $F - 1$ degrees of freedom would exactly cancel, and the double phase space average would collapse to a single phase space average for all degrees of freedom except the one involving operator $\hat{B}$.

To implement this idea — but without making any approximations about separability of degrees of freedom — suppose that operator $\hat{B}$ in the correlation function involves only a few (perhaps collective) degrees of freedom, as is often the case. For example, if $B(\mathbf{q})$ is a function of one collective variable $s(\mathbf{q})$ (e.g., a collective reaction coordinate),

$$B(\mathbf{q}) = B(s(\mathbf{q})), \tag{10}$$

then the FB-IVR result for the correlation function is given by[44]

$$C_{AB}(t) = \int_{-\infty}^{\infty} dp_s\, \tilde{B}(p_s)\, (2\pi\hbar)^{-F} \int d\mathbf{p}_0 \int d\mathbf{q}_0\, C(\mathbf{p}_0, \mathbf{q}_0; p_s)$$

$$\times \langle \mathbf{p}_0, \mathbf{q}_0 | \hat{A}(\beta) | \mathbf{p}_0', \mathbf{q}_0' \rangle e^{iS(\mathbf{p}_0, \mathbf{q}_0; p_s)/\hbar}, \tag{11}$$

where here the coherent state (Herman–Kluk) IVR has been used. $(\mathbf{p}_0, \mathbf{q}_0)$ in Eq. (11) are the initial conditions for a trajectory that is evolved to time t in the usual way, but here the momentum vector undergoes the following *momentum jump*,

$$\mathbf{p}_t \rightarrow \mathbf{p}_t + \frac{\partial s(\mathbf{q}_t)}{\partial \mathbf{q}_t} p_s, \tag{12}$$

and the trajectory is then propagated back to time 0; $(\mathbf{p}_0', \mathbf{q}_0')$ is the final phase point of this forward-backward trajectory, S the classical action along it, and C the Herman–Kluk pre-exponential factor; $\tilde{B}(p_s)$ is the (1-dimensional) Fourier transform of $B(s)$.

This FB-IVR result thus involves only a 1d integral (over the "jump parameter" p_s) in addition to a *single* phase space average over initial conditions (which one recalls is also required for the standard classical calculation), and is perhaps the simplest result of all that is capable of describing quantum coherence. The contribution to the FB action integral S [the phase of the integrand of Eq. (11)] from all the degrees of freedom that are coupled only weakly to the motion of the collective variable $s(\mathbf{q})$ will largely cancel, so that the integrand of Eq. (11) is much less oscillatory than the double phase space average of the full SC-IVR expression; i.e., much of the oscillatory structure of the double phase space average that would have cancelled numerically has been eliminated analytically by this FB approach, by combining the forward and backward time evolution operators into one effective forward-backward propagator.

Figures 3 and 4 show the results[48] of a calculation which illustrates this FB-IVR approach, namely, a model of real time molecular structure (i.e., a time-dependent radial distribution function). The specific model is a Morse potential (with parameters corresponding to the B-state of I_2) coupled to a harmonic bath (modeling the environmental degrees of freedom, e.g., a cluster, a liquid, etc.). $|\phi\rangle$ is the ground vibrational state of the diatomic in the ground electronic state, which becomes (upon Franck–Condon excitation) the initial vibrational wavefunction in the B-state. The time-dependent radial distribution function — i.e., the probability distribution of the diatomic coordinate at time t — is given by the correlation function of Eq. (1), where operators $\hat{A}(\beta)$ and $\hat{B}$ are

$$\hat{A}(\beta) = |\phi\rangle\langle\phi|e^{-\beta\hat{H}_b} , \tag{13a}$$

$$\hat{B}(r) = \delta(r - \hat{r}) , \tag{13b}$$

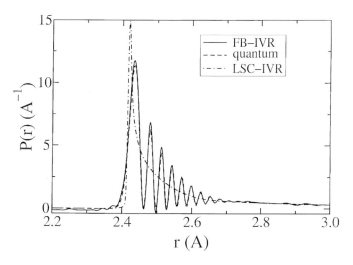

Fig. 3. Probability distribution of the vibrational coordinate of I_2 (modeled as a 1d Morse oscillator) at time $t = 192$ fsec ($\sim 1\frac{1}{4}$ vibrational periods after excitation). The dashed line and solid line (almost indistinguishable) are the exact quantum and forward-backward SC-IVR (FB-IVR) results, respectively, and the dash-dot line the results of linearized approximation to the SC-IVR (LSC-IVR).

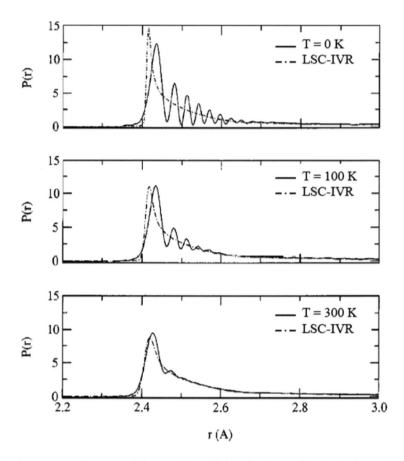

Fig. 4. Same quantity as Fig. 3, but with the addition of a harmonic bath that is coupled to the Morse oscillator (I_2); T is the temperature of the harmonic bath. The solid line is the result of the FB-IVR, and the dash-dot line that of the LSC-IVR.

where $\hat{H}_b$ is the Boltzmann operator for the harmonic bath. The correlation $C_{AB}(t)$ is then $P_t(r)$, the probability distribution of the diatomic coordinate at time t, i.e., the radial distribution function of I_2. Figure 3 shows this for a time of 192 fsec (about $1\frac{1}{4}$ vibrational periods of I_2) for the isolated diatomic (i.e., no coupling to the bath), and one sees very pronounced coherence structure (due to the fact that the initial state is a coherent superposition of many different vibrational eigenstates of the B-state); the FB-IVR result is essentially indistinguishable from that of the exact quantum calculation (which is easy for the isolated diatom case). Also shown is the result of the LSC-IVR/classical Wigner calculation, which shows none of the coherence structure. Figure 4 then shows $P_t(r)$ with coupling to the bath, for several values of the bath temperature T. For $T = 0$, the result is essentially the same as the isolated molecule result of Fig. 3, i.e., the bath is "frozen out". But as T is increased, the coherence structure is progressively quenched (or "de-coherred") by coupling to the bath, and by the time it has increased to 300 K the coherence features have mostly disappeared (for the assumed coupling strength), and in this case one sees that the LSC-IVR does an excellent job in describing $P_t(r)$. So just as one would expect, when quantum coherence features are averaged out, classical mechanics (which is effectively what the LSC-IVR gives) works well. This is not a surprising result. The point of this example is to show that semiclassical theory is able to simulate these coherence effects (and the extent to which they are quenched) in systems with many degrees of freedom. This model system is of course a simple one, but the nature of the calculation for a realistic model of a large molecular system would be essentially the same (though the computational time for each trajectory would of course be greater for a more complicated potential energy surface).

2.3. *Calculating the Hessian along a Trajectory*

Though the FB-IVR approach described above largely solves the problem of the oscillatory integrand, it still requires the monodromy matrices and thus the Hessian of the potential [Eq. (5)] along the trajectory. Having to explicitly calculate $K(q)$ along the trajectory q_t would be a major increase in computational complexity for a large molecular system with a very complicated potential energy function. Very recently, though, we have found a way to generate the Hessian along the trajectory without explicitly calculating the matrix of second derivatives of the potential.[49]

We avoid explicit calculation of $\mathbf{K}$ by differentiating one of Hamilton's equations

$$\dot{\mathbf{p}}_t = -\frac{\partial V(\mathbf{q}_t)}{\partial \mathbf{q}_t} \tag{14}$$

with respect to time, which (for a Cartesian Hamiltonian) gives

$$\ddot{\mathbf{p}}_t = -\frac{\partial^2 V}{\partial \mathbf{q}_t, \partial \mathbf{q}_t} \cdot \dot{\mathbf{q}}_t = -\frac{\mathbf{K}(\mathbf{q}_t) \cdot \mathbf{p}_t}{m}. \tag{15}$$

Along a trajectory it is quite easy to obtain $\ddot{\mathbf{p}}_i = \ddot{\mathbf{p}}(t_i)$ at time grid point i by finite difference,

$$\ddot{\mathbf{p}}(t_i) \cong \frac{\left[\mathbf{p}(t_{i+1}) + \mathbf{p}(t_{i-1}) - 2\mathbf{p}(t_i)\right]}{\Delta t^2}, \tag{16}$$

so knowing $\ddot{\mathbf{p}}(t_i)$ and $\mathbf{p}(t_i)$ at times $\{t_i\}$, one can use Eq. (15) to determine $\mathbf{K}$ (assuming it to be approximately constant for a short time interval). To see how this works, it is useful to write Eq. (15) in component notation

$$m\ddot{p}_{k,i} = -\sum_{k'} K_{k,k'}\, p_{k',i}, \tag{17}$$

where k and $k' = 1,\ldots,F$ (the number of degrees of freedom), and the matrix $p_{k,i} = p_k(t_i)$, etc. If one uses F time points $\{t_i\}$, then $\{\ddot{p}_{k,i}\}$ and $\{p_{k,i}\}$ are square matrices, so that one can solve the matrix equation, Eq. (17), to obtain

$$K_{k,k'} = -m \sum_i \ddot{p}_{k,i}(\mathbf{p}^{-1})_{i,k'}, \tag{18}$$

where $\mathbf{p}^{-1}$ is the inverse matrix of $\{p_{k,i}\}$.

As described above, one needs F time points $\{t_i\}$, a large number for a large molecular system, and this would largely invalidate our assumption that $\mathbf{K}$ is approximately constant for a short time interval. But this is not the case. Thus consider Eq. (17): for *fixed* k, the left-hand side is a vector (index i), $K_{k,k'}$ is a vector (index k'), and $p_{k',i}$ is a matrix. For *fixed* k, the index k' takes on only a small number of values, because the force constant matrix is highly banded (along its diagonal), i.e., it is only non-zero for a small number of k' values (for *fixed* k). Thus only a small number of time values $\{t_i\}$ are necessary to make $\{p_{k',i}\}$ a square matrix (and thus invertible). The matrix $\mathbf{p}^{-1}$ in Eq. (18) is the inverse of this "small square" matrix. Since one needs only a small number of time steps for each inversion, the approximation of a constant $\mathbf{K}$ matrix should be reasonable for that time interval. This calculation does have to be carried out for every value of k, which takes on F values, meaning that the overall procedure is linear in the number of degrees of freedom, i.e., the same order as the trajectory calculation itself.

3. Concluding Remarks

Ways of handling the additional complications of a semiclassical dynamics calculation, compared to the corresponding classical treatment, are now in hand, and a number of examples of increasing complexity have been carried out which demonstrate these capabilities. The simplest, most approximate version of semiclassical initial value methods — its linearized approximation, which leads to the classical Wigner model — is able to provide a good description of tunneling effects in chemical reactions and zero point energy effects in dynamical processes. The forward-backward IVR goes further and is able also to describe true quantum coherence effects in the dynamics, thus showing when these effects are present or when they are averaged out (quenched, or "de-cohered"). Both of these approaches are more difficult to apply than a standard classical MD simulation, but not extraordinarily so.

There are many cases, of course, for which quantum mechanical aspects of the dynamics will not be significant, but one may not always know this *a priori*; and if one is only able to perform classical MD simulations there is no way to know whether such effects are present or not. In other situations, e.g., dynamical processes which involve H atom motion, one can be fairly certain that quantum effects will indeed be important. In these cases the ability to still use classical MD methodology, but as input to a semiclassical treatment, should provide a useful enhancement of theoretical capabilities.

Acknowledgments

This work was supported by the Director, Office of Science, Office of Basic Energy Sciences, Chemical Sciences, Geosciences, and Biosciences Division, U.S. Department of Energy under Contract No. DE-AC02-05CH11231, by the National Science Foundation Grant No. CHE-0345280 and by the Office of Naval Research Grant Nos. N00014-07-1-0586 and N00014-05-1-0457.

References

1. Ford KW, Wheeler JA. (1959) Semiclassical description of scattering. *Ann Phys* **7:** 259–287.
2. Berry MV, Mount KE. (1972) Semiclassical approximations in wave mechanics. *Rept Prog Phys* **35:** 315–397.
3. Miller WH. (1970) Semiclassical theory of atom-diatom collisions: Path integrals and the classical S-matrix. *J Chem Phys* **53:** 1949–1959.
4. Miller WH. (1970) The classical S-matrix: Numerical application to inelastic collisions. *J Chem Phys* **53:** 3578–3587.
5. Miller WH. (1970) The classical S-matrix: A more detailed study of classically forbidden transitions in inelastic collisions. *Chem Phys Lett* **7:** 431–435.
6. Marcus RA. (1970) Extension of the WKB method to wave functions and transition probability amplitudès (S-matrix) for inelastic or reactive collisions. *Chem Phys Lett* **7:** 525–532.
7. Marcus RA. (1971) Theory of semiclassical transition probabilities (S-matrix) for inelastic and reactive collisions. *J Chem Phys* **54:** 3965–3979.
8. Miller WH. (1974) Classical-limit quantum mechanics and the theory of molecular collisions. *Adv Chem Phys* **25:** 69–177.
9. Miller WH. (1975) The classical S-matrix in molecular collisions. *Adv Chem Phys* **30:** 77–136.
10. Miller MH, McCurdy CW. (1978) Classical trajectory model for electronically non-adiabatic collision phenomena: A classical model for electronic degrees of freedom. *J Chem Phys* **69:** 5163–5173.

11. Meyer HD, Miller WH. (1979) A classical analog for electronic degrees of freedom in non-adiabatic collision processes. *J Chem Phys* **70:** 3214–3223.

12. Miller WH. (1991) Comment on: Semiclassical time evolution without root searches. *J Chem Phys* **95:** 9428–9430.

13. Herman MF, Kluk E. (1984) A semiclassical justification for the use of non-spreading wavepackets in dynamics calculations. *Chem Phys* **91:** 27–34.

14. Heller EJ. (1991) Cellular dynamics: A new semiclassical approach to time-dependent quantum mechanics. *J Chem Phys* **94:** 2723–2729.

15. Heller EJ. (1991) Reply to Comment on: Semiclassical time evolution without root searches: Comments and perspective. *J Chem Phys* **95:** 9431–9432.

16. Kay KG. (1994) Integral expressions for the semiclassical time-dependent propagator. *J Chem Phys* **100:** 4377–4392.

17. Kay KG. (1994) Numerical study of semiclassical initial value methods for dynamics. *J Chem Phys* **100:** 4432–4445.

18. Capolieti G, Brumer P. (1994) Semiclassical propagation: Phase indices and the initial-value formalism. *Phys Rev* **A50:** 997–1018.

19. Tannor DJ, Garaschuk S. (2000) Semiclassical calculation of chemical reaction dynamics via wavepacket correlation functions. *Ann Rev Phys Chem* **51:** 553–600.

20. Kay KG. (2005) Semiclassical initial value treatments of atoms and molecules. *Ann Rev Phys Chem* **56:** 255–280.

21. Miller WH. (2001) The semiclassical initial value representation: A potentially practical way for adding quantum effects to classical molecular dynamics simulations. *J Phys Chem* **A105:** 2942–2955.

22. Miller WH. (2005) Quantum dynamics of complex molecular systems. *Proc Natl Acad Sci* **102:** 6660–6664.

23. Miller WH. (2006) Including quantum effects in the dynamics of complex (i.e., large) molecular systems. *J Chem Phys* **125:** 132305.1-8.

24. Berne BJ, Harp CD. (1970) On the calculation of time correlation functions. *Adv Chem Phys* **17:** 63–228.

25. Sun X, Miller WH. (1997) Mixed semiclassical-classical approaches to the dynamics of complex molecular systems. *J Chem Phys* **106:** 916–927.

26. Wang H, Sun X, Miller WH. (1998) Semiclassical approximations for the calculation of thermal rate constants for chemical reactions in complex molecular systems. *J Chem Phys* **108:** 9726–9736.

27. Imre K, Ozizmir E, Rosenbaum M, Zweifel PF. (1967) Wigner method in quantum statistical mechanics. *J Math Phys* **8:** 1097–1108.

28. Heller EJ. (1976) Wigner phase space method: Analysis for semiclassical applications. *J Chem Phys* **65:** 1289–1298.

29. Heller EJ, Brown RC. (1976) Errors in the Wigner approach to quantum dynamics. *J Chem Phys* **75:** 1048–1050.

30. Lee HW, Scully MO. (1980) A new approach to molecular collisions: Statistical quasiclassical method. *J Chem Phys* **73:** 2238–2242.

31. Liao JL, Pollak E. (1998) A new quantum transition state theory. *J Chem Phys* **108:** 2733–2743.

32. Poulsen JA, Nyman G, Rossky PJ. (2003) Practical evaluation of condensed phase quantum correlation functions: A Feynman–Kleinert variational linearized path integral method. *J Chem Phys* **119:** 12179–12193.

33. Miller WH. (1974) Quantum mechanical transition state theory and a new semiclassical model for reaction rate constants. *J Chem Phys* **61:** 1823–1834.

34. Miller WH, Schwartz SD, Tromp JW. (1983) Quantum mechanical rate constants for bimolecular reactions. *J Chem Phys* **79:** 4889–4898.

35. Yamamoto T, Wang H, Miller WH. (2002) Combining semiclassical time evolution and quantum Boltzmann operator to evaluate reactive flux correlation function for thermal rate constants of complex systems. *J Chem Phys* **116:** 7335–7349.

36. Frantsuzov P, Neumaier A, Mandelshtam VA. (2003) Gaussian resolutions for equilibrium density matrices. *Chem Phys Lett* **381:** 117–122.

37. Frantsuzov PA, Mandelshtam VA. (2004) Quantum statistical mechanics with Gaussians: Equilibrium properties of van der Waals clusters. *J Chem Phys* **121:** 9247–9256.

38. Liu J, Miller WH. (2007) Linearized semiclassical initial value time correlation functions using the thermal Gaussian approximation: Applications to condensed phase systems. *J Chem Phys* **127:** 114506.1-10.

39. Shi Q, Geva E. (2004) A semiclassical generalized quantum master equation for an arbitrary system-bath coupling. *J Chem Phys* **120:** 10647–10658.

40. Bonella S, Coker DF. (2005) Chemical theory and computation special feature: Linearized path integral approach for calculating nonadiabatic time correlation functions. *Proc Natl Acad Sci* **102:** 6715–6719.

41. Makri N, Nakayama A, Wright NJ. (2004) Forward-backward semiclassical simulation of dynamical processes in liquids. *J Theo Comp Chem* **3:** 391–417.

42. Cao J, Voth GA. (1993) A new perspective on quantum time correlation functions. *J Chem Phys* **99:** 10070–10073.

43. Craig IR, Manolopoulos DE. (2004) Quantum statistics and classical mechanics: Real time correlation functions from ring polymer molecular dynamics. *J Chem Phys* **121:** 3368–3373.

44. Miller WH. (1998) Quantum and semiclassical theory of chemical reaction rates. *Faraday Disc Chem Soc* **110:** 1–21.

45. Sun W, Miller WH. (1999) Forward-backward initial value representation for semiclassical time correlation functions. *J Chem Phys* **110:** 6635–6644.

46. Wang H, Thoss M, Miller WH. (2000) Forward-backward initial value representation for the calculation of thermal rate constants for reactions in complex molecular systems. *J Chem Phys* **112:** 47–55.

47. Makri N, Thompson K. (1998) Semiclassical influence functionals for quantum systems in anharmonic environments. *Chem Phys Lett* **291:** 101–109.

48. Wang H, Thoss M, Sorge K, Gelabert R, Gimenez X, Miller WH. (2001) Semiclassical description of quantum coherence effects and their quenching: A forward-backward initial value representation study. *J Chem Phys* **114:** 2562–2571.

49. Ananth N, Miller WH. (2007) Work in progress.

Biographies

David Baltimore

David Baltimore served as President of the California Institute of Technology for nine years, then in 2006 he was appointed President Emeritus and the Robert Andrews Millikan Professor of Biology. Awarded the Nobel Prize at the age of 37 for research in virology, Baltimore has profoundly influenced national science policy on such issues as recombinant DNA research and the AIDS epidemic.

Born in New York City, he received his B.A. in Chemistry from Swarthmore College in 1960 and a Ph.D. in 1964 from Rockefeller University, where he returned to serve as President from 1990–91 and faculty member until 1994.

For almost 30 years, he was a faculty member at MIT. While his early work was on poliovirus, in 1970 he identified the enzyme reverse transcriptase in tumor virus particles, thus providing strong evidence for a process of RNA to DNA conversion. Baltimore and Howard Temin (with Renato Dulbecco, for related research) shared the 1975 Nobel Prize in Physiology or Medicine for their discovery, which provided the key to understanding the life-cycle of HIV. In the following years, he has contributed widely to the understanding of cancer, AIDS and the molecular basis of the immune response. His present research focuses on control of inflammatory and immune responses as well as on the use of gene therapy methods to treat HIV and cancer in a program called "Engineering Immunity."

He played an important role in creating a consensus on national science policy regarding recombinant DNA research. He served as founding director of the Whitehead Institute for Biomedical Research at MIT from 1982 until 1990. He co-chaired the 1986 National Academy of Sciences committee on a National Strategy for AIDS and was appointed in 1996 to head the National Institutes of Health AIDS Vaccine Research Committee.

His numerous honors include the 1999 National Medal of Science. He was elected to the National Academy of Sciences in 1974 and is a foreign member of both the Royal Society of London and the French Academy of Sciences. For 2007/8, he is President of the AAAS. He has published more than 600 peer-reviewed articles.

Carlos J. Bustamante

Carlos J. Bustamante received a Ph.D. in biophysics in 1981 from the University of California, Berkeley. Since 1994, Dr. Bustamante has held an appointment as a Howard Hughes Medical Institute Investigator. In 1998, he became the director for the Advance Microscopies Department at Lawrence Berkeley National Laboratory, and a Professor of Physics as well as Molecular and Cell Biology at Berkeley. His research interests include single molecule manipulation methods and their application to investigate various biochemical processes: torque measurements on single DNA molecules, reversible folding of single RNA molecules by force, and the mechanochemistry of nucleic-acid binding molecular motors. He was nominated as America's Best in *Time* magazine (2001), received the Biological Physics Prize of the American Physical Society (2002), and he accepted the Alexander Hollaender Award in Biophysics from the National Academy of Science (2004). He has also received the Richtmyer Memorial Lecture Award by the American Association of Physics Teachers (2005), and a Doctor *honoris causa* by the University of Chicago (2005). Dr. Bustamante has given well over 400 presentations and lectures and has published over 200 papers in several journals such at PNAS, Nature, and Cell. He currently serves as a member of the Science Advisory Board of the Searle Scholars Program. He is an elected member of the National Academy of Sciences, the Science Advisory Committee, and the Board of Directors of the Burroughs–Wellcome Fund. He also holds several other advisory roles within the University of California, Berkeley and the larger scientific community.

Christopher M. Dobson

Christopher M. Dobson received his doctorate from the University of Oxford in 1976, having worked on the application of NMR spectroscopy as a means of defining the structures and dynamics of proteins in solution. After a short period as a Research Fellow he moved to Harvard University as an Assistant Professor of Chemistry. In 1980 he returned to the University of Oxford first as a Lecturer and later as Professor and Director of the Oxford Centre for Molecular Sciences. His research interests increasingly focused on defining the mechanism of protein folding and more recently on understanding the consequences of misfolding particularly in terms of its relationship to human disease. In 2001 he moved to the University of Cambridge as John Humphrey Plummer Professor of Chemical and Structural Biology. Since October 2007, he has become the Master of St. John's College, Cambridge.

Leroy Hood

Leroy Hood received his M.D. from Johns Hopkins in 1964 and his Ph.D. from Caltech in 1968. He spent several years at the NIH Cancer Institute before moving to Biology at Caltech. He spent 10 years as Chairman of Biology before moving in 1992 to the University of Washington as founder and Chairman of the cross-disciplinary Department of Molecular Biotechnology (MBT). In 2000, he co-founded the Institute for Systems Biology in Seattle. Dr. Hood's research has focused on immunology, the study of a variety of diseases including cancer and technology development.

Dr. Hood was awarded the Lasker Prize in 1987 for his studies on the mechanism of immune diversity, the 2002 Kyoto Prize in Advanced Technology for the development of the five different instruments, the 2003 Lemelson — MIT Prize for Innovation and Invention — for the development of the DNA sequencer, the 2004 Biotechnology Heritage Award for life-long achievements in biotechnology, and the 2006 Heinz Award in Technology, the Economy and Employment for his extraordinary breakthroughs in biomedical science at the genetic level. In 2007 he was elected to the Inventors Hall of Fame (for the automated DNA sequencer). Dr. Hood has received 15 honorary degrees from Institutions such as Johns Hopkins, UCLA, and Whitman College. He has published more than 600 peer-reviewed papers, received 14 patents, and has co-authored textbooks in biochemistry, immunology, molecular biology, and genetics, and is just finishing a text book on systems biology.

Dr. Hood is a member of the National Academy of Sciences, the American Philosophical Society, the American Academy of Arts and Sciences, the Institute of Medicine and the National Academy of Engineering. Indeed, Dr. Hood is one of 7 (of more than 6000 members) scientists elected to all three academies (NAS, NAE and IOM). Dr. Hood has also played a role in founding or cofounding more than 12 biotechnology companies, including Amgen, Applied Biosystems, Systemix, Darwin and Rosetta.

Noel S. Hush

Noel S. Hush, Australian-born with dual British citizenship, earned both his bachelor's and master's degrees in chemistry from the University of Sydney. He was appointed a lecturer in physical chemistry at the University of Manchester in the U.K. in 1950, and later joined the faculty at the University of Bristol from 1955 until 1971, when he returned to Australia to the University of Sydney as the Foundation Professor of Theoretical Chemistry. He served as Head of the department from 1971 until 1989 and continues to maintain a vigorous research program at the University of Sydney.

His interests are principally in application of quantum mechanics to elucidation of electronic structures and dynamics, and range widely in this area. A particular emphasis has been on fundamental understanding of thermal and optical electron transfer dynamics in condensed phases or at interfaces, including electron exchange in solution, charge separation in photosynthesis and charge transfer at electrode/solution interfaces. This is connected with the phenomenon of mixed valency, ubiquitous in physics, chemistry and biology, in which the extent of electron localization has a key role. His current research interests focus on an area of nanotechnology known as molecular electronics, which embodies many of the concepts of the earlier electron transfer theory, and collaborates with University of Sydney colleague, Jeffrey Reimers, in theoretical studies of molecular circuitry and the design of molecular devices, such as ultra-dense random access memories. He is an Officer of the Order of Australia, a Fellow of the Australian Academy of Science, a Fellow of the Royal Society of London, a Fellow of the Royal Australian Chemical Institute and a Foreign Member of the American Academy of Arts and Sciences. Awards he has received include a Doctorate of Science from Manchester University, the Flinders Medal of the Australian Academy of Science, the Centenary Medal of the Royal Society of Chemistry, the inaugural David Craig Medal of the Australian Academy of Science, the Physical Chemistry Medal of the Royal Australian Chemical Society and the Australian Federation Medal.

Christof Koch

Christof Koch was born in 1956 in the American Midwest. He grew up in Holland, Germany, Canada, and Morocco, where he graduated from the Lycèe Descartes in 1974. He studied Physics and Philosophy at the University of Tübingen in Germany and was awarded his Ph.D. in Biophysics in 1982.

After four years at MIT, Dr. Koch joined Caltech in 1986, where he is the Lois and Victor Troendle Professor of Cognitive and Behavioral Biology. He lives in Pasadena, and loves to run and to climb.

The author of three hundred scientific papers and journal articles, and several books, Dr. Koch studies the biophysics of computation, and the neuronal basis of visual perception, attention, and consciousness. Together with Francis Crick, he is one of the pioneers of the neurobiological approach to consciousness.

Roger D. Kornberg

Roger D. Kornberg is Winzer Professor in Medicine in the Department of Structural Biology at Stanford University. In his doctoral research, he demonstrated the diffusional motions of lipids in membranes, termed flip-flop and lateral diffusion. He was a postdoctoral fellow and member of the scientific staff at the Laboratory of Molecular biology in Cambridge, England from 1972–5, where he discovered the nucleosome, the basic unit of DNA coiling in chromosomes. He moved to his present position in 1978, where his research has focused on the mechanism and regulation of eukaryotic gene transcription. Notable findings include the demonstration of the role of nucleosomes in transcriptional regulation, the establishment of a yeast RNA polymerase II transcription system and the isolation of all the proteins involved, the discovery of the Mediator of transcriptional regulation, the development of two-dimensional protein crystallization and its application to transcription proteins, and the atomic structure determination of an RNA polymerase II transcribing complex. Kornberg's awards include the 2001 Welch Award and the 2006 Nobel Prize in Chemistry (unshared). Kornberg's closest collaborator has been his wife, Dr. Yahli Lorch. They have three children, Guy, Maya, and Gil.

Roderick MacKinnon

Roderick MacKinnon is currently John D. Rockefeller Jr. professor in the laboratory of Molecular Neurobiology and Biophysics at Rockefeller University and Investigator of the Howard Hughes Medical Institute. His research aims to understand the molecular mechanisms of a class of integral membrane proteins known as ion channels. By catalyzing the rapid and selective flow of inorganic ions, such as potassium and chloride, across cell membranes, these proteins generate electrical signals in cells. Among their many biological functions, ion channels control the pace of the heart, regulate hormone secretion and generate the electrical impulses underlying information transfer in the nervous system. Central questions in the field of mechanistic ion channel studies include: How do their pores discriminate between very similar ions such as sodium and potassium, and how does neurotransmitter binding or a change in a cell's membrane voltage control the gating (opening and closing) process? These questions are being addressed through functional and structural studies of potassium and other ion channels. Dr. MacKinnon is a member of the U.S. National Academy of Sciences and the recipient of numerous scientific awards including the 1999 Albert Lasker Basic Medical Research Award, the 2000 Lewis S. Rosenstiel Award for Distinguished Research Award, the 2001 Gairdner Foundation International Award, the 2003 Louisa Gross Horwitz Prize, and the 2003 Nobel Prize in Chemistry.

J. Andrew McCammon

J. Andrew McCammon is the Joseph E. Mayer Chair Professor of Theoretical Chemistry and Distinguished Professor of Pharmacology at UCSD, and is an Investigator of the Howard Hughes Medical Institute. He received his B.A. from Pomona College, and his Ph.D. in chemical physics from Harvard University, where he worked with John Deutch. In 1976–78, he developed the computer simulation approach to protein dynamics in Martin Karplus's lab at Harvard. He joined the University of Houston as Assistant Prof. of Chemistry in 1978, and became the M.D. Anderson Chair Prof. of Chemistry in 1981. He moved to UCSD in 1995. He has invented theoretical methods for accurately predicting and interpreting molecular recognition, the rates of reactions, and other properties of chemical systems. In addition to their fundamental interest, these methods play a growing role in the design of new drugs and other materials. He is the author with Stephen Harvey of *Dynamics of Proteins and Nucleic Acids* (Cambridge University Press), and is the author or co-author of more than 500 publications on subjects in theoretical chemistry and biochemistry. About 50 of his former students have tenured or tenure-track positions at leading universities or research institutes. In the 1980s, he guided the establishment of the computer-aided drug discovery program of Agouron Pharmaceuticals (now Pfizer Global Research and Development, La Jolla Laboratories), and contributed to the development of the widely prescribed HIV-1 protease inhibitor, Viracept. His group's studies of HIV-1 integrase flexibility contributed to the discovery of the first in a new class of antiviral drugs by Merck & Co., named Raltegravir in 2007. He received the first George Herbert Hitchings Award for Innovative Methods in Drug Design from the Burroughs Wellcome Fund in 1987. In 1995, he received the Smithsonian Institution's Information Technology Leadership Award for Breakthrough Computational Science, sponsored by Cray Research. He is a Fellow of the American Academy of Arts and Sciences, the American Assoc. for the Advancement of Science, the American Physical Society, and the Biophysical Society.

William H. Miller

William H. Miller was born in Kosciusko, Mississippi, in 1941, and grew up in Jackson. He received a B.S. in Chemistry from Georgia Tech (1963) and a Ph.D. in Chemical Physics from Harvard (1967). During 1967–69 he was a Junior Fellow in Harvard's Society of Fellows, the first year of which was spent as a NATO postdoctoral fellow at the University of Freiburg (Germany). He joined the chemistry department of the University of California, Berkeley, in 1969 and has been Professor since 1974, serving as Department Chairman from 1989 to 1993 and becoming the Kenneth S. Pitzer Distinguished Professor in 1999.

Professor Miller's research has dealt with essentially all aspects of molecular collision theory and chemical reaction dynamics. The more significant of his contributions include a comprehensive semiclassical scattering theory (the classical S-matrix) of inelastic and reactive scattering processes, the reaction path Hamiltonian for describing polyatomic reactions, the S-matrix Kohn variational method for state-to-state reactive scattering, and a rigorous quantum theory of chemical reaction rates (and its semiclassical limit, the "instanton" model) that generalizes transition state theory. Most recently his efforts have focused on developing the initial value representation (IVR) of semiclassical theory into a practical way of adding quantum effects to classical molecular dynamics simulations of chemical processes.

Professor Miller is a member of the International Academy of Quantum Molecular Sciences (1985), the National Academy of Sciences (1987), and the American Academy of Arts and Sciences (1993). His awards include the Annual Prize of the International Academy of Quantum Molecular Sciences (1974), the E. O. Lawrence Memorial Award (1985), the Irving Langmuir Award in Chemical Physics (1990), the American Chemical Society Award in Theoretical Chemistry (1994), the Hirschfelder Prize in Theoretical Chemistry (1996), the Ira Remsen Award (1997), the Spiers Medal of the Royal Society of Chemistry (1998), and the Peter Debye Award in Physical Chemistry (2003).

Michele Parrinello

Michele Parrinello is known for his many technical innovations in the field of atomistic simulations and for a wealth of interdisciplinary applications ranging from materials science to chemistry and biology. Together with Roberto Car he introduced ab-initio molecular dynamics, also known as the Car–Parrinello method, marking the beginning of a new era both in the area of electronic structure calculations and in molecular dynamics simulations. He is also known for the Parrinello–Rahman method, which allows crystalline phase transitions to be studied by molecular dynamics. More recently he has introduced metadynamics for the study of rare events and the calculation of free energies. For his work he has been awarded many prizes and honorary degrees. He is a member of numerous academies and learned societies, including the German Berlin–Brandenburgische Akademie der Wissenschaften, the British Royal Society and the Italian Accademia Nazionale dei Lincei, which is the major academy in his home country of Italy. Born in Messina in 1945, he got his degree from the University of Bologna and is currently professor of Computational Sciences at ETH in Switzerland.

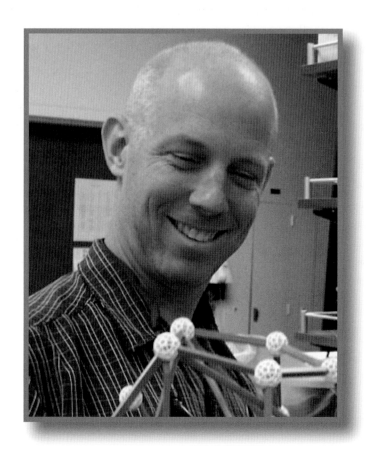

Rob Phillips

Rob Phillips is a professor of Applied Physics at the California Institute of Technology in Pasadena, California where he currently serves as the Option Representative for Biochemistry and Molecular Biophysics. Phillips received his Ph.D. in condensed matter physics at Washington University in 1989. Prior to the privilege of a life in science, he spent seven years of travel, self-study and work as an electrician.

Work in his group centers on physical biology of the cell, the use of physical models to explore biological phenomena and the construction of experiments designed to test them. Some of the key areas of interest include the physics of genome management such as how viruses and cells physically manipulate DNA as part of their standard repertoire during their life cycles and how the physical properties of lipid bilayers are tied to the behavior of ion channels.

Over the last seven years, Phillips has been working with Professor Jané Kondev (Brandeis University) and Professor Julie Theriot (Stanford University) on a book entitled "Physical Biology of the Cell" to be published by Garland Science.

Stephen R. Quake

Stephen R. Quake studied physics (B.S. 1991) and mathematics (M.S. 1991) at Stanford University before earning his doctorate in physics from Oxford University (1994) as a Marshall scholar. He then spent two years as a post-doc in Nobel Laureate Steven Chu's group at Stanford developing techniques to manipulate single DNA molecules with optical tweezers. In 1996 Quake joined the faculty of the California Institute of Technology, where he rose through the ranks and was ultimately appointed the Thomas and Doris Everhart Professor of Applied Physics and Physics. Quake moved back to Stanford in 2004 to help launch a new department in Bioengineering.

Quake's interests lie at the nexus of physics, biology and biotechnology. Over the past half decade, he has focused on understanding the basic physics and biological applications of microfluidic technology. His group pioneered the development of Micro-fluidic Large Scale Integration (LSI), demonstrating the first integrated microfluidic devices with thousands of mechanical valves. This technology is helping to pave the way for large scale automation of biology at the nanoliter scale, and he and his students have been exploring applications of "lab on a chip" technology in functional genomics, genetic analysis, and protein design. Throughout his career, Quake has also been active in the field of single molecule biophysics; he has focused on precision measurements on single molecules, and in 2003 his group demonstrated the first successful single molecule DNA sequencing experiments.

Quake received "Career" and "First" awards from the National Science Foundation and National Institutes of Health in 1997, was named a Packard Fellow in 1999, was in the inaugural class of NIH Director's Pioneer Awards in 2004, and in 2005 was selected as an investigator of the Howard Hughes Medical Institute. His contributions to the development of new biotechnology at the interface between physics and biology have been recognized by recent awards from the MIT Technology Review Magazine, Forbes, and Popular Science. He is a founder and scientific advisory board chair of Fluidigm, Inc. and Helicos Biosciences, Inc.

Douglas C. Rees

Douglas C. Rees is the Roscoe Gilkey Dickinson Professor of Chemistry at the California Institute of Technology and an Investigator of the Howard Hughes Medical Institute. He received his B.S. in Molecular Biophysics and Biochemistry from Yale College, working with C.W. Slayman, and his Ph.D. in Biophysics from Harvard University, where he conducted his graduate research in protein crystallography with W.N. Lipscomb. Following a postdoctoral appointment at the University of Minnesota studying nitrogenase with J.B. Howard, Dr. Rees joined the faculty of the Department of Chemistry and Biochemistry at UCLA until moving to Caltech in 1989. He is a member of the American Academy of Arts and Sciences and the National Academy of Sciences. The research interests of the Rees group emphasize the general area of structural bioenergetics, using crystallographic and functional approaches to characterize metalloproteins and membrane proteins such as nitrogenase and ABC transporters that mediate ATP-dependent energy transduction processes.

John Meurig Thomas

John Meurig Thomas is Honorary Professor of Solid-State Chemistry at the Department of Materials Science, Cambridge. A graduate of the University of Wales (Swansea), he taught and researched at Bangor and Aberystwyth for 20 years before he became Head of Physical Chemistry at Cambridge I3 in 1978, and Director of the Royal Institution of Great Britain, London (1986–91). He later returned to pursue his research in Cambridge, and was Master of Peterhouse from 1993–2002.

A pioneer of solid-state chemistry and an expert on the design, synthesis and *in situ* characterization of heterogeneous catalysts, his work on single-site solid catalysts has produced a strategy for the creation of new inorganic catalysts, many of which have been exploited commercially. He has assembled numerous solid catalysts for environmentally benign processes, such as the regio-, shape-, and enantio-selective conversions of organic molecules in air or oxygen.

His early work focused on chemical and electronic consequences of imperfections in solids and this prompted him to adapt various forms of electron microscopy for the retrieval of chemical information. Many new families of structures have been discovered by him. In recognition of his investigations in geo-chemistry (involving the deployment of microscopy, NMR and synchrotron radiation) a new mineral, Meurigite, was named in his honour.

The recipient of many awards (Davy Medal, Faraday Medal, the Willard Gibbs and Linus Pauling Medals), he also has several honorary doctorates in the U.K., U.S., Australia, Europe, Africa and Japan. A foreign Fellow of the American Philosophical Society and of many national academies, he was chairman of CHEMRAWN (Chemical Research Applied to World Needs) of IUPAC, 1988–92. In 1991 he was knighted by Queen Elizabeth II "for services to chemistry and the popularization of science."

David A. Tirrell

David A. Tirrell is the Ross McCollum–William H. Corcoran Professor and Chairman of the Division of Chemistry and Chemical Engineering at the California Institute of Technology. After earning the B.S. in Chemistry at MIT, Tirrell enrolled in the Department of Polymer Science and Engineering at the University of Massachusetts, where he was awarded the Ph.D. in 1978 for work done under the supervision of Otto Vogl. After a brief stay with Takeo Saegusa at Kyoto University, Tirrell accepted an assistant professorship in the Department of Chemistry at Carnegie–Mellon University in the fall of 1978.

Tirrell returned to Amherst in 1984 and served as Director of the Materials Research Laboratory before moving to Caltech in 1998. He has been a Visiting Professor at the University of Queensland, at the Institut Charles Sadron in Strasbourg, at the University of Wisconsin, and at the Institut Curie in Paris. He was Editor of the *Journal of Polymer Science* from 1988 until 1999, and has chaired the Gordon Research Conferences on *Polymers in Biosystems and on Chemistry of Supramolecules and Assemblies.*

Tirrell's contributions to macromolecular chemistry have been recognized in a variety of ways, including his election to the American Academy of Arts and Sciences and the National Academy of Sciences.

Alexander Varshavsky

Alexander Varshavsky is Smits Professor of Cell Biology at the California Institute of Technology (Caltech). He moved to Caltech in 1992, after 15 years at MIT. Dr. Varshavsky's early work resulted in the 1978–1980 discoveries of the first exposed region in chromosomes and the first (multicatenanes-based) pathway of chromosome segregation. In the 1980s, a set of interconnected discoveries by Dr. Varshavsky revealed the biological significance of the ubiquitin system, by showing that it plays a strikingly broad, previously unsuspected part in cellular physiology. Ubiquitin conjugation was demonstrated, by Dr. Varshavsky and coworkers, to be required for the protein degradation *in vivo*, for cell viability and also, more specifically, for the cell cycle, DNA repair, protein synthesis, transcriptional regulation and stress responses. He also discovered the first degradation signals in short-lived proteins (including the N-end rule), the first physiological substrate of the ubiquitin system, the first polyubiquitin chains, the first specific ubiquitin ligase, and the subunit selectivity of ubiquitin-mediated proteolysis. The latter is a fundamental capability of the ubiquitin system that underlies the cell cycle, gene transcription, cell differentiation and many other processes. Dr. Varshavsky and coworkers invented major methods in biochemistry and genetics, including the ubiquitin fusion technique, in 1986; the chromatin immunoprecipitation (ChIP) assay, in 1988; and the first split-protein (splitubiquitin) assay for *in vivo* protein interactions, in 1994. Dr. Varshavsky is a member of the National Academy of Sciences, the American Academy of Arts and Sciences, the American Philosophical Society, and foreign member of the Academia Europaea and the European Molecular Biology Organization. Among his awards are the Gairdner Award (1999), the Lasker Award (2000), the Sloan Prize in Cancer Research (2000), the Wolf Prize (2001), the Max Planck Award (2001), the Horwitz Prize (2001), the Merck Award (2001), the Wilson Medal (2002), the Stein & Moore Award (2005), the Gagna Prize (2006), the March of Dimes Prize in Developmental Biology (2006), and the Griffuel Prize in Cancer Research (2006).

George M. Whitesides

George M. Whitesides was born August 3, 1939 in Louisville, Kentucky. He received an A.B. degree from Harvard University in 1960 and a Ph.D. from the California Institute of Technology (with J.D. Roberts) in 1964. He was a member of the faculty of MIT from 1963 to 1982. He joined the Department of Chemistry of Harvard University in 1982, and was Department Chairman 1986–89, and Mallinckrodt Professor of Chemistry from 1982–2004. He is now the Woodford L. and Ann A. Flowers University Professor.

Awards: American Chemical Society (ACS) Award in Pure Chemistry (1975)–Arthur C. Cope Award (ACS) (1995)–Defense Advanced Research Projects Agency Award for Significant Technical Achievement (1996)–U.S. National Medal of Science (1998)–Von Hippel Award (Materials Research Society) (2000)–Kyoto Prize for Advanced Technology (Inamori Foundation, Japan) (2003)–Paracelsus Prize (Swiss Chemical Society) (2004)–Dan David Prize in Future Science (Dan David Foundation, Israel) (2005)–Welch Award (2005)–UAA-Dhirumbhai Ambani Award (National Academy of Science, India) (2006)–Priestley Medal (American Chemical Society) (2007).

Memberships and Fellowships: American Academy of Arts and Sciences, National Academy of Sciences, National Academy of Engineering, American Philosophical Society, Fellow of the American Association for the Advancement of Science, Fellow of the Institute of Physics, New York Academy of Sciences, World Technology Network, Foreign Fellow of the Indian National Academy of Science, Honorary Member of the Materials Research Society of India, Honorary Fellow of the Chemical Research Society of India, Royal Netherlands Academy of Arts and Sciences, and Honorary Fellow of the Royal Society of Chemistry (U.K.).

Present research interests include: physical and organic chemistry, materials science, biophysics, complexity and emergence, surface science, microfluidics, optics, self-assembly, micro- and nanotechnology, science for developing economies, catalysis, energy production and conservation, origin of life, rational drug design, and cell-surface biochemistry.

Peter G. Wolynes

Peter G. Wolynes was born in Chicago, Illinois April 21, 1953. He graduated with an A.B. from Indiana University in 1971 and received a Ph.D. in Chemical Physics from Harvard University in 1976. He spent most of the year as a postdoctoral fellow with John Deutch at the Massachusetts Institute of Technology. He then spent from Fall 1976 to 1979 as an Assistant Professor in the Chemistry department at Harvard. In 1980 he moved to the University of Illinois, eventually becoming the Center for Advanced Study Professor of Chemistry, Physics and Biophysics. In 2000 he moved to the Department of Chemistry and Biochemistry at the University of California, San Diego.

Wolynes has been a visiting scholar for extended periods at the Institute for Theoretical Physics (UCSB), the Institute for Molecular Science (Okazaki, Japan) and the Ecole Normale Superieure (Paris, France). He was a Fogarty Scholar-in-Residence at the National Institute of Health and the Himshelwood Lecturer at Oxford.

Wolynes' work across the spectrum of theoretical chemistry and biochemistry has been recognized by the 1986 ACS Award in Pure Chemistry, the 2000 Peter Debye Award for Physical Chemistry of the ACS and the Fresenius Award. He received an honorary Doctor of Science from Indiana University in 1988 and was elected to both the National Academy of Sciences and the American Academy of Arts and Sciences in 1991. He is a fellow of the American Association for the Advancement of Science and the American Physical Society and a Fellow of the Biophysical Society. Most recently, Wolynes has been elected to the American Philosophical Society, the German Academy of Sciences "Leopoldina," and Foreign Member, the Royal Academy.

Wolynes holds the Francis Crick Chair in the Physical Sciences at UCSD and in addition to continuing his work on many body chemical physics, protein folding and structure prediction, he is also studying stochastic aspects of cell biology.

Ahmed H. Zewail

Ahmed H. Zewail is the Linus Pauling Chair Professor of Chemistry and Professor of Physics. For ten years, he has been the Director of the NSF Laboratory for Molecular Sciences (LMS) at the California Institute of Technology (Caltech). Currently, he is the Director of the Physical Biology Center at Caltech.

Dr. Zewail was awarded the 1999 Nobel Prize for his development of the field of femtochemistry, making possible discoveries of phenomena on the femtosecond timescale. At present, the focus of his research group is mainly on the development of four-dimensional microscopy for visualization in the four dimensions of space and time, and the understanding of complexity of chemical and biological transformations.

Among the other honors he received are the Albert Einstein World Award (2006), Benjamin Franklin Medal (1998), Robert A. Welch Award (1997), Leonardo da Vinci Award (1995), Wolf Prize (1993), and the King Faisal Prize (1989). From Egypt he was awarded the Order of the Grand Collar of the Nile, the highest state honor. Postage stamps have been issued to honor his contributions to science and humanity. He holds honorary degrees in the sciences, arts, philosophy, law, medicine, and humane letters from some thirty universities around the world, and is an elected member of the National Academy of Sciences, the American Philosophical Society, and the American Academy of Achievement. He is also a member of the European Academy of Arts, Sciences, and Humanities, the Royal Society of London, the Pontifical Academy, the French Academy, the Russian Academy, and the Royal Swedish Academy of Sciences.

Dr. Zewail serves on several boards of international institutions of higher learning. He has been a Chief Editor of Chemical Physics Letters (1991–2007) and is now the Honorary Advisory Editor. He was a member of the Board of Directors of TIAA-CREF (2004–2007), and since 2002 has been a member of the Scientific Advisory Board of the Welch Foundation.